1976

Contemporary Organic Chemistry

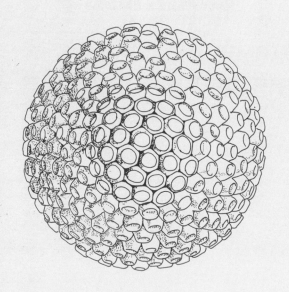

Contemporary Organic Chemistry

Molecules, Mechanisms, and Metabolism

Marion H. O'Leary

Associate Professor of Chemistry
University of Wisconsin, Madison

McGraw-Hill Book Company

New York / St. Louis / San Francisco / Auckland / Düsseldorf / Johannesburg
Kuala Lumpur / London / Mexico / Montreal / New Delhi / Panama
Paris / São Paulo / Singapore / Sydney / Tokyo / Toronto

Contemporary
Organic
Chemistry

1 2 3 4 5 6 7 8 9 0   KPKP   7 9 8 7 6 5

This book was set in Times Roman by York Graphic Services, Inc.
The editors were Robert H. Summersgill and David Damstra;
the designer was J. Paul Kirouac, A Good Thing, Inc.;
the production supervisor was Leroy A. Young.
The drawings were done by Vantage Art, Inc.
Kingsport Press, Inc., was printer and binder.

Library of Congress Cataloging in Publication Data

O'Leary, Marion H
Contemporary organic chemistry.

Includes index.
1. Chemistry, Organic.   I. Title.
QD251.2.043   547   75-9792
ISBN 0-07-047694-2

for Sandra

Contents

Preface

Most students who enroll in a short organic chemistry course are not chemistry majors but students who need a background in organic chemistry to prepare for other subjects. A large number of these students are majors in the life sciences—biology, premedicine, nursing, pharmacy, and related fields. The teaching of organic chemistry has not always reflected the particular needs and interests of these students, who have often been presented with a shortened version of the organic chemistry course taught to chemistry majors, with no attention paid to their interest in the life sciences.

My purpose in writing this text was to present organic chemistry in a manner more relevant to the needs and interests of these students without sacrificing the rigor which should be a part of the introductory course. The quantitative aspects of organic chemistry—orbital theory, electronic structure of molecules, reaction mechanisms, and the like—are the stuff of which a good introductory chemistry course is made, and these features should not be compromised merely in order to teach a more "relevant" course. However, my thesis in this book is that these basic features can be used to explain the chemical reactions which occur in living systems just as they explain the reactions which occur in the laboratory.

The unique features of this book reflect my own prejudice about introducing nonmajor students to organic chemistry. Because of the central place of stereochemistry in organic chemistry, and because of the difficulty many students have with stereochemistry, I have included many problems which require the use of molecular models. My own belief in the importance of understanding bonding and reaction mechanisms leads me to treat these subjects in more detail than is usual in introductory textbooks. Finally, I believe that biochemistry and organic chemistry are really two separate aspects of the same subject, and I have integrated biochemical examples at appropriate places throughout the text, rather than collecting them in a separate chapter at the end.

I would like to thank the many people who provided support and encouragement for this project. My friend Peter Guthrie sometimes understood my objectives better than I. A number of reviewers helped to focus and clarify my ideas. My own students in introductory organic chemistry courses were firm, but tolerant, and forced me to say what needed to be said. Bill Newmann and Gary Niehaus asked many thoughtful questions. My own research students accepted my absences from the lab with equanimity and gave me static when static was due; particular thanks to Steve Koontz, Gil Gilbert, and Tony Young. Karen Rulland typed and retyped my drafts and never lost patience when I changed my mind. My wife Sandra relieved me of many burdens. And last, my thanks to my children, who without really understanding all the fuss about my writing a book nonetheless contributed in their own way to its completion.

Marion H. O'Leary

To the Student

CONTEMPORARY ORGANIC CHEMISTRY

Organic chemistry is a complex science built on a small number of basic principles of structure and reactivity. These basic principles can also be used to understand the chemical processes which go on in living systems. My purpose in writing this book was to present these principles in a way which will make possible an understanding of both organic chemistry and the chemistry of life.

MOLECULES

The first principles of organic chemistry, considered in the first few chapters of this book, are those of structural chemistry: how electrons form bonds in organic molecules and how those bonds give rise to the characteristic shapes of organic compounds. Because of the importance of understanding molecular shapes, I have included throughout these chapters problems which require the use of molecular models. I urge that you interrupt your reading of the text and do these problems as they occur.

MECHANISMS

The second group of principles are those which govern chemical reactions. These principles are used to derive reaction mechanisms which are detailed descriptions of how reactions occur. Because of the importance of reaction mechanisms, I have discussed them at length in the text. Study these carefully. If you understand mechanisms, you will understand reactions.

METABOLISM

Many of the most fascinating reactions in organic chemistry are in the field of metabolism, the chemical reactions which occur in living systems and thus make possible the existence of life. Although these reactions often seem quite complex, they are no different in principle from the simple organic reactions whose mechanisms we discuss. I have included a large number of examples from living systems because they illustrate very clearly the principles of structures and mechanisms. Study these carefully. They will increase your understanding and your appreciation of organic chemistry.

HOW TO STUDY

Here are some useful hints for learning organic chemistry: Learn the material on molecular structures and work all problems which require the use of molecular models. Whenever you look at a new reaction, study the mechanism, then test your understanding of the reaction by trying to write the mechanism yourself. A number of common themes run through the reaction mechanisms given in this book; when studying a new reaction, it is useful to review the mechanisms of earlier reactions of the same type. Do the in-text problems as they occur; they are intended to test your understanding of the material which immediately precedes them. Use your understanding of organic reactions to explain how the examples from living systems work. When you finish a chapter, do the problems at the end of the chapter. The answers to many of these are given at the end of the book.

Organic chemistry can be fun. I enjoyed writing this book, and I hope you will enjoy reading it and working with it.

Marion H. O'Leary

Contemporary Organic Chemistry

Introduction

The living things which inhabit the earth are diverse. A butterfly and an elephant may seem to have little in common, but the chemical principles which describe the life of a butterfly are the same as those which describe that of an elephant, or that of a tree, or even that of a human being.

Chemistry provides a description of the world of observable things in terms of atoms and molecules. To a chemist, the diversity of life is all the more intriguing because of the basic chemical unity which underlies these different forms.

The purpose of this book is to present organic chemistry in a manner which will lead to an appreciation of the chemistry of life. Although we will never reach the point where we can describe in detail the chemistry of a butterfly or an elephant (such descriptions are the province of biochemistry), we will consider in detail the chemical principles which make such a description possible.

1-1 A BIT OF HISTORY

Until the first part of the nineteenth century, all of chemistry was divided into two parts—*organic chemistry* (the chemistry of compounds derived from living things) and *inorganic chemistry* (the chemistry of compounds derived from minerals). Organic compounds were intimately associated with life and were believed to be endowed with some kind of *vital force,* which made them different from inorganic compounds. This vital force provided the key to the existence of life.

However, in 1828 the German chemist Friedrich Wöhler succeeded in preparing an organic compound (urea) from an inorganic compound (ammonium cyanate)

ammonium cyanate   urea

Wöhler demonstrated that his synthetic urea was indistinguishable from the urea isolated from a living organism. Thus, the association of a vital force with organic compounds was proved false.

Following Wöhler, organic chemistry developed into a manageable science governed by demonstrable chemical principles. The leading role of carbon in organic chemistry was established. Chemical bonds and chemical reactions became the central focus of the subject.

However, Wöhler's work also created a schism which was to last for over a hundred years. Following Wöhler, scientists who were interested in the more complex aspects of the chemistry of life began to develop a discipline of their own: biochemistry. They, too, studied structures and reactions, but the structures and reactions which they studied were much more complex than those being studied by organic chemists. Biochemists learned to identify the structures of such important compounds as proteins, carbohydrates, and fats, and they began to study the ability of living systems to carry out reactions of these complex substances. These transformations seemed to be somehow different from the simple reactions studied by organic chemists, and biochemists continued to explain their results in terms of a vital force.

The demise of the vital-force theory began early in the twentieth century, when scientists showed that many of the chemical reactions carried out by living organisms could also be carried out by cellular extracts from which all life was obviously gone. The catalytic power of these extracts was attributed to unknown entities called *enzymes*. The nature of these mysterious enzymes and the question of whether they themselves were endowed with a vital force was finally settled in 1926 by the American biochemist J. B. Sumner, who prepared an enzyme in pure, crystalline form (figure 1-1). Oddly enough, it was the enzyme *urease*, which catalyzes the reaction of urea with water

$$H{-}N{-}\overset{\displaystyle O}{\underset{\displaystyle H}{\overset{\|}{C}}}{-}\underset{\displaystyle H}{N}{-}H + H_2O \xrightarrow{\text{urease}} O{=}C{=}O + 2H{-}\overset{\displaystyle H}{N}{-}H$$

urea   water   carbon   ammonia
                              dioxide

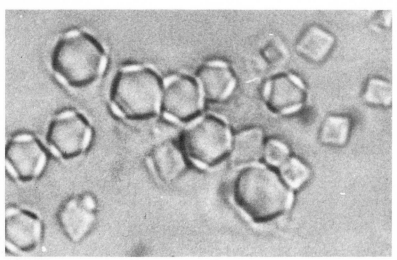

Purification and crystallization of this enzyme played a key part in the demise of the vital-force theory. [W. N. Fishbein and K. Nagarajan, Archives of Biochem- *istry and Biophysics, **144:** 700 (1971); reproduced with permission of the authors and Academic Press, Inc.]*

Figure 1-1
Crystals of the enzyme urease.

Sumner showed that urease is a protein—a large molecule made up of amino acids (chapter 12)—and that it does not possess any kind of vital force. This demonstration signaled the end of a very productive era in biochemistry and at the same time marked the beginning of a new era, in which biochemistry and organic chemistry could be reunited into a consistent and coherent partnership.

A MODERN VIEW

Organic chemistry is the chemistry of carbon. Among the hundred-odd atoms of the periodic table, carbon is unique in the variety and complexity of the compounds which it forms. Its ability to form stable single, double, and triple bonds to itself is particularly significant.

The chemistry of living things is called *biochemistry*. Because almost all compounds which occur in living things contain carbon, there is a large overlap between biochemistry and organic chemistry. For example, a compound synthesized by an organic chemist may turn out to be a useful drug. A molecule isolated from a living organism by a biochemist may prove to be of interest in studies of organic structures and reactions. The border line between the two fields is indistinct, and the crossover from one to the other has been of great benefit to both.

1-2 VITAMIN B_{12}: A CASE STUDY

Progress in understanding how living systems work has been marked by interaction among organic chemists, biochemists, physicians, and others. The history of vitamin B_{12} provides an interesting example of this cooperation.

Pernicious anemia, first described in the mid-nineteenth century by Thomas Addison, an English physician, is characterized by weakness and fatigue and often leads to death. Subsequent to Addison's report, this condition became well known, but little progress was made in alleviating it until 1920, when G. H. Whipple in California showed that raw liver can be used to treat anemia in dogs. In 1926, George R. Minot and William P. Murphy in Boston showed that pernicious anemia in humans can also be treated with liver. Whipple, Minot, and Murphy received the 1934 Nobel prize for their discoveries.

The search for the anti-pernicious anemia factor of liver was started by biochemists shortly after Whipple's discovery, but little progress was made in the next few years. The total amount of the factor in liver is extremely small, and it often decomposes when purification is attempted. In addition, there was no convenient chemical or biological test for its presence except for its effect on patients with pernicious anemia.

Finally, in 1948, the isolation of the pure, dark-red crystalline compound, by then called vitamin B_{12}, was reported by K. Folkers and his collaborators at the Merck Research Laboratories in the United States, and shortly thereafter by E. Lester Smith and L. F. J. Parker at the Glaxo Laboratories in England. Isolation required the expenditure of very large amounts of money and manpower and the use of massive amounts of material. It is not surprising, therefore, that the isolation was accomplished in industrial laboratories, which were able to commit the vast resources needed for this momentous effort.

Isolation of pure vitamin B_{12} opened a new era in biochemistry. Studies of the structure of the new compound began immediately, but it soon became clear that vitamin B_{12} is a very complex substance (its empirical formula is $C_{63}H_{88}O_{14}N_{14}PCo$). Investigations by a large number of organic chemists were unsuccessful in determining the structure. In 1954 the structure was determined by Dorothy C. Hodgkin in England by the use of x-ray crystallography, a feat for which she received the 1964 Nobel prize. The structure of vitamin B_{12} is shown in figure 1-2.

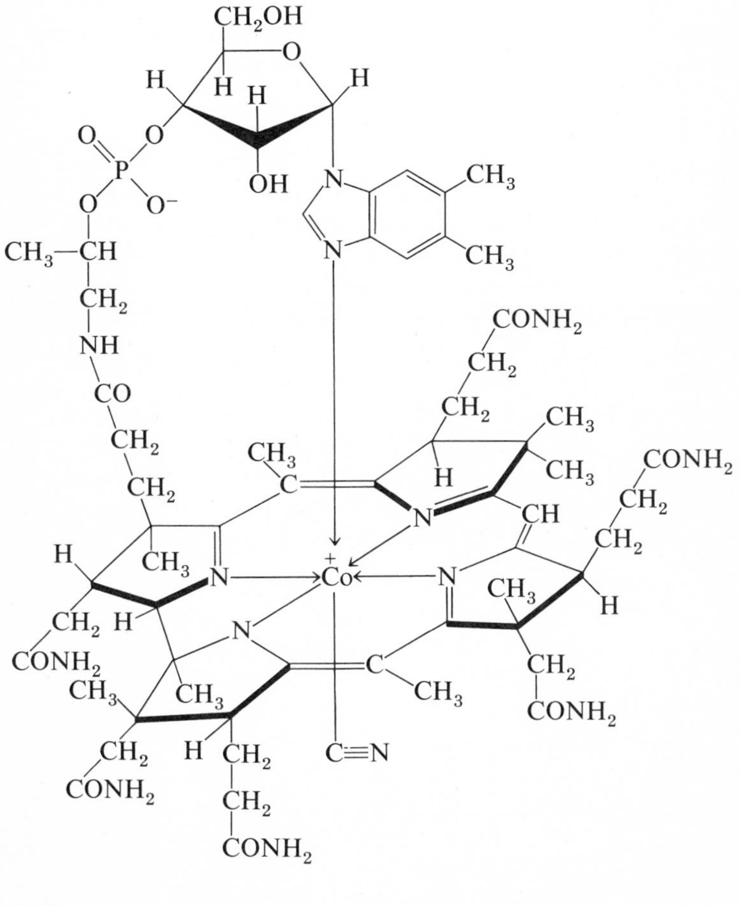

Figure 1-2
Vitamin B_{12}.

The chemical reactions of vitamin B_{12} were studied during this period by a number of scientists, most notably by Sir Alexander Todd, in England (he won the 1957 Nobel prize). These chemical and structural studies made possible the complete chemical synthesis of vitamin B_{12}. The mammoth effort required for the synthesis was directed by R. B. Woodward in the United States (he won the Nobel prize in 1965) and Albert Eschenmoser in Switzerland. The synthesis was completed in 1972 after an eleven-year effort involving a hundred scientists from nineteen different countries.

Current studies of vitamin B_{12} center on the detailed molecular mechanism by which it functions. This research involves physiologists, biochemists, organic chemists, inorganic chemists, and other scientists.

1-3 WHAT ORGANIC CHEMISTRY IS ABOUT

Organic chemistry is a vast field. Its practitioners study a wide variety of subjects, from very simple molecules such as methane, CH_4, to such complex molecules as vitamin B_{12}. For our purposes, we will divide organic chemistry into two principal subdivisions, *structures* and *reactions*.

STRUCTURES Organic chemistry is concerned largely with covalent bonds, that is, bonds in which one or more pairs of electrons are shared between adjacent atoms. Covalent bonds impart characteristic shapes to organic molecules, and one important part of our study will be molecular shape and its effects on the properties of molecules.

We will frequently consider particular substructures of molecules containing a small number of atoms arranged in a particular way. Such a substructure is called a *functional group;* its properties are largely independent of the molecule in which it occurs. For example, the properties of the functional group —O—H, called the *hydroxyl group,* are very similar whether it occurs in a simple compound like ethyl alcohol (the "alcohol" of alcoholic beverages)

$$
\begin{array}{ccc}
 & H & H \\
 & | & | \\
H- & C- & C-O-H \\
 & | & | \\
 & H & H
\end{array}
$$

ethyl alcohol

or in a complex molecule like pyridoxine (vitamin B_6)

pyridoxine

A chemical reaction is the transformation of one structure into another **REACTIONS** by a change in bonds. Until early in the twentieth century a reaction was characterized only by the starting materials, the products, and the reaction conditions. Under those circumstances, the study of organic chemistry becomes largely a matter of memorizing a long catalog of chemical reactions.

Fortunately, organic chemistry has progressed far beyond that point. We can now understand organic reactions in terms of *reaction mechanism,* the description of the pathway by which a conversion takes place and the enumeration of the structural factors responsible for the chemical reaction.* Thus described, organic chemistry becomes a study of chemical principles, from which chemical reactions can be understood. This book uses this approach of principles and mechanisms, rather than simply presenting countless reactions to be memorized.

1-4   WHAT BIOCHEMISTRY IS ABOUT

Except for its greater complexity, biochemistry is like organic chemistry; the same physical laws govern both fields. Thus, we are able to move from principles of organic chemistry to examples from biochemistry without any inconsistency.

As in organic chemistry, structures and reactions are central to the understanding of biochemistry. However, we must also consider some features which are peculiar to biochemistry, in particular, enzymes (biological catalysts) and metabolism (the total chemistry of an organism).

---

*If a chemical reaction is considered to be like a journey, say, from Madison, Wisconsin, to New York City, then a reaction mechanism is a description of the route of that journey, say, by car from Madison to Chicago and then by air nonstop to New York. Understanding the route (the mechanism) makes it easier to predict the destination (the product).

STRUCTURES   Simple compounds like those in organic chemistry also occur in biochemistry, but complex compounds containing thousands of atoms are common as well. In chapters 12 and 13 we will consider some of these complex structures and will see that in spite of their complexity they obey the same chemical laws as simple organic compounds.

ENZYMES   Most chemical reactions which occur in living systems operate under the influence of *enzymes:* specific biological catalysts which increase the rates of reactions and thus control the chemistry of the system. Although we do not consider the structures of enzymes in detail until chapter 12, we will have many occasions to examine the effects of these essential substances.

METABOLISM   Even the simplest living thing depends for its existence on a very large group of chemical reactions. These reactions must provide for survival, growth, reproduction, and every other characteristic of life. The totality of chemical transformations which an organism is capable of carrying out is called its *metabolism.* Some important aspects of metabolism are shown in figure 1-3.

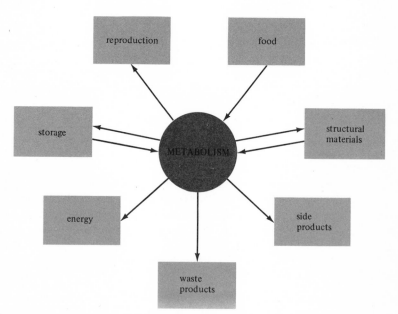

Figure 1-3
Metabolism.

Even the very simplest organism is capable of carrying out several thousand chemical transformations needed for survival. Almost all these transformations are catalyzed by enzymes. The whole collection of chemical transformations is called the organism's metabolism.

Even though the metabolism of even the simplest creature may seem hopelessly complex, metabolism can be divided into large groups of enzyme-catalyzed reactions. Each group, in turn, can be divided into a number of individual chemical reactions which can be explained in terms of principles of organic chemistry. This reduction of life to chemical terms is one of the most fascinating and rewarding features of modern chemistry.

SUGGESTED READINGS

Goodman, M., and F. Morehouse: *Organic Molecules in Action,* Gordon and Breach, New York, 1973. A very readable account of the relationship of organic chemistry to biology, medicine, and other natural sciences.

Hendrickson, J. B., D. J. Cram, and G. S. Hammond: *Organic Chemistry,* 3d ed., McGraw-Hill, New York, 1970. An excellent comprehensive introduction to organic chemistry.

Hill, J. W.: *Chemistry for Changing Times,* 2d ed., Burgess, Minneapolis, 1974. Highly readable and highly relevant account of the relationship between chemistry and society.

Mahler, H. R., and E. H. Cordes: *Biological Chemistry,* 2d ed., Harper and Row, New York, 1971. One of the best comprehensive introductory texts on biochemistry.

Morrison, R. T., and R. N. Boyd: *Organic Chemistry,* 3d ed., Allyn and Bacon, Boston, 1973. A popular and clearly written introductory textbook.

Organic Chemistry of Life: Readings from Scientific American, Freeman, San Francisco, 1973. Interesting articles on a variety of subjects in the usual engaging *Scientific American* style.

Roberts, J. D., R. Stewart, and M. C. Caserio: *Organic Chemistry,* Benjamin, Menlo Park, Calif., 1971. An excellent introductory textbook.

Watson, J. D.: *The Molecular Biology of the Gene,* 2d ed., Benjamin, Menlo Park, Calif., 1970. An eminently readable introduction to the border line between biology and chemistry.

White, A., P. Handler, and E. L. Smith: *Principles of Biochemistry,* 5th ed., McGraw-Hill, New York, 1973. An excellent comprehensive introduction to biochemistry, with emphasis on its medical aspects.

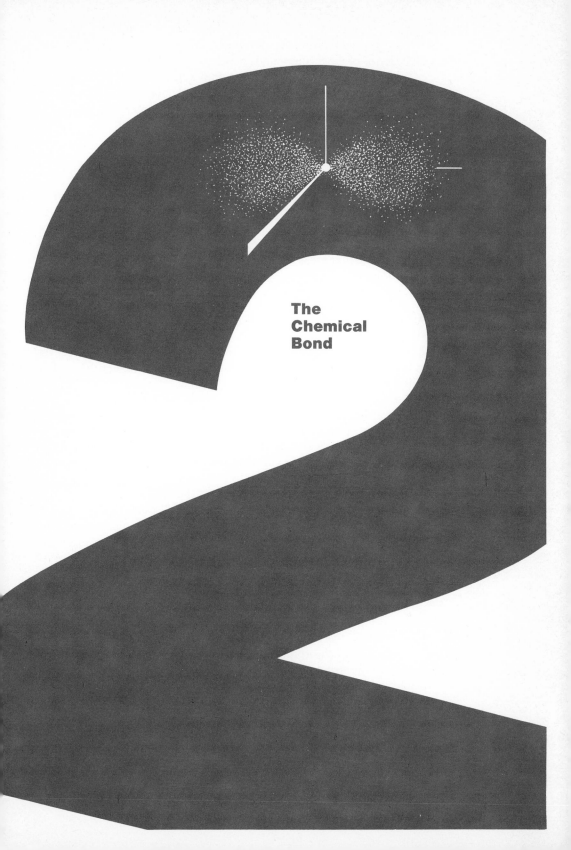

**The
Chemical
Bond**

2

The variety and complexity of organic chemistry arise from the great diversity of organic compounds which exist and the variety of chemical reactions which they undergo. Before we survey the properties and reactions of organic molecules, we must understand the various types of chemical bonds which occur in them. This chapter will provide us with that understanding.

2-1 ATOMIC STRUCTURE

An atom is composed of two parts: the *nucleus* and the *electrons*. The nucleus is small but heavy. The electrons are very light and move rapidly about the nucleus. Their motion is so rapid that it is not possible to define their positions or their paths. Instead, we can speak only of particular areas about the nucleus where the electrons are likely to be found.

ORBITALS   An *orbital* is the region of space around the nucleus where a particular electron is likely to be found. Orbitals are usually represented as clouds whose density corresponds to the probability of finding the electron at that particular point; the higher the density, the higher the probability that the electron is there.

The appearances and properties of various kinds of orbitals are derived from quantum mechanics. Orbitals are of two classes, *atomic orbitals,* which define the locations of electrons in isolated atoms, and *molecular orbitals,* which define the locations of electrons in molecules. An orbital of either class can hold only two electrons.

Atomic orbitals are designated by two labels; a number, the *principal quantum number,* indicates the energy of the orbital, and a letter, the *orbital type,* indicates the shape of the orbital. The energy of an orbital of a particular type (or, more properly, the energy which an electron would have if it were in that orbital) increases with increasing principal quantum number. Many types of orbital exist, but only two are of great interest in organic chemistry, *s* orbitals and *p* orbitals. Seldom will we have occasion to discuss the more complex *d* orbitals.

All atoms have the same set of orbitals available for electrons; different types of atoms have different numbers of electrons to go in these orbitals. The first nine orbitals which an atom can use are shown in figure 2-1. Each orbital can accommodate two electrons. Defining the electronic structure of an atom is simply a matter of placing pairs of electrons in orbitals of successively higher energy until no more electrons remain to be placed.

s ORBITALS   The hydrogen atom, with a single proton in the nucleus and a single electron, is the simplest atom. This electron is placed in the 1*s* orbital.

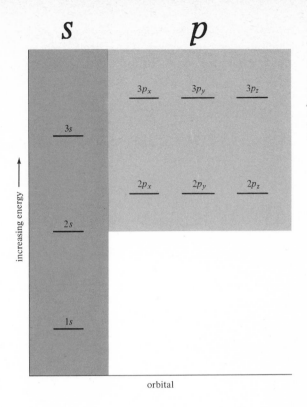

s   p

increasing energy →

3p_x 3p_y 3p_z

3s

2p_x 2p_y 2p_z

2s

1s

orbital

Figure 2-1
Atomic orbitals.
*Energy increases
toward the top of the
chart. Atoms are
constructed by
placing each electron
in the lowest
available unfilled
orbital.*

The *s* orbitals are the simplest, being symmetrical about the nucleus. That is, if you stood at the nucleus looking out toward the orbital, it would appear the same no matter in what direction you looked.

A hydrogen atom is shown in figure 2-2. The electron occupies the 1*s* orbital.

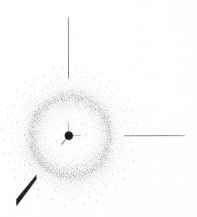

Figure 2-2
The hydrogen atom.
*The nucleus is at the
center, and the
electron occupies
a spherically
symmetrical σ
orbital.*

In order to define the structure of an atom, we must list the orbitals which its electrons occupy. This is called the *electronic configuration* of the atom. Each orbital in figure 2-1 can accommodate two electrons. We construct the electronic configuration of an atom by putting pairs of electrons into successive orbitals until no more electrons remain to be placed.

The second atom in the periodic table is helium, which has two electrons. These two electrons occupy the $1s$ orbital, and we write the electronic configuration of helium as

He: $1s^2$

The exponent 2 indicates the presence of two electrons in the $1s$ orbital. Helium has a completely filled $1s$ orbital, and no other orbitals of principal quantum number 1 remain to be filled. As a result, helium is the first *inert gas*. Its electronic configuration is extremely stable, and helium has no inclination to participate in electron exchanges with other atoms: helium is chemically inert.

Figure 2-3 gives the configurations of the first eighteen atoms of the periodic table. All atoms past helium (those of atomic number 3 and higher) have a filled $1s$ orbital. Since a filled orbital does not participate in chemical bonding, only one atom, hydrogen, uses its $1s$ orbital for bonding.

The next two atoms in the periodic table, lithium and beryllium, have filled $1s$ orbitals and one or two electrons in the $2s$ orbital. All

Figure 2-3
Electronic
configurations of
some atoms.
Filled shells are
shaded. The
electrons outside of
these shells are
called **valence**
electrons *because*
they are responsible
for the formation of
chemical bonds.

H	$1s$		
He	$1s^2$	inert gas	
Li	$1s^2$	$2s$	
Be	$1s^2$	$2s^2$	
B	$1s^2$	$2s^2$	$2p$
C	$1s^2$	$2s^2$	$2p^2$
N	$1s^2$	$2s^2$	$2p^3$
O	$1s^2$	$2s^2$	$2p^4$
F	$1s^2$	$2s^2$	$2p^5$
Ne	$1s^2$	$2s^2$	$2p^6$	inert gas
K	$1s^2$	$2s^2$	$2p^6$	$3s$
Mn	$1s^2$	$2s^2$	$2p^6$	$3s^2$

atoms beyond beryllium have filled $2s$ orbitals and one or more electrons in the $2p$ orbitals.

Boron is the first atom which has electrons in one or more of the three $2p$ orbitals. The $2p$ orbitals are of higher energy than the $2s$ orbital and have different shapes.

The three $2p$ orbitals are all of the same size, shape, and energy, but they differ in spatial orientation about the nucleus. Each p orbital consists of two halves, or *lobes,* of equal size. The lobes are roughly spherical and are arranged on both sides of the nucleus, as shown in figure 2-4. The three $2p$ orbitals have their principal axes mutually perpendicular.

The atoms lithium through fluorine have filled $1s$ shells and partially filled $2s$ or $2p$ orbitals. The $2s$ and $2p$ orbitals of these atoms are said to be the *bonding orbitals* because they participate in bond formation when these atoms form molecules. The last element in this row of the periodic table, neon, has all its $2s$ and $2p$ orbitals filled, and neon is unable to form chemical bonds.

2-2 MOLECULAR STRUCTURE

A seemingly infinite variety of molecules confronts the student of chemistry. This variety, however, is not the result of chaos; instead, two principal driving forces control the formation of molecules from atoms:

1 In forming a molecule, each atom attempts to achieve an *inert gas configuration.*

2 Each atom tries to remain *electrically neutral.*

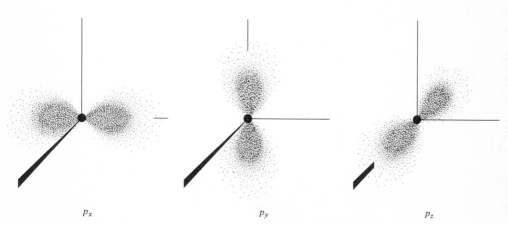

p_x p_y p_z

All three orbitals have the same shape and the same energy, but they are differently oriented in space. The   *principal axes of the three orbitals are mutually perpendicular.*   **Figure 2-4
The three 2p orbitals.**

Chemical bonds are of two types, *ionic* and *covalent*. Atoms involved in ionic bonds are able to meet the first of the above conditions but not the second; atoms involved in covalent bonds are able to meet both conditions.

IONIC BONDS   When an ionic compound is formed from atoms, electrons are transferred from one atom to another. By means of such a transfer, each atom achieves an inert gas configuration. The resulting chemical bond is an *ionic bond,* an electrostatic attraction between ions of unlike sign.

The formation of lithium fluoride from lithium atoms and fluorine atoms illustrates how ionic molecules are formed from atoms. Lithium and fluorine have the electronic configurations

Li: $1s^2 2s$ and   F: $1s^2 2s^2 2p^5$

A lithium fluoride molecule is formed from these two atoms by the transfer of a single electron from lithium to fluorine. By this transfer, lithium becomes positively charged and achieves the inert gas configuration of helium; simultaneously, fluorine becomes negatively charged and achieves the inert gas configuration of neon

Li^+: $1s^2$ and F^-: $1s^2 2s^2 2p^6$

Molecular structures can also be represented by *Lewis dot structures,* in which electrons in the outer shell are indicated by dots. The formation of lithium fluoride from a lithium atom and a fluorine atom is shown as

$$Li\cdot \; + \; \cdot \overset{..}{\underset{..}{F}}: \; \longrightarrow \; Li^+ \quad :\overset{..}{\underset{..}{F}}:^-$$

 atoms   molecule

PROBLEM 2-1   Give Lewis dot structures of the following ionic molecules: (*a*) $CaCl_2$; (*b*) Na_2S; (*c*) HCl.

COVALENT   *Covalent bonds* are bonds in which electrons are shared between atoms.
BONDS   The simplest molecule containing a covalent bond is the hydrogen molecule, H_2. When the hydrogen molecule is formed from two hydrogen atoms, each atom donates its electron to form a covalent bond

$$H\cdot \; + \; \cdot H \; \longrightarrow \; H:H$$

 atoms   molecule

Covalent bonds always involve pairs of electrons.

The covalent bond serves as a slightly rubbery molecular glue which holds nuclei firmly together. As a result, molecules which are held together by covalent bonds have well-defined geometries and bond lengths.

The unique thing about covalent bonds is that the shared electrons are counted by both nuclei toward achievement of an inert gas configuration. Thus, both hydrogen atoms in the hydrogen molecule act as though they had achieved the inert gas configuration of helium.

The shared electrons in a covalent bond are no longer localized around one nucleus but may be in the vicinity of either nucleus. The orbitals which contain shared electrons are called *molecular orbitals* and are given Greek letters for identification to distinguish them from atomic orbitals, which are identified by Latin letters.

MOLECULAR ORBITALS

Molecular orbitals are formed from atomic orbitals by a process akin to simple addition. The two atomic orbitals are combined, and this combination is a molecular orbital. The formation of the hydrogen molecule by the combination of hydrogen atomic orbitals is shown in figure 2-5. The molecular orbital formed in this case is a σ (sigma) molecular orbital. This is the simplest and most common type of molecular orbital and one that we will see in virtually every molecule. Like all orbitals, this σ orbital can hold two electrons. The electron density in the orbital is highest in the region between the two nuclei.

The fluorine molecule, F_2, is formed from two fluorine atoms by sharing a pair of electrons in a σ molecular orbital. A fluorine atom needs one electron in order to achieve an inert gas configuration. Combining two fluorine atoms by sharing one electron from each gives both atoms inert gas configurations

$$:\overset{..}{\text{F}}\cdot \; + \; \cdot\overset{..}{\underset{..}{\text{F}}}: \; \longrightarrow \; :\overset{..}{\underset{..}{\text{F}}}:\overset{..}{\underset{..}{\text{F}}}:$$

atoms   molecule

The shared pair of electrons in the fluorine molecule is in a σ

Figure 2-5
The hydrogen molecule.
The σ molecular orbital is formed by the combination of a 1s orbital from each of the two hydrogen atoms.

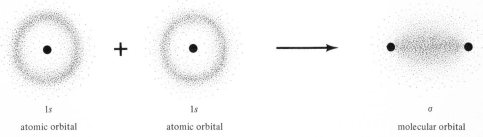

1s   1s   σ

atomic orbital   atomic orbital   molecular orbital

molecular orbital. This σ orbital is similar to that in the hydrogen molecule, but it differs in one important respect because it is formed by the combination of two *p* orbitals, one on each atom, rather than two *s* orbitals (figure 2-6).

COVALENT VALENCE   The ground rules for forming molecules containing covalent bonds are simple: electron-pair bonds are formed by using one electron from each of· the two atoms. Additional bonds can be formed until every atom has achieved an inert gas configuration. Each covalent bond contributes one new electron toward achievement of this configuration. Thus, the number of bonds an atom needs in order to achieve an inert gas configuration is equal to the number of electrons the atom needs to achieve this configuration. This characteristic number of bonds is called the *covalent valence* of the element.

We have already seen the workings of this concept for hydrogen and fluorine; both need one covalent bond in order to achieve an inert gas configuration. That is, hydrogen and fluorine each have a covalent valence of 1. Therefore hydrogen and fluorine can also form a molecule HF, or

$$H : \overset{\cdot\cdot}{\underset{\cdot\cdot}{F}} :$$

hydrogen fluoride

Covalent valences of a number of elements important for organic chemistry are given in table 2-1. Phosphorus and sulfur do not always conform to these values; the covalent valences given are for the lowest oxidation states of these two elements (-3 and -2, respectively) and do not apply to higher oxidation states.

The values given in table 2-1 apply only to cases where the atom is uncharged. Charged atoms do not follow these rules. Oxygen, for example, has two bonds in the neutral water molecule, H_2O, but three bonds in the positively charged hydronium ion, H_3O^+, and one bond in the negatively charged hydroxide ion, OH^-.

Covalent valences can be calculated quite easily. The covalent valence of carbon, for example, can be derived as follows. Carbon has

Figure 2-6
The fluorine molecule.
The σ molecular orbital is formed by the combination of a 2p orbital from each of the two fluorine atoms.

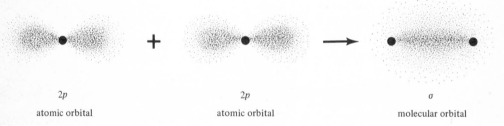

2p   2p   σ

atomic orbital   atomic orbital   molecular orbital

		Table 2-1
H	1	*Table 2-1*
C	4	*Covalent Valences*
N	3	*of a Number of*
O	2	*Important Atoms*
F	1	
Si	4	
P*	3	
S*	2	
Cl	1	
Br	1	
I	1	

*In its lowest oxidation state.

four electrons in its outer shell and therefore needs four more electrons to achieve an inert gas configuration. The simplest way to do this is to make electron-pair bonds with each of four hydrogen atoms

$$\cdot \overset{\cdot}{\underset{\cdot}{C}}\cdot + 4\cdot H \longrightarrow H\colon\overset{\overset{\displaystyle H}{\cdot\cdot}}{\underset{\underset{\displaystyle H}{\cdot\cdot}}{C}}\colon H$$

  atoms   molecule

Each bond is made with one electron from carbon and one electron from hydrogen. Thus, the simplest compound of carbon and hydrogen which obeys our valence rules is CH_4. This is, in fact, the simplest organic compound, *methane.*

Use the rules of covalent valence to construct a compound containing silicon and chlorine. Draw a Lewis structure of the compound.   **PROBLEM 2-2**

Do the same for phosphorus and hydrogen.   **PROBLEM 2-3**

Not all covalent species contain equal numbers of protons and electrons; some covalent species are charged. Examples include the hydronium ion, H_3O^+, the sulfate ion, SO_4^{2-}, and the ammonium ion, NH_4^+. In these cases each individual atom still maintains its inert gas configuration of electrons, but the number of protons in the molecule as a whole is not equal to the number of electrons, and one or more of the atoms in the ion is said to possess a *formal charge.* The charge can be placed on the correct atom by the following procedure: **FORMAL CHARGE**

1 Draw the Lewis structure.

2 Assign every electron in the molecule (for bookkeeping purposes) to a particular atom by use of the following rules:

a For each bond, assign half the electrons to one of the bonded atoms and half to the other.

b Assign unshared electrons to the atom to which they are attached.

3 Atoms which have a number of assigned electrons *equal to their number of valence electrons* (figure 2-3), that is, electrons in their outer shell, are neutral; those having more or less electrons than their number of valence electrons are negative or positive.

By this procedure we can confirm that the atoms in water are uncharged:

$$H:\overset{..}{\underset{..}{O}}:H$$

water

Each hydrogen has the requisite one electron, and oxygen has the requisite six. Thus, all atoms are uncharged. By the same procedure, we see that the hydronium ion has a positive charge on oxygen; hydroxide ion has a negative charge on oxygen

$$H:\overset{..}{\underset{..}{O}}\overset{+}{:}H \qquad H:\overset{..}{\underset{..}{O}}:^-$$
$$H$$

hydronium ion   hydroxide ion

PROBLEM 2-4   Assign a formal charge to each atom in the following molecules: (a) NH_4^+; (b) NO_2^-; (c) HCO_2^-.

2-3 COVALENT BONDS TO CARBON

We showed in section 2-2 that carbon is tetravalent and that the simplest compound of carbon is methane, CH_4. We must now consider the orbitals which carbon uses to make methane.

In its outer shell, carbon has an *s* orbital and three *p* orbitals which might be used for bonding to other atoms. However, because of their different shapes and energies, *s* orbitals and *p* orbitals differ in their ability to form bonds, and use of these orbitals for bonding would be inefficient. Instead, carbon undergoes a quantum-mechanical mixing process called *hybridization* before bonding occurs.

HYBRIDIZATION   Hybridization is the process of transforming a number of dissimilar atomic orbitals into a new set of *hybrid orbitals,* which are identical in all respects except orientation. Hybridization occurs because hybrid orbitals are better able to form covalent bonds than *s* or *p* orbitals are. It must be understood at the outset that hybridization is a process of quantum mechanics and is not a chemical reaction. One cannot observe the conversion of one kind of orbital into another. This hybridization process is mainly a means by which we can understand the relationship between molecular orbitals and various kinds of atomic orbitals.

Carbon forms hybrid orbitals from s and p orbitals in different ways according to the type of bonding. The formation of the sp^3 hybrid, which is used to form four single bonds to other atoms, is the simplest and will be considered first.

To form sp^3 hybrid orbitals, the $2s$ orbital and three $2p$ orbitals are first added quantum-mechanically, and then this combination is divided into *four equivalent sp^3 hybrid orbitals,* as shown in figure 2-7. The four sp^3 hybrid orbitals formed all have the same energy and the same shape; thus, they all have the same ability to form bonds.

The two lobes of a hybrid orbital are decidedly unlike in size. As a result of this distention, sp^3 hybrid orbitals can overlap very efficiently with orbitals of other atoms, and stronger bonds are formed from such hybrid orbitals than would be formed from s or p orbitals.

The arrangement of the sp^3 hybrid orbitals in space about the nucleus is different from that of the p orbitals. The sp^3 hybrid orbitals are arranged in the most symmetrical possible array, the *tetrahedral* arrangement, so called because the four orbitals point toward the corners of a regular tetrahedron whose center is at the carbon nucleus. *GEOMETRY*

It is these four sp^3 hybrid orbitals which are used to construct molecular orbitals in molecules such as methane. The four σ molecular orbitals in methane are each constructed by combination of an sp^3 hybrid orbital with a $1s$ orbital of a hydrogen. Four such orbitals constitute the molecular orbitals of methane (figure 2-8).

The geometry of methane is a result of the geometry of the hybrid orbitals from which it is formed: methane is tetrahedral (figure 2-8). In fact, carbon is always tetrahedral when it is singly bonded to four other atoms, except where the formation of rings distorts the tetrahedral geometry (chapter 3).

Methane is a highly symmetrical molecule, and all the carbon-hydrogen bonds are exactly alike. They all have the same length, and the angle between any two carbon-hydrogen bonds is 109.5°. It is not possible to distinguish any one bond from the others.

Build a molecular model of methane. Draw pictures of the molecule from three different angles. *PROBLEM 2-5**

---

*Understanding the geometry of organic compounds is both important and difficult. To facilitate this understanding, many problems in early chapters require the use of molecular models. It is strongly recommended that the student purchase a set of molecular models and stop and solve these problems whenever they occur in the text.

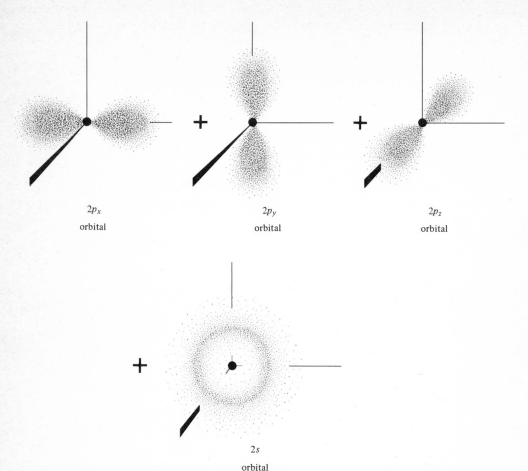

$2p_x$ orbital $2p_y$ orbital $2p_z$ orbital

$2s$ orbital

Figure 2-7 *sp³ hybridization.* The 2s atomic orbital can be combined with the three 2p atomic orbitals to form four equivalent sp³ hybrid orbitals which have the same shape and energy. The principal axes of the four hybrid orbitals point toward the corners of a regular tetrahedron and are therefore 109.5° apart.

CARBON-CARBON SINGLE BONDS

Carbon forms strong bonds to other carbon atoms. *Ethane* contains one carbon-carbon bond:

$$H : \overset{\overset{\displaystyle H}{..}}{\underset{\underset{\displaystyle H}{..}}{C}} : \overset{\overset{\displaystyle H}{..}}{\underset{\underset{\displaystyle H}{..}}{C}} : H \qquad \text{or} \qquad H - \overset{\overset{\displaystyle H}{|}}{\underset{\underset{\displaystyle H}{|}}{C}} - \overset{\overset{\displaystyle H}{|}}{\underset{\underset{\displaystyle H}{|}}{C}} - H$$

ethane

The carbon-carbon single bond consists of a σ molecular orbital, which is formed from an sp^3 hybrid orbital of each of the two carbon atoms.

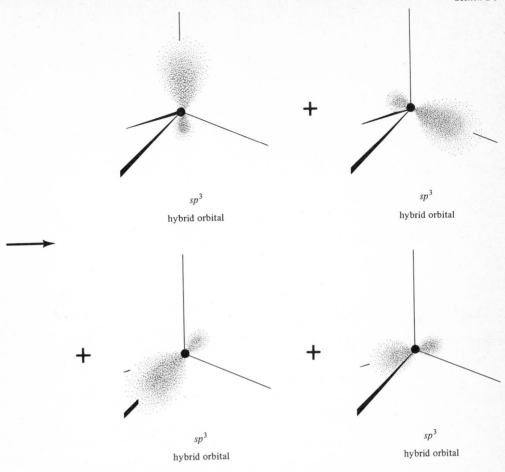

This process of forming carbon-carbon single bonds can be continued indefinitely:

propane

butane

decane

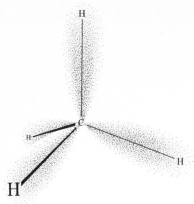

Figure 2-8
Methane.

Each σ bond in methane is formed by the combination of a 1s orbital of a hydrogen atom with an sp³ hybrid orbital of carbon.

Like the hybrid orbitals with which it was formed, methane is tetrahedral. All H—C—H bond angles are 109.5°.

In fact, the well-known plastic *polyethylene* is a collection of just such chains of carbon atoms, but the chains are many thousands of atoms long.*

Organic chemistry is complex because it includes the structures and properties of several million compounds containing a large number of different arrangements of atoms. But at the same time organic chemistry is simple because the properties of a particular bond are largely independent of the molecule in which that bond occurs. For example, the carbon-carbon bonds in polyethylene are σ bonds formed by the combination of two sp^3 hybrid orbitals, and the properties of these bonds can be adequately understood by considering small molecules such as ethane, propane, and butane.

PROBLEM 2-6   Build a molecular model of decane and use it to answer the following question: Is the picture of decane above an accurate representation of the molecule? Why or why not?

CARBON-CARBON
DOUBLE
BONDS

Whenever carbon shares one pair of electrons with another atom, a σ molecular orbital is used to make the covalent bond. Carbon can also share two or three pairs of electrons with another atom, thereby forming a double or triple bond. The simplest such compound is *ethylene:*

---

*Although this compound contains no carbon-carbon double bonds, it is named as a derivative of ethylene, which has a carbon-carbon double bond, because it is made by the *poly*merization of *ethylene.*

H $\cdot$ $\cdot$ H $\quad\quad$ H $\quad\quad\quad$ H

$\quad\quad$ $\overset{..}{C}::\overset{..}{C}$ $\quad$ or $\quad$ $\overset{\diagdown}{}C=C\overset{\diagup}{}$

H $\cdot$ $\quad$ $\cdot$ H $\quad\quad$ H $\quad\quad\quad$ H

$\quad\quad$ ethylene

Four electrons are involved in the double bond, and so two molecular orbitals are required. Here the hybridization of carbon is different from that when only single bonds are formed.

When a carbon atom is linked by single bonds to four other atoms, the hybridization is sp^3, with four equivalent orbitals available for bonding. The hybridization of a carbon atom which participates in a double bond occurs by mixing the $2s$ orbital of carbon with only *two* of the three $2p$ carbon orbitals, resulting in the formation of three equivalent sp^2 hybrid orbitals. One p orbital is not included, but is left intact to form a π molecular orbital. This hybridization process is shown in figure 2-9.

The sp^2 hybrid orbitals have shapes similar to those of sp^3 hybrid orbitals, but they are arranged differently about the nucleus. The axes of the three orbitals are all in the same plane and at 120° angles to each other (figure 2-9). The $2p$ orbital which is left over has its principal axis perpendicular to the plane of the three hybrid orbitals. This p orbital is involved in the formation of the second half of the double bond.

The carbon-hydrogen bonds in ethylene are σ bonds made from an s orbital of hydrogen and an sp^2 hybrid orbital of carbon. The carbon-carbon double bond is made in two parts: the first is a σ bond made from an sp^2 hybrid orbital on each carbon; the second is a π molecular orbital formed by sideways overlap of the unhybridized p orbitals at the two carbon atoms (figure 2-10).

The π molecular orbitals differ from σ molecular orbitals because they are formed by side-to-side overlap of orbitals rather than end-to-end overlap, as σ orbitals are. As a result, the electron density in a π molecular orbital is not concentrated in the region immediately be-tween the two nuclei, but lies farther from the nucleus. The π orbital is composed of two separate parts, or *lobes,* one above and one below the plane of the nuclei. However, notwithstanding its two-part con-struction, this is still only one orbital and it is capable of holding only a single pair of electrons.

To achieve proper overlap of the two p orbitals, all the atoms in ethylene must lie in the same plane. Draw a picture showing that overlap could not be achieved if the two ends of the molecule were rotated 90° out of line. **PROBLEM 2-7**

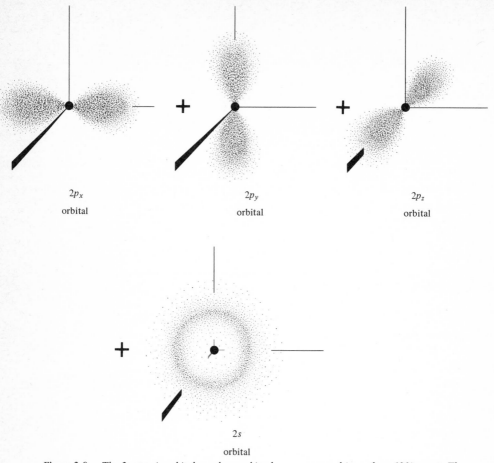

$2p_x$
orbital

$2p_y$
orbital

$2p_z$
orbital

$2s$

orbital

Figure 2-9
sp^2 hybridization.
The 2s atomic orbital can be combined with two 2p atomic orbitals to form three equivalent sp^2 hybrid orbitals which have the same shape and energy. The principal axes of the three hybrid orbitals are arranged in a plane 120° apart. The unhybridized p orbital has its principal axis perpendicular to the plane of the three hybrid orbitals.

CARBON-CARBON
TRIPLE
BONDS

A triple bond is formed by sharing six electrons between two atoms. The simplest compound with a carbon-carbon triple bond is acetylene:

H:C:::C:H   or   H—C≡C—H

acetylene

The hybridization of carbon required for formation of a triple bond is different from that for a single or double bond. In order to form a triple bond, carbon hybridizes by combining the $2s$ orbital with a single $2p$ orbital to form two equivalent sp hybrid orbitals. Of course,

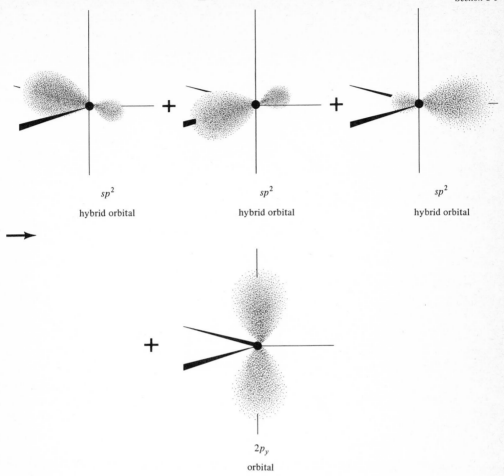

sp^2

hybrid orbital

sp^2

hybrid orbital

sp^2

hybrid orbital

$2p_y$

orbital

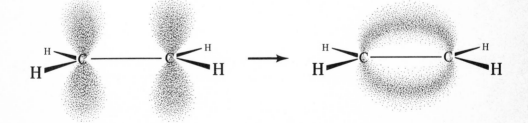

p

atomic orbitals

π

molecular orbital

This bond is formed by sideways overlap of two p orbitals. The π orbital has two parts, one above and one below the plane of the nuclei. The lines represent σ bonds.

Figure 2-10
Formation of the π bond in ethylene.

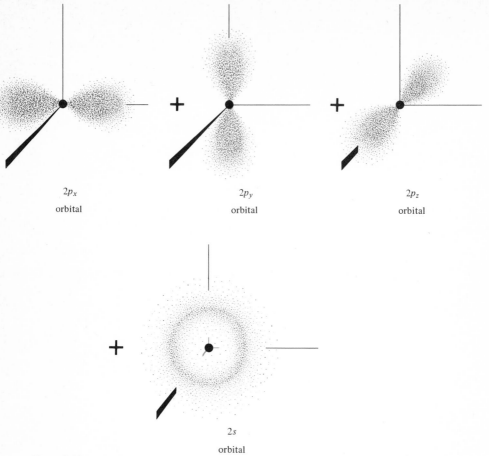

$2p_x$
orbital

$2p_y$
orbital

$2p_z$
orbital

$2s$
orbital

Figure 2-11
***sp* hybridization.** Combination of the 2s atomic orbital with a single 2p atomic orbital forms two equivalent sp hybrid orbitals which have the same shape and energy. The principal axes of the two orbitals point in opposite directions from the nucleus. The two p orbitals which are left unhybridized have their principal axes perpendicular to each other and perpendicular to the axes of the two hybrid orbitals.

there are two *p* orbitals left over. This hybridization is shown in figure 2-11.

The *sp* hybrid orbitals are similar in shape to the hybrid orbitals we have considered previously, but they are arranged differently about the nucleus. These two hybrid orbitals are on opposite sides of the nucleus, 180° apart (figure 2-11). As a result, the two σ bonds which are formed using these orbitals lie along a straight line.

The molecular orbitals in acetylene are formed like those in ethylene. Each carbon-hydrogen bond is formed by the combination of a 1*s* orbital of hydrogen with an *sp* hybrid orbital of carbon. The carbon-carbon triple bond is formed in two parts: a σ molecular orbital is

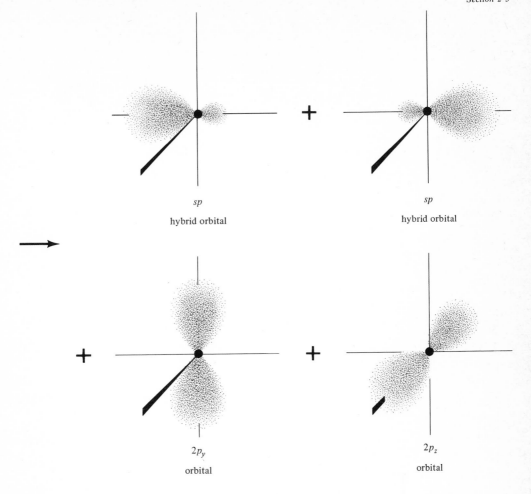

sp

hybrid orbital

sp

hybrid orbital

$2p_y$

orbital

$2p_z$

orbital

formed by the combination of an sp hybrid orbital on each carbon, and *two* π orbitals are formed by the side-to-side overlap of a p orbital at each carbon atom (figure 2-12). The formation of two π orbitals in acetylene 90° apart results in the overall formation of π electron density in a cylinder which is symmetrical about the internuclear axis. Because of the linear geometry of the sp hybrid orbitals, all four atoms in acetylene lie in a straight line.

Carbon is a very versatile element. It can form strong bonds to hydrogen, it can form strong single, double, and triple bonds to carbon, and it can form strong bonds to many other atoms. Such bonds occur in a wide variety of organic molecules.

    Carbon can form single bonds to all the elements listed in table 2-1.

BONDS BETWEEN CARBON AND OTHER ATOMS

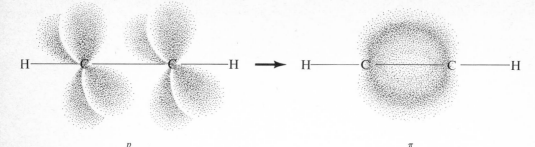

p

atomic orbitals

π

molecular orbitals

Figure 2-12
Formation of the π
bonds in acetylene.
These bonds are
formed by sideways
overlap of p orbitals.
The lines represent σ
bonds.

In each case the bond between carbon and the other atom is a σ bond formed from an sp^3 hybrid orbital of carbon and an orbital of the other atom. The oxygen, chlorine, and nitrogen atoms in the following molecules all have unshared electrons:

$$\begin{array}{ccc}
\text{H} & \text{H} \quad \text{H} & \text{H} \\
| & | \quad | & | \\
\text{H—C—Ö—H} & \text{H—C—N—H} & \text{H—C—Cl:} \\
| & | \quad & | \\
\text{H} & \text{H} & \text{H}
\end{array}$$

methanol   methylamine   chloromethane

Carbon can also form double bonds with atoms other than carbon; for example,

$$\begin{array}{cc}
\text{H}_{\diagdown} & \text{H}_{\diagdown} \quad _{\diagup}\text{H} \\
\quad \text{C=O:} & \quad \text{C=N} \\
\text{H}^{\diagup} & \text{H}^{\diagup}
\end{array}$$

formaldehyde   formaldehyde imine

Such double bonds are invariably made up of a σ bond and a π bond, just as in ethylene.

Carbon can also form triple bonds to nitrogen:

$$\text{H—C}\equiv\text{N:} \qquad \begin{array}{c} \text{H} \\ | \\ \text{H—C—C}\equiv\text{N:} \\ | \\ \text{H} \end{array}$$

hydrogen   acetonitrile
cyanide

ELECTRO-
NEGATIVITY

Ethane, CH_3—CH_3, has a σ bond connecting two carbon atoms. The two carbon atoms share the electrons in this bond equally; neither carbon has a greater affinity for the electrons than the other. However, when carbon forms bonds to other atoms, this equality no longer holds. Some atoms have a greater ability than others to attract the electrons. This ability to attract electrons in a bond is called *electronegativity*. Atoms with a greater ability than carbon to attract electrons are said to be more *electronegative* than carbon.

The electronegativities of atoms vary with their positions in the periodic table. They are greatest for atoms in the upper right corner of the table and least for those in the lower left. Thus, nitrogen, oxygen, fluorine, and chlorine all have a greater electronegativity than carbon.

Because of the electronegativity of oxygen the carbon-oxygen bond in a compound such as methanol, CH_3—OH, is *polar;* that is, the electrons in the σ bond are not shared equally between oxygen and carbon. Oxygen, being more electronegative, attracts the electrons more strongly than carbon does, making the carbon end of the bond electron-poor and the oxygen end electron-rich. This slight electron imbalance has many important consequences in organic chemistry.

In the preceding sections we have surveyed the different modes by which carbon can form bonds with carbon and with other atoms. These possibilities are summarized in table 2-2.

REVIEW OF BONDING TO CARBON

2-4 TYPES OF MOLECULES IN ORGANIC CHEMISTRY

In this chapter we have seen a variety of organic compounds, some containing only carbon and hydrogen, and some containing oxygen, nitrogen, chlorine, or other atoms as well. By using the information in this chapter we can describe the bonding and hybridization of the atoms in many of these compounds.

We have not yet seen all the types of molecules with which this book will be concerned, but the following discussion outlines all the simple types of compounds. The remainder of this book serves to expand this outline.

Carbon Bonding	Hybridization	Geometry	Bond Angle
four single bonds	sp^3	tetrahedral	109.5°
two single bonds and one double bond	sp^2	planar	120°
one single bond and one triple bond	sp	linear	180°

*Table 2-2
Summary of Types of Bonding in Carbon Compounds*

Although the concept of oxidation levels is not usually considered to be as central in the chemistry of carbon as it is in the chemistry of other elements, it remains a useful (though somewhat arbitrary) basis for classifying organic structures.

Oxidation numbers apply only to particular carbon atoms, and not all carbons in a molecule need have the same oxidation number. The oxidation number of a carbon atom is defined as the number of atoms other than carbon and hydrogen bound to that carbon *plus* the number of π bonds to that carbon:

Oxidation number = number of bound atoms other than C and H
+ number of π bonds

The structures of some simple organic compounds of various types are arranged according to oxidation number in table 2-3. The beauty of this table as an aid in organizing organic chemistry is that it enables us to classify nearly all organic reactions into one of two types:

1 Oxidations and reductions, which represent horizontal movements in table 2-3

Table 2-3
Important Functional Classes In Organic Chemistry

Atoms Present	Oxidation Level				
	0	*1*	*2*	*3*	*4*
C, H	H–C–C–H (alkane)	C=C (alkene)	H–C≡C–H (alkyne)		
C, H, Cl		H–C–Cl (alkyl chloride)	H–C–Cl (dichloride)	H–C–Cl (trichloride)	Cl–C–Cl (tetrachloride)
C, H, O		H–C–OH (alcohol)	H–C–C–H (aldehyde)	H–C–C–OH (acid)	O=C=O (carbon dioxide)
C, H, N		H–C–NH₂ (amine)	C=N (imine)	H–C–C≡N (nitrile)	

2 Nonoxidative functional-group interchanges, which represent vertical movements in table 2-3

We will see that this organization frequently serves to remind us of simple relationships between seemingly dissimilar classes of compounds.

Table 2-3 can also be used to explain the organization of the next several chapters in this book. Chapter 3 deals with hydrocarbons at all three oxidation levels. Chapter 4 deals with certain aspects of stereochemistry. Chapters 5 and 6 deal with halogen, oxygen, and nitrogen compounds of oxidation level 1. Chapter 7 deals with oxygen compounds at oxidation level 2. Chapters 9 and 10 deal with all compounds at oxidation levels 3 and 4. Subsequent chapters deal with more complex compounds.

2-8 Give the covalent valence of (*a*) hydrogen; (*b*) carbon; *PROBLEMS* (*c*) oxygen; (*d*) nitrogen; (*e*) chlorine.

2-9 For each of the following molecules, draw a Lewis dot structure, indicate the hybridization of each carbon atom, and identify polar bonds: (*a*) CH_4; (*b*) C_2H_4; (*c*) CH_2Cl_2; (*d*) CH_2O; (*e*) CH_5N.

2-10 Give the electronic configuration of (*a*) F^-; (*b*) Ne; (*c*) K^+.

2-11 Draw a picture of each of the following: (*a*) an *s* orbital; (*b*) a *p* orbital; (*c*) a σ bond; (*d*) a π bond.

2-12 Use the rules of covalent valence to construct a compound containing (*a*) hydrogen and sulfur; (*b*) hydrogen, nitrogen, and oxygen; (*c*) hydrogen, carbon, and sulfur.

2-13 Propene has the molecular structure CH_3—CH=CH_2. (*a*) Draw a picture of the molecular orbitals in propene and label each orbital as σ or π; (*b*) indicate the hybridization of each carbon atom; (*c*) indicate what all bond angles are.

2-14 Do the same for allene, CH_2=C=CH_2. Is the molecule planar? Why or why not?

2-15 The covalent valence of carbon is 4. That being so, why doesn't C≡C exist?

2-16 $NaOCH_3$ has both ionic and covalent bonds. (*a*) Draw a Lewis structure of it and indicate clearly which bonds are ionic and which are covalent. (*b*) Do the same for CH_3NH_3Cl.

2-17 Assign formal charges in each of the following structures: (*a*) $(CH_3)_3NO$, total charge 0; (*b*) CH_3O, total charge -1; (*c*) CH_3CO_2, total charge -1; (*d*) $(CH_3)_4N$, total charge $+1$. *Caution:* Do not attempt to do this problem without first drawing correct Lewis structures.

2-18 The nitrogen atom in ammonia, NH_3, might have any of three possible hybridizations. What are they? For each case, what would the geometry of ammonia be?

2-19   Diimide, H—N=N—H, might have either of two geometries, depending on the hybridization of the nitrogen atoms. Describe these two possible geometries and hybridizations.

2-20   Methylene, CH_2 (no charge), is not a stable molecule, but it is observed as an unstable intermediate in a number of chemical reactions. Although the lifetime of a methylene molecule is considerably less than a second, it has been possible by indirect methods to deduce that methylene exists in at least two different configurations. Draw two possible configurations, indicating the hybridization of carbon and the kind of orbital holding each electron.

SUGGESTED READINGS   Most general chemistry textbooks contain a complete description of atomic structure and chemical bonding.

Hertz, W.: *The Shape of Carbon Compounds,* Benjamin, Menlo Park, Calif., 1964. Good introduction to structure and bonding in organic compounds.

3

Hydrocarbons

Carbon-hydrogen and carbon-carbon bonds are present in almost every molecule we will consider. In this chapter we survey the structures and properties of *hydrocarbons,* compounds containing only carbon and hydrogen. Our survey has three parts: classes of hydrocarbons, isomerism, and reactions of hydrocarbons.

ISOMERS   Much of this chapter will be concerned with *isomers;* that is, compounds which have the same empirical formula but different structures. Several types of isomerism are possible; some are obvious and easy to understand, but some are very subtle.

One of the most important steps at the beginning of any course on organic chemistry is learning to perceive structures of molecules correctly. There is no substitute for the use of molecular models to achieve this. Be sure to work all the problems that require molecular models as you read the chapter.

3-1 ALKANES

Alkanes are hydrocarbons which contain no double or triple bonds. As a result, alkanes contain the greatest number of hydrogen atoms compared with the number of carbon atoms of any type of hydrocarbon, and they are called *saturated.*

Saturated hydrocarbons which contain no rings (such compounds are called *acyclic*) have the empirical formula C_nH_{2n+2}, where n can be any number. The first two members of the family ($n = 1$ and $n = 2$), called *methane* and *ethane,* have the structures

methane   ethane

There is only one compound C_3H_8. Because of the tetrahedral geometry of the saturated carbon atom, the two representations

propane   propane

are the same molecule.

36

Build four models of methane. Set them in a line on a table in front **PROBLEM 3-1**
of you, with each molecule in the orientation shown in figure 3-2.
In the first model, remove the left front hydrogen; in the second,
remove the right front hydrogen; in the third, the top hydrogen; and
in the fourth, the rear hydrogen. Show that the four seemingly different molecules you have made are indistinguishable. Do the same for
the hydrogens of ethane. Is a similar demonstration possible with
propane?

Use molecular models to show that there is only one compound **PROBLEM 3-2**
CH_2Cl_2. Draw pictures of it from two different angles.

There are two compounds with empirical formula C_4H_{10}: **STRUCTURAL**
ISOMERISM

normal butane   isobutane
(*n*-butane)

These two compounds are *structural isomers:* separate and distinct
compounds which have the same empirical formula but which have
different atoms connected to different atoms.

There are three structural isomers having the empirical formula
C_5H_{12}:

n-pentane   isopentane   neopentane

Although other structures can be written which appear at first to be
different, every such structure is actually the same as one of these
three.

The number of structural isomers rises rapidly as the number of carbon atoms increases. There are 5 isomers of C_6H_{14}, 9 of C_7H_{16}, 18 of C_8H_{18}, 75 of $C_{10}H_{22}$, and 366,319 of $C_{20}H_{42}$.

PROBLEM 3-3   Draw two compounds with the empirical formula C_5H_{12} which look different from those shown above and use molecular models to show that they are not really different.

PROBLEM 3-4   Draw all the structural isomers of C_6H_{14}. Use molecular models to help confirm that all your structures are different and that you have not missed any isomers.

CONDENSED STRUCTURAL FORMULAS   Structural formulas like those above, in which every covalent bond is represented by a line, are cumbersome for any but the simplest compounds. It is usual to condense these structural formulas by writing the hydrogen atoms next to the carbon atom to which they are attached. Thus, the three structural isomers of pentane become:

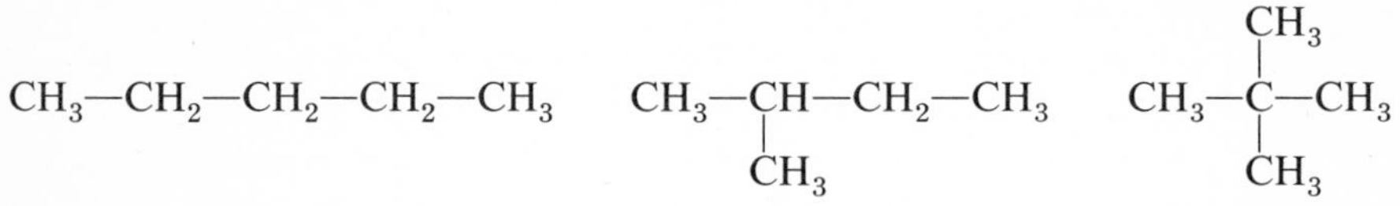

$$CH_3-CH_2-CH_2-CH_2-CH_3 \qquad CH_3-\underset{\underset{CH_3}{|}}{CH}-CH_2-CH_3 \qquad CH_3-\underset{\underset{CH_3}{|}}{\overset{\overset{CH_3}{|}}{C}}-CH_3$$

$$\textit{n}\text{-pentane} \qquad\qquad\qquad \text{isopentane} \qquad\qquad\qquad \text{neopentane}$$

HOW MANY ISOMERS?   Recognizing whether or not two structures are identical is an important and difficult problem. One sure way to decide is to build molecular models of both structures and compare them, but this approach is frequently impractical. When a decision must be made from written structures, the best procedure is to start by finding the longest chain of carbon atoms in the molecule and then comparing the positions of atoms or groups of atoms attached to this chain. In fact, this is the basis of the most important method for naming organic compounds, as we shall now see.

NOMENCLATURE   The structural theory of organic chemistry has now been in existence for slightly over a hundred years. In that time, the naming of organic compounds has progressed from a haphazard state to a very orderly one.

   The best and most systematic method for naming organic compounds is the method adopted by the International Union of Pure and Applied Chemistry (IUPAC). For the most part, we will use the IUPAC system. However, many common organic compounds are known by *trivial names,* which have no relation to the IUPAC system but which have been widely used for a long time. Many compounds isolated from nature are called by their trivial names because these are much simpler than the IUPAC names. In this book we use both IUPAC names and trivial names.

The IUPAC names of the straight-chain (sometimes called *normal*) saturated hydrocarbons are given in table 3-1. These names form the basis of the IUPAC system of nomenclature for all types of compounds.

Branched-chain saturated hydrocarbons are named in the following way:

1 Find the longest continuous chain of carbon atoms in the molecule. This is the *main chain,* and the final part (called the *root*) of the name of the compound is the name of the straight-chain hydrocarbon containing that number of carbon atoms.

2 Number the carbon atoms in the main chain sequentially starting from the end that will give the carbon groups attached to the chain the smaller numbers.

3 Carbons which are not part of the main chain are identified as *substituents* on the main chain. The names of substituent groups (arranged in alphabetical order) and the numbers identifying their positions on the chain become prefixes on the name of the main chain. The names of some important substituent groups are given in table 3-2.

4 The presence of two identical groups is indicated by the prefix *di-* in front of the name of the group. Three identical groups are indicated by the prefix *tri-*, four by *tetra-*, etc.

Some examples of the names of saturated hydrocarbons are given in figure 3-1.

Empirical Formula	Structure	Name
CH_4	CH_4	methane
C_2H_6	CH_3CH_3	ethane
C_3H_8	$CH_3CH_2CH_3$	propane
C_4H_{10}	$CH_3CH_2CH_2CH_3$	butane
C_5H_{12}	$CH_3CH_2CH_2CH_2CH_3$	pentane
C_6H_{14}	$CH_3CH_2CH_2CH_2CH_2CH_3$	hexane
C_7H_{16}	$CH_3CH_2CH_2CH_2CH_2CH_2CH_3$	heptane
C_8H_{18}	$CH_3CH_2CH_2CH_2CH_2CH_2CH_2CH_3$	octane
C_9H_{20}	$CH_3CH_2CH_2CH_2CH_2CH_2CH_2CH_2CH_3$	nonane
$C_{10}H_{22}$	$CH_3CH_2CH_2CH_2CH_2CH_2CH_2CH_2CH_2CH_3$	decane

Table 3-1
The Names of the Straight-Chain Saturated Hydrocarbons

Table 3-2
Some Important Alkyl Groups

Structure	Name	Comment
CH_3—	methyl	
CH_3CH_2—	ethyl	
$CH_3CH_2CH_2$—	*n*-propyl	*n*- (or *normal*) means straight chain
CH_3CH— $\quad\vert$ $\quad CH_3$	isopropyl	*iso*- means that the chain is branched one carbon from the end
$CH_3CH_2CH_2CH_2$—	*n*-butyl	
CH_3CH_2CH— $\qquad\vert$ $\qquad CH_3$	*s*-butyl	*s*- (or *secondary*) means that the carbon by which the group is attached to the main chain has two other carbons attached to it
CH_3CHCH_2— $\quad\vert$ $\quad CH_3$	isobutyl	
$\qquad CH_3$ $\qquad\vert$ CH_3C— $\qquad\vert$ $\qquad CH_3$	*t*-butyl	*t*- (or *tertiary*) means that the carbon by which the group is attached to the main chain has three other carbons attached to it
$\qquad CH_3$ $\qquad\vert$ CH_3CCH_2— $\qquad\vert$ $\qquad CH_3$	neopentyl	*neo*- means that the center carbon of the group is attached to four other carbons
 (benzene ring)—	phenyl	a benzene ring (section 3-7) attached by one carbon

PROBLEM 3-5 Name each of the following compounds:

$$\text{(a)} \quad CH_3CH_2\underset{\underset{CH_3}{\vert}}{C}HCH_2CH_3$$

$$\text{(b)} \quad CH_3\underset{\underset{CH_2CH_3}{\vert}}{C}H-\underset{\underset{CH_2CH_3}{\vert}}{C}CH_3$$

with CH_3CH_2 above and CH_2CH_3 groups

$$\text{(c)} \quad CH_3CH_2\underset{\underset{CH_2CH_3}{\vert}}{C}HCH_2CH_3$$

3-2 WHAT MOLECULES LOOK LIKE

The letter and line formulas used in the previous section to represent molecules are good because they are convenient. At the same time,

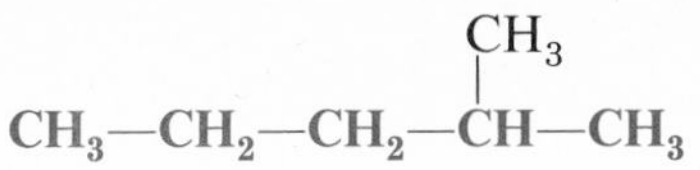

$$CH_3-CH_2-CH_2-\overset{\overset{\displaystyle CH_3}{|}}{CH}-CH_3$$

2-methylpentane

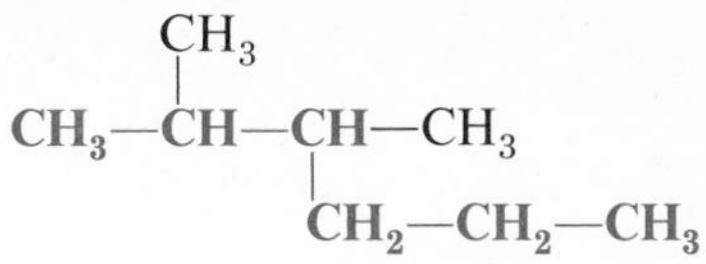

$$CH_3-\overset{\overset{\displaystyle CH_3}{|}}{CH}-\overset{\underset{\displaystyle CH_2-CH_2-CH_3}{|}}{CH}-CH_3$$

2,3-dimethylhexane

$$CH_3-\overset{\overset{\displaystyle CH_3}{|}}{CH}-CH-CH_2-CH_2-CH_3$$
$$CH_3-CH_2-\overset{|}{CH}-CH_3$$

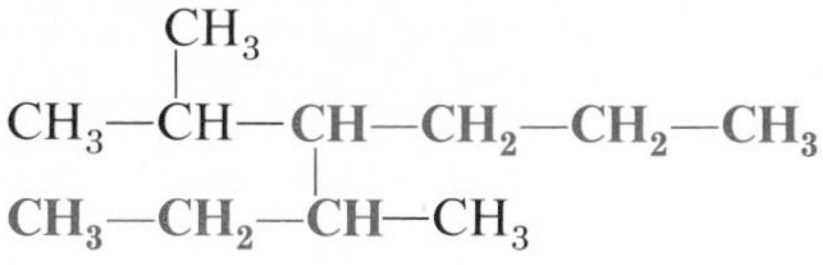

4-isopropyl-3-methyl-
heptane

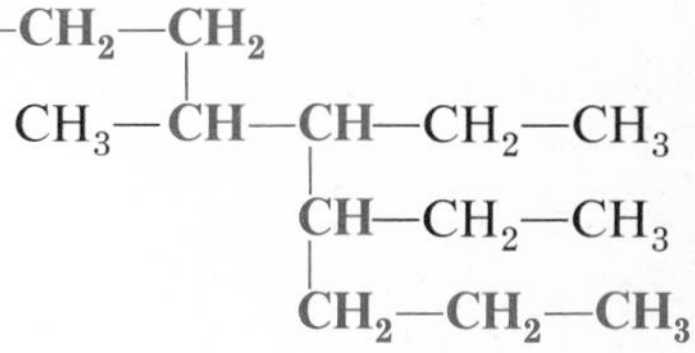

$$CH_3-\overset{\overset{\displaystyle CH_3-CH_2-CH_2}{|}}{CH}-CH-CH_2-CH_3$$
$$\overset{|}{CH}-CH_2-CH_3$$
$$\overset{|}{CH_2}-CH_2-CH_3$$

4,5-diethyl-6-methylnonane

Figure 3-1
Names of some
saturated
hydrocarbons.

they are bad because molecules really do not look like that; molecules are three-dimensional and occupy space.

When an accurate picture of a molecule is required, molecular models must be used. Such models are of two types, *framework* molecular models and *space-filling* models, each having its own advantages and disadvantages. Framework molecular models show clearly the spatial relationships between nuclei and the positions of bonds. However, they do not adequately show the electron density in the molecule; this is better done with space-filling models. Framework and space-filling models of methane and ethane are shown in figure 3-2.

The spatial arrangement of atoms around the tetrahedral carbon atoms of molecules like methane and ethane is best understood by studying molecular models. It is clear from such study that all the hydrogen atoms in methane are indistinguishable; only one compound can be obtained by replacing a hydrogen atom of methane with some other atom. A similar situation holds with ethane.

THE TETRAHEDRAL CARBON

Conformational isomers are structures which are different but which can be made identical by the rotation of one or more σ bonds. Such isomerism is not possible for methane, but it is possible for ethane and almost all other organic compounds. As a result, conformational isomerism is an important phenomenon in organic chemistry.

The two carbon atoms of ethane are connected by a σ bond. Since this bond is roughly egg-shaped, the two ends of the ethane molecule can rotate with respect to each other without any change in the

CONFORMATIONAL ISOMERISM IN ETHANE

42

METHANE

H

H —— C —— H

H

structure

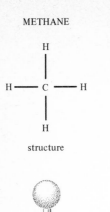

framework molecular model

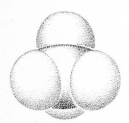

space–filling model

ETHANE

H   H

H —— C ————— C —— H

H   H

structure

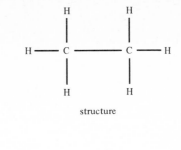

framework molecular model

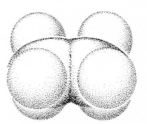

space–filling model

Figure 3-2
Structures and
models of methane
and ethane.

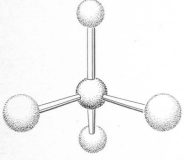

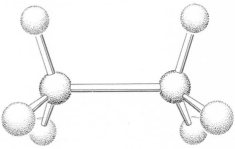

carbon-carbon bond.* Thus, an ethane molecule can change itself by rotation of the carbon-carbon single bond. The different-looking structures which can be produced by this rotation are called *conformational isomers* and are shown in figure 3-3. Although a large number of conformations for the ethane molecule are possible, we will consider only the two most important conformations, the *eclipsed* conformation, in which the two ends of the molecule are precisely aligned, and the

---

*We will shortly see that a single bond is very different from a double bond in this respect. Because of the presence of the π bond, a double bond is not free to rotate in this way.

ECLIPSED CONFORMATION

STAGGERED CONFORMATION

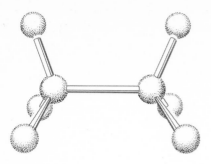

side view

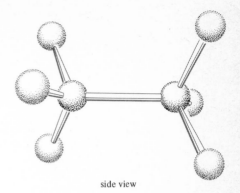

side view

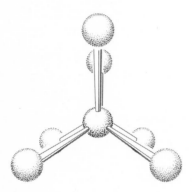

end view

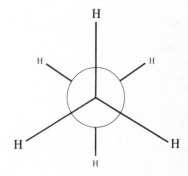

end view

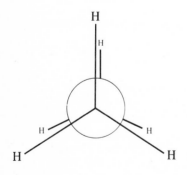

Newman projection

Newman projection

*Rotation of the carbon-carbon σ bond produces two slightly different structures called **conformational isomers**. Newman*

projections are often used to represent conformational isomers.

**Figure 3-3
Conformations of ethane.**

staggered conformation, in which the two ends are rotated as far as possible away from each other.

The rotation of carbon-carbon single bonds occurs rapidly (many thousands of times each second), and it is not possible to separate molecules of ethane in the eclipsed conformation from molecules in the staggered conformation. That is, eclipsed ethane and staggered ethane are not different compounds but rapidly interconverting forms of the same compound. This fact makes conformational isomers different from other kinds of isomers which we will study; only conformational isomers are incapable of being separated from each other.

The energies of the eclipsed and staggered conformations of ethane differ slightly. The eclipsed form is the less stable of the two by about 3 kilocalories/mole, approximately one-thirtieth of the energy of a typical carbon-hydrogen bond. Though modest, this energy difference is sufficient to ensure that more than 99 percent of any collection of ethane molecules will exist in the staggered conformation at any instant.

NEWMAN PROJECTIONS

The best way to achieve accurate representation of conformational isomers of compounds like ethane is to use Newman projections. A Newman projection is a schematic view of a compound looking down one carbon-carbon bond, as shown in figure 3-3. The front carbon is represented as the intersection point of the three front carbon-hydrogen bonds. The back carbon is represented as a circle from which the three back carbon-hydrogen bonds emanate. It is possible to use Newman projections to show conformations in molecules more complex than ethane, but a Newman projection can usually show the conformation of only one bond at a time.

CONFORMATIONAL ISOMERISM IN LARGER MOLECULES

Propane has two carbon-carbon single bonds, each of which can rotate and give rise to conformational isomerism. Butane has three carbon-carbon bonds and many possibilities for conformational isomerism.

Because of the tetrahedral geometry of the carbon atoms in butane, the carbon backbone of this molecule is not straight but bent, as shown in figure 3-4. The conformation about the center carbon-carbon bond of butane is particularly interesting because it illustrates many points about conformational isomerism which will be important in future discussions. Four principal conformations of butane are shown in figure 3-4. Although the two terminal carbons seem very far apart when butane is drawn on paper, in certain conformations they are close enough for their electron clouds to get in each other's way. The most favorable conformation is the one in which these two carbons are a maximum distance apart.

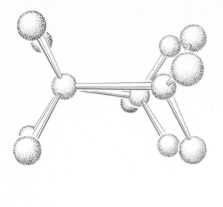

eclipsed: high energy

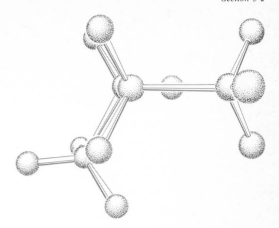

eclipsed : high energy

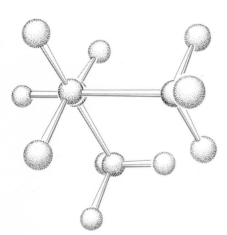

staggered : low energy

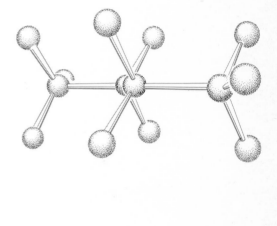

staggered: low energy

Only conformations about the center carbon-carbon bond are shown. The first conformation, in which the two terminal methyl groups are eclipsed with each other, and the second conformation, in which methyl groups are eclipsed with hydrogens, are of very high energy. The third and fourth conformations, which are entirely staggered, are of much lower energy than the first two. The fourth conformation is of slightly lower energy than the third. A complete description of the conformation of butane would also include a discussion of conformations of the two terminal carbon-carbon bonds.

**Figure 3-4
Conformations of
butane.**

PROBLEM 3-6   Build a model of hexane and use the information in figure 3-4 to place it in its most stable conformation.

PROBLEM 3-7   Draw four Newman projections of Cl—CH_2—CH_2—Cl corresponding to those of figure 3-4 and arrange them in order of increasing energy.

3-3   ALKENES

An *alkene* is a compound with a carbon-carbon double bond. The carbon-carbon double bond is composed of a σ bond and a π bond connecting two sp^2 hybridized carbon atoms (section 2-3). The rigidity of the π bond in alkenes makes a new type of isomerism possible which is not possible for acyclic alkanes.

Alkenes, also called *olefins*, are the first compounds we consider which are *unsaturated;* that is, they contain one or more π bonds. As a result, they can react with hydrogen in the presence of an appropriate catalyst to form compounds which are saturated (section 3-9).

GEOMETRICAL ISOMERISM   The sp^2 hybrid orbitals are arranged in a plane at 120° angles. The p orbital used to make the π molecular orbital is above and below this plane. As a result, the two carbons in a carbon-carbon double bond and all four atoms attached to those two carbons lie in the same plane, as shown in figure 3-5.

The π molecular orbital consists of two parts, called *lobes,* one above and one below the plane of the double bond. These two lobes result from overlap of the p orbitals on either side of the plane of the double bond. If the two ends of the double bond were rotated with respect to each other, this overlap would be destroyed. As a result, rotation does not occur; double bonds are completely rigid. This rigidity leads to a new kind of isomerism: *geometrical isomerism,* occasionally called *cis-trans isomerism*. There are two compounds called 2-butene:

trans-2-butene   *cis*-2-butene

These compounds are, respectively, trans (meaning across) and cis (meaning same side) isomers. These two isomers are separate compounds. They have different physical properties, and they have no tendency to undergo interconversion; a bottle of *cis*-2-butene will remain forever the cis isomer and will never turn into the trans isomer unless a catalyst is added.

Geometrical isomers are a type of *stereoisomers:* separable compounds which have the same atoms connected to the same atoms, but

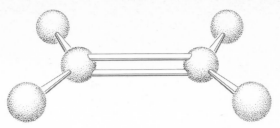

Because of the geometry of the two sp^2 carbon atoms and because of the existence of the π bond, this molecule is planar. The π bond prevents the rotation of the carbon-carbon double bond.

Figure 3-5
Geometry of ethylene.

with one or more of those atoms differently arranged in space. Other stereoisomers include cis-trans isomers of cyclic compounds (section 3-4) and optical isomers (chapter 4).

Not all olefins show geometrical isomerism, even though the double bond is always rigid. In order for an olefin to show geometrical isomerism, the two substituents on each carbon atom must be different. Examples of olefins which do not show geometrical isomerism include

$$CH_3CH_2 \quad H$$
$$C=C$$
$$H \quad\quad H$$
1-butene

$$CH_3 \quad H$$
$$C=C$$
$$CH_3 \quad H$$
2-methylpropene

$$CH_3CH_2 \quad CH_3$$
$$C=C$$
$$H \quad\quad CH_3$$
2-methyl-2-pentene

Build a model of propene and use it to help explain why the different conformations produced by rotation of the carbon-carbon single bond have similar energies, unlike the conformations of ethane.

PROBLEM 3-8

The IUPAC names of alkenes are based on the IUPAC names of alkanes. Alkene names differ from alkane names because it is necessary to indicate the position and geometry of the double bond. Simple alkenes are named by the following rules:

NOMENCLATURE OF ALKENES

1 Find the longest continuous chain of carbon atoms containing the double bond. This is the *main chain,* and the root of the name of the compound is the name of the straight-chain hydrocarbon having that number of carbon atoms (table 3-1) with the alkane ending *-ane* replaced by the alkene ending *-ene.*

2 Number the carbon atoms in the main chain sequentially, starting at the end nearer the double bond. Use the number of the lower-numbered carbon atom of the double bond to indicate the position

of that bond. Place this number immediately before the name of the main chain.

3 Locate other substituents on the carbon chain, and assign names and numbers to them exactly as for alkanes. Substituent names are arranged in alphabetical order.

4 A compound with two double bonds is called a *diene,* and two numbers are used to indicate the positions of the double bonds. A compound with three double bonds is a *triene,* etc.

5 If the compound is capable of having cis and trans isomers, the prefix *cis-* or *trans-* is placed before the name, according to whether the main chain crosses the double bond cis or trans.

The use of this nomenclature system is illustrated in figure 3-6.

PROBLEM 3-9 Name the following compounds:

(a)
$$CH_3CH_2 \diagdown_{} \diagup H$$
$$C{=}C$$
$$CH_3CH_2 \diagup \diagdown CH_3$$

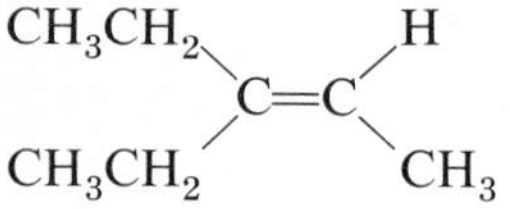

(b)
$$H \diagdown_{} \diagup CH_2CH_3$$
$$C{=}C$$
$$CH_3 \diagup \diagdown CH_3$$

(c)
$$CH_3CH_2CH_2 \diagdown_{} \diagup H$$
$$C{=}C$$
$$CH_3CH \diagup \diagdown CH_3$$
$$|$$
$$CH_3$$

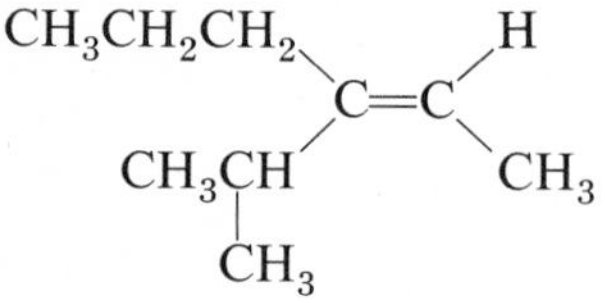

CIS-TRANS ISOMERIZATION; VISION Nature recognizes molecules by their shapes much as we recognize articles of clothing by their shapes. Just as we are able to choose a properly fitting coat for a cold day, nature is able to choose the proper

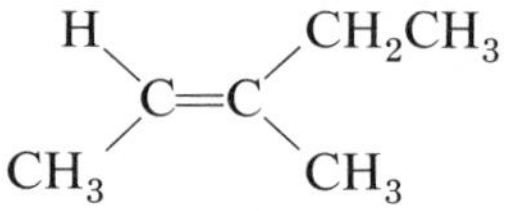

Figure 3-6
The names of some alkenes.

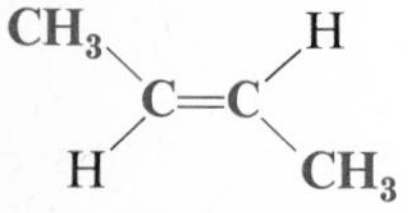

$$CH_3 \diagdown_{} \diagup H$$
$$C{=}C$$
$$H \diagup \diagdown CH_3$$
trans-2-butene

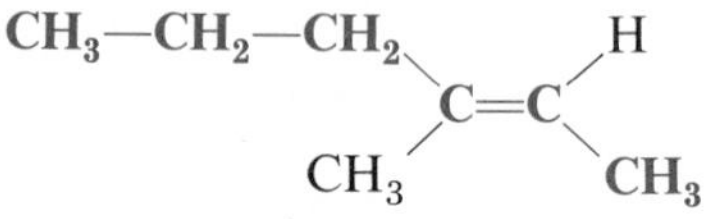

$$CH_3{-}CH_2{-}CH_2 \diagdown_{} \diagup H$$
$$C{=}C$$
$$CH_3 \diagup \diagdown CH_3$$
trans-3-methyl-2-hexene

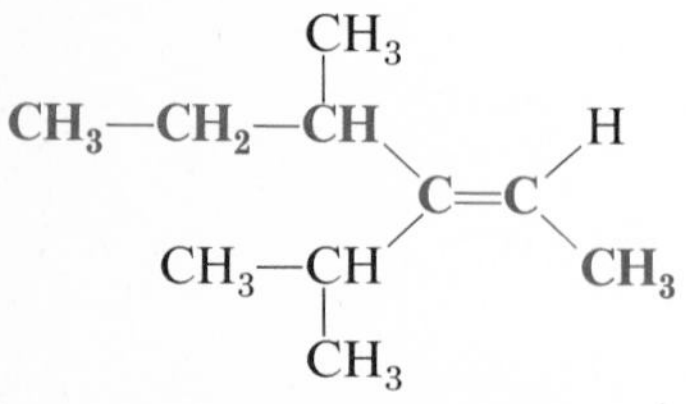

$$CH_3$$
$$|$$
$$CH_3{-}CH_2{-}CH \diagdown_{} \diagup H$$
$$C{=}C$$
$$CH_3{-}CH \diagup \diagdown CH_3$$
$$|$$
$$CH_3$$
trans-3-isopropyl-4-methyl-2-hexene

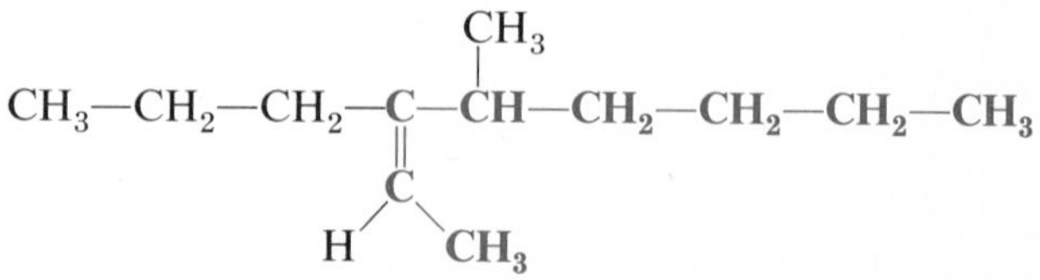

$$CH_3$$
$$|$$
$$CH_3{-}CH_2{-}CH_2{-}C{-}CH{-}CH_2{-}CH_2{-}CH_2{-}CH_3$$
$$\|$$
$$C$$
$$H \diagup \diagdown CH_3$$
cis-4-methyl-3-*n*-propyl-2-octene

substance for each of the multitude of tasks necessary for survival. For example, this ability to recognize shapes is the key to the operation of the eye.

The human eye contains two types of light-sensitive cells; the *rods* are sensitive to small amounts of light but have no color discrimination, and the *cones* are less sensitive but can distinguish colors. The rods contain a purple substance called *rhodopsin,* which is made up of a large molecule (actually a protein, chapter 12) called *opsin* attached to a small molecule called *retinal.* Retinal has five carbon-carbon double bonds. When retinal is attached to opsin, all these double bonds are in the trans configuration except one. The shape of this isomer is exactly complementary to the shape of opsin, and the two stick very tightly together (figure 3-7).

When light strikes rhodopsin, the energy of the light causes the cis double bond of retinal to become a trans double bond. This new geometrical isomer of retinal has a different shape from its predecessor and can no longer stick to the opsin. The retinal-opsin complex falls apart, causing the message that light has been detected to be sent to the brain. At this point a series of chemical reactions takes place leading ultimately back to the cis isomer, which promptly binds to opsin again and forms the light-sensitive pigment rhodopsin.

Each rod cell in the eye contains many molecules of rhodopsin, and under most circumstances only part of the retinal and opsin molecules in the rod cell are in the form of the light-sensitive rhodopsin. Dark adaptation involves conversion of nearly all the molecules into the rhodopsin form. Retinal is produced in the body from vitamin A; a deficiency of this vitamin causes night blindness.

3-4 CYCLIC HYDROCARBONS

Many organic compounds contain three or more atoms arranged in a ring. They are called *cyclic compounds* to distinguish them from *acyclic compounds,* which do not contain rings. In the following sections we will discuss some of the more important cyclic compounds containing only carbon atoms in the ring. In later chapters we will consider cyclic compounds containing other atoms in the ring.

Because of the tetrahedral geometry of the saturated carbon atom, cyclopentane is a rather rigid molecule, and all its carbon atoms are constrained to lie nearly in the same plane. The structure of cyclopentane is shown in figure 3-8. The representation of cyclopentane as an unadorned pentagon is quick and easy, but not very informative. The Haworth projection of cyclopentane is much more useful for showing structure and geometry.

CYCLOPENTANE

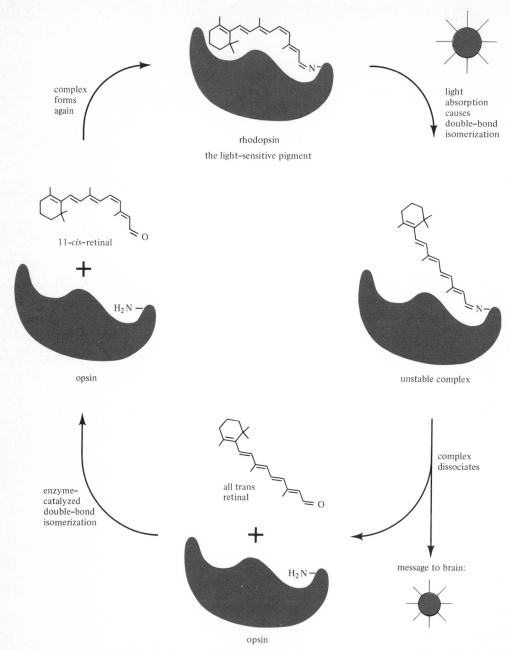

complex
forms
again

rhodopsin
the light-sensitive pigment

light
absorption
causes
double-bond
isomerization

11-*cis*-retinal

+

opsin

unstable complex

enzyme-
catalyzed
double-bond
isomerization

all trans
retinal

+

opsin

complex
dissociates

message to brain:

Figure 3-7
The chemistry of
vision.

The light-sensitive pigment in the rods is rhodopsin, a tight complex of opsin and the cis isomer of retinal. When light is absorbed, the cis isomer becomes a trans isomer, thereby changing the shape of retinal and causing it to dissociate from

opsin. Dissociation causes a message to be sent to the brain ("I see light!"), after which a sequence of chemical reactions changes the trans double bond back to cis and rhodopsin is again formed.

TOP VIEW   SIDE VIEW

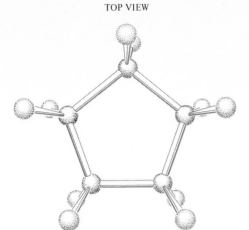

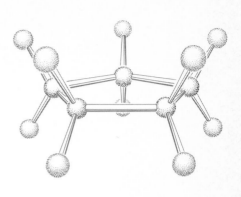

molecular model   molecular model

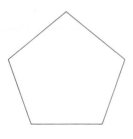

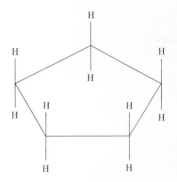

schematic drawing   Haworth projection

Molecular models show that cyclopentane   *can be used to show stereochemistry. The*   **Figure 3-8**
has its carbon atoms all very near the   *lower part of the ring is assumed to be*   **Cyclopentane.**
same plane, with hydrogens above and   *toward the viewer, and the upper part of*
below that plane. The Haworth projection   *the ring is away from the viewer.*

    All the carbon atoms of cyclopentane are indistinguishable; each
is bonded to one hydrogen above the plane of the ring and one below
the plane of the ring. All ten hydrogens of cyclopentane are indistinguishable, and the replacement of any one of those hydrogens by a
—CH_3 produces methylcyclopentane. There is only one methylcyclopentane

methylcyclopentane

PROBLEM 3-10 Draw four different Haworth projections for methylcyclopentane with the —CH₃ group appearing at different places on the ring. Demonstrate by molecular models that all these pictures represent the same compound.

GEOMETRICAL ISOMERISM There are two different compounds called 1,2-dichlorocyclopentane (figure 3-9), which differ according to whether the two chlorines are on the same side (cis) or on opposite sides (trans) of the plane of the carbon atoms.* These isomers, like the cis and trans isomers of olefins, are *geometrical isomers,* exist separately, and have distinct properties.

PROBLEM 3-11 Use molecular models to determine how many isomers there are of 1,3-dichlorocyclopentane. Draw Haworth projections of these compounds.

CYCLIC OLEFINS Cyclic compounds are able to accommodate double bonds easily. Cyclopentene, like cyclopentane, has all of its carbon atoms nearly in the same plane. The six hydrogen atoms on saturated carbons are above and below the plane of the ring, as in cyclopentane, but the two hydrogens on the unsaturated carbons are in the plane of the ring (figure 3-10).

PROBLEM 3-12 Cyclopentene can exist only if the carbon-carbon double bond is cis. Use molecular models to estimate the number of carbons needed in a cyclic unsaturated compound in order for the double bond to be trans.

CYCLOHEXANE The bond angles at an sp^3-hybridized carbon atom are 109.5° whenever possible. In acyclic compounds, this optimum angle can usually be attained without difficulty, but in cyclic compounds the additional constraint imposed by the ring may make this angle difficult to achieve.

Cyclopentane can achieve 109° bond angles by being nearly planar. Cyclohexane, however, would have 120° bond angles if it were planar; as a result, cyclohexane exists in a nonplanar configuration.

Unlike cyclopentane, cyclohexane shows conformational isomerism. The two most important conformations, the *boat* and *chair* confor-

---

*There is actually a third isomer as well; it is the mirror image of the trans isomer shown here. It is also a trans isomer and is an *optical isomer,* or *enantiomer,* of the structure shown here. This type of isomer will be considered more fully in chapter 4.

CIS ISOMER

TRANS ISOMER

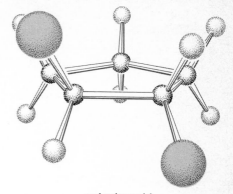

molecular model

molecular model

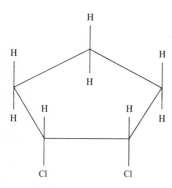

Haworth projection

*There are two geometrical isomers of
1,2-dichlorocyclopentane, the cis isomer,
in which the two chlorines are on the*

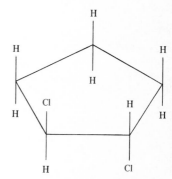

Haworth projection

*same side, and the trans isomer, in which
they are on opposite sides.*

**Figure 3-9
1,2-Dichloro-
cyclopentane.**

**Figure 3-10
Cyclopentene.** *The
two hydrogens on
the double bond are
in the plane of the
ring.*

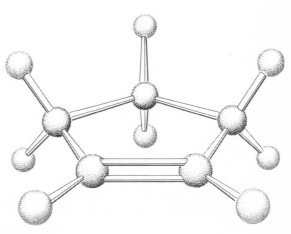

mations, are shown in figure 3-11. The boat form of cyclohexane has two kinds of unfavorable interactions which make it *less stable* than the chair form. This difference in stability is about 6 kilocalories/mole (about 7 percent of the energy of a carbon-carbon single bond), but this energy difference is sufficient for cyclohexane to exist more than 99.99 percent in the chair form at room temperature.

Two factors make the boat form of cyclohexane of higher energy than the chair form: (1) one pair of hydrogens in the boat form are quite close together and form a "flagpole" or "bow-stern" interaction, thereby trying to push each other apart; (2) many hydrogens on adjacent carbons are in conformations corresponding to the unfavorable eclipsed conformation of ethane.

Figure 3-11
The boat and chair forms of cyclohexane. The chair form is the more stable of the two because it lacks eclipsing interactions and flagpole interactions.

BOAT FORM

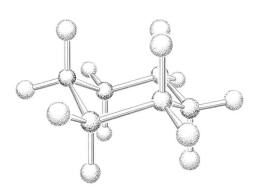

CHAIR FORM

side view

side view

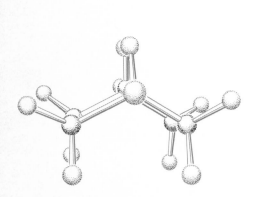

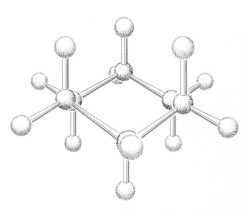

end view

end view

The chair form of cyclohexane has neither of these unfavorable interactions. The bow-stern interaction is absent because of the way the ring is arranged. The eclipsing interactions are absent because the conformations around all carbon-carbon bonds in the chair form are staggered.

Build a molecular model of cyclohexane in the boat conformation. Identify the flagpole interactions. Look down the molecule from bow to stern and draw a picture of the molecule which shows the eclipsing interactions. Build a model of the chair form and draw a corresponding picture showing that the eclipsing interactions are absent.

PROBLEM 3-13

The hydrogen atoms in the chair form of cyclohexane can be divided into two groups, *axial* hydrogens and *equatorial* hydrogens (figure 3-12). Axial hydrogens lie distinctly above and below the ring, three hydrogens on alternate carbons above the ring and three on the other carbons below the ring. The axes of all the *axial* carbon-hydrogen bonds are parallel. The equatorial hydrogens lie neither above nor below the carbon ring but are outside its periphery.

The chair form of a cyclohexane ring can exist in two indistinguishable conformations, the hydrogens which are axial in one conformation becoming equatorial in the other. The interchange of conformations occurs by twisting one end of the cyclohexane ring downward and the other end upward. By that flip, the two conformations are interconverted. The interconversion of the two chair forms of cyclohexane occurs quite readily—about 10^6 times per second at room temperature.

CHAIR-CHAIR INTERCONVERSIONS

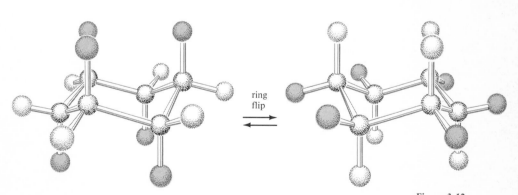

ring flip

There are equal numbers of axial and equatorial hydrogens. The axial hydrogens, shown in pink in the left structure, become equatorial when the ring is flipped to the other chair form.

Figure 3-12 Interconversion of hydrogen types in cyclohexane.

Chlorocyclohexane can exist in two different chair conformations, one with an axial chlorine and one with an equatorial chlorine (figure 3-13). Like cyclohexane itself, chlorocyclohexane can flip from one chair form to the other very rapidly; this flip interconverts axial and equatorial chlorine atoms. The two chair conformations of chlorocyclohexane, unlike those of cyclohexane, are not of equal energy. When the chlorine is axial, it is quite close to the two axial hydrogens on the same side of the ring. This *1,3 diaxial interaction* makes the axial conformation slightly higher in energy than the equatorial conformation, and at room temperature chlorocyclohexane exists about 30 percent in the axial conformation and 70 percent in the equatorial conformation. This energy difference between axial and equatorial conformations is greater when groups larger than a chlorine atom are substituted on the cyclohexane ring.

PROBLEM 3-14 With the help of molecular models, draw two different chair conformations for each of the two geometrical isomers of 1,2-dimethylcyclohexane.

PROBLEM 3-15 Build a model of cyclohexene and consider what kind of conformation this molecule might adopt.

---

*OTHER CYCLIC
COMPOUNDS* Compounds containing five- and six-membered rings are common in organic chemistry, but a wide variety of other ring sizes and types also occur. A number of examples are given in figure 3-14.

Cyclic compounds with fewer than five atoms in the ring cannot achieve the desired 109.5° bond angles at carbon. Such compounds are of higher energy than compounds which have proper bond angles

*Figure 3-13
Chlorocyclohexane.
The chlorine
changes from axial
to equatorial when
the ring flips from
one chair form to
the other.*

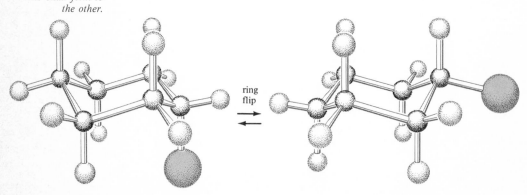

ring
flip

axial conformation   equatorial conformation

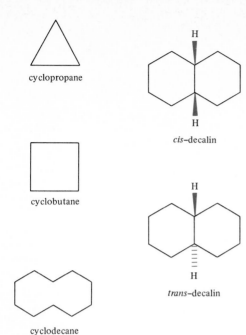

*Figure 3-14
Some cyclic
saturated
hydrocarbons.*

cyclopropane

cis–decalin

norbornane

cyclobutane

adamantane

trans–decalin

cyclodecane

and are said to be *strained*. That such compounds exist at all must
be attributed to the ability of carbon to accommodate bent bonds
without the loss of an exorbitant amount of bond energy. Syntheses,
structures, and chemistry of small rings have been of great interest
to organic chemists in recent years. Cyclopropane, with 60° internal
bond angles, is the most strained of the simple cyclic hydrocarbons.
Most of the reactions of cyclopropane result in opening the strained
three-membered ring. Cyclobutane, with 90° internal bond angles, is
less reactive than cyclopropane.

Rings with more than six carbons are usually quite flexible and do
not show such well-defined conformations as cyclohexane does. Compounds containing rings with thirty or more members have been
prepared.

Many important compounds have more than one ring. Often the
presence of additional rings severely limits the possibilities for conformational isomerism. Decalin, $C_{10}H_{18}$, is made up of two cyclohexane
rings which share two adjacent carbons. Two geometrical isomers of
decalin exist; they differ in whether the connection between the rings
is cis or trans. The cyclohexane rings in both geometrical isomers exist
in chair conformations. Norbornane and adamantane are highly symmetrical structures which are completely rigid and cannot accommodate conformational isomerism.

PROBLEM 3-16   Draw all the isomers of dimethylcyclopropane.

3-5 ALKYNES

An *alkyne* is a hydrocarbon having one or more carbon-carbon triple bonds. Because the carbon-carbon triple bond is linear, no geometrical isomerism is possible in alkynes. Structural isomerism, however, is still possible. There is only one ethyne (usually called acetylene) and one propyne

$$H-C\equiv C-H \qquad H-C\equiv C-CH_3$$

    acetylene   propyne
    (ethyne)

but there are two butynes

$$H-C\equiv C-CH_2-CH_3 \qquad CH_3-C\equiv C-CH_3$$

       1-butyne   2-butyne

and two pentynes. As usual in organic chemistry, the number of isomers rises rapidly as the number of carbon atoms increases.

The names of alkynes are derived just like those of alkenes, except that the ending *-yne* appropriate for alkynes is substituted for the *-ene* of alkenes. The cis and trans prefixes used for alkenes are not necessary for alkynes because of their linear geometry.

PROBLEM 3-17   Give the structures and names of all alkynes having empirical formula (a) C_5H_8; (b) C_6H_{10}.

3-6 ALIPHATIC HYDROCARBONS IN NATURE; TERPENES

A large and varied collection of hydrocarbons occurs in small amounts in nature, particularly in plants. Many of these compounds appear to have been synthesized from a five-carbon precursor which has the carbon structure of *isoprene,* 2-methyl-1,3-butadiene,

$$CH_2=C\begin{matrix} \diagup CH_3 \\ \diagdown CH=CH_2 \end{matrix}$$

    isoprene

The reactions by which these compounds are made will not be considered here. The most striking thing about these compounds is that despite an apparently great diversity of structures, they can usually be shown to consist of isoprene units put together in various ways.

A few examples are shown in figure 3-15. Compounds which are derived from isoprene are called *terpenes*. Not all terpenes are hydrocarbons; terpene alcohols, aldehydes, ketones, and other classes of compounds exist. Examples of complex terpenes are also shown in figure 3-15.

The ultimate in complex terpenes is natural rubber, a long-chain hydrocarbon made up of a large number of isoprene units put together in a head-to-tail fashion. The carbon-carbon double bonds are all in the cis configuration. A section of the structure of natural rubber is shown in figure 3-15.

3-7 AROMATIC HYDROCARBONS

The compounds considered so far in this chapter are all *aliphatic* compounds; each can be adequately represented by a single Lewis structure, with each pair of electrons in the molecule either held by a single atom or shared between a particular pair of atoms. In this section we will consider the structures of *aromatic compounds:* compounds in which some of the pairs of electrons cannot be assigned to definite locations between particular pairs of atoms. Such compounds cannot be represented by single Lewis structures and must be represented as composites, or *hybrids,* of several contributing structures.

BENZENE

The simplest and best-known aromatic compound is *benzene,* a planar molecule with empirical formula C_6H_6. The carbon atoms in benzene are sp^2-hybridized and form a regular hexagon with a hydrogen attached to each carbon. The fourth bond to each carbon atom is a π bond to an adjacent carbon, but it is impossible to specify which carbon is doubly bonded to which. Two structures for benzene must be considered:

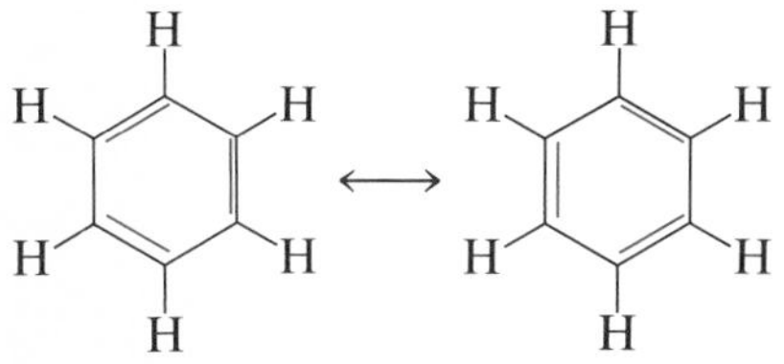

benzene

Neither structure by itself adequately represents the structure of benzene; instead, benzene is an average, or *hybrid,* of the two structures. Each carbon-carbon bond is neither a single bond nor a double bond, but something halfway between.

This averaging, or mixing, of several bond structures is called *resonance,* and the two structures above are called *resonance structures.*

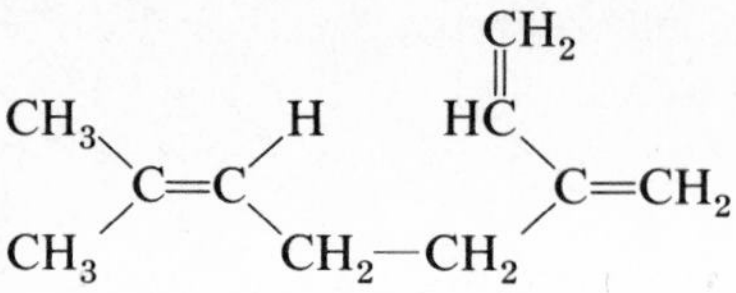

myrcene
from oil of bay, oil of hops

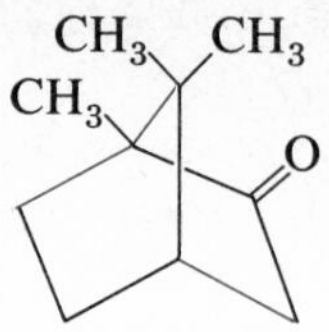

camphor
long used for medicinal purposes

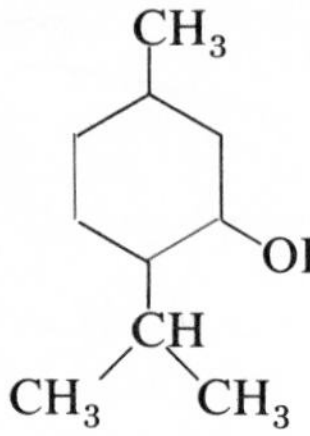

menthol
from peppermint oil

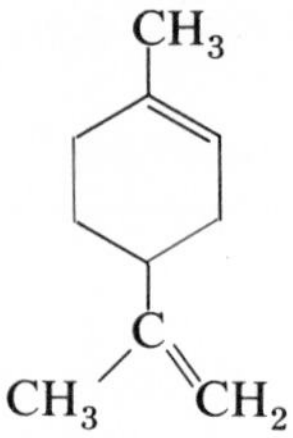

limonene
from oil of lemon, oil of orange

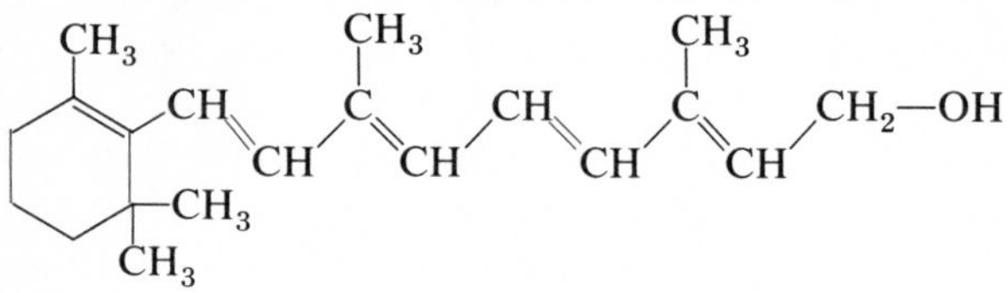

vitamin A
natural precursor of retinal

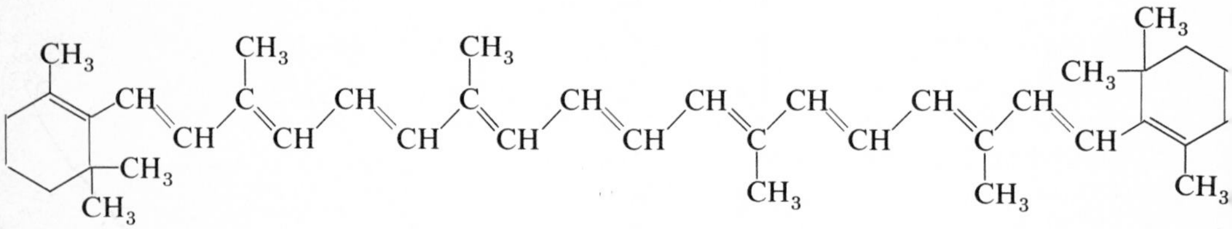

β-carotene
from carrots

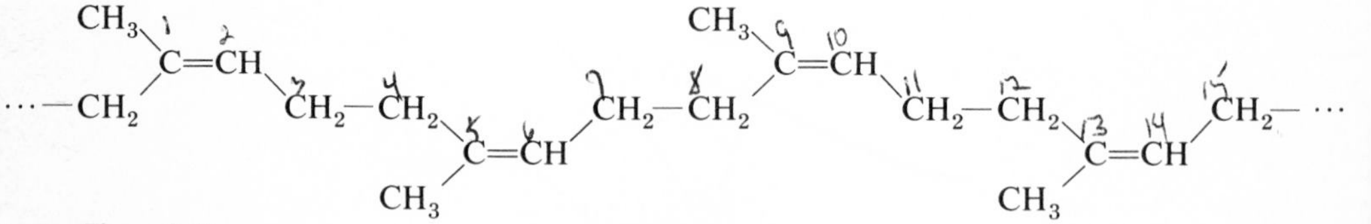

Figure 3-15
Some terpenes.
Position of
connection of
isoprene units is
indicated by colored
bonds.

natural rubber
a very long chain molecule made
up of several thousand of the units
shown here

Resonance structures are always connected by double-headed arrows like the one used here. These arrows are used in organic chemistry *only* to indicate resonance; structures connected by such arrows are always resonance structures, and the actual structure of the molecule in question is a combination, or *hybrid,* of the structures so connected.

The structure of benzene is sometimes written

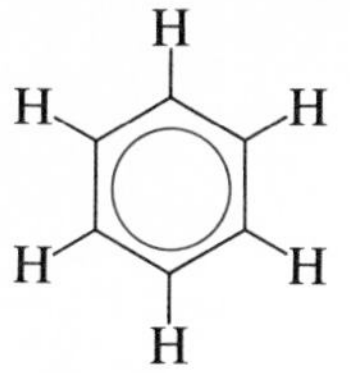

to indicate the uniformity of the carbon-carbon bonds, but this representation is less informative than the resonance structures shown above.

Although the two resonance forms above appear to be the same molecule simply rotated by 60°, such a rotation is not permissible in writing resonance structures. When resonance structures are written, the nuclei are kept in place; they may not exchange places. Resonance structures are not real; they must be combined and essentially averaged in order to convey the true structure of an aromatic molecule. Resonance structures are necessary because ordinary valence-bond structures do not represent the structures of aromatic molecules adequately.

Writing resonance forms of aromatic molecules is subject to rigid rules:

WRITING RESONANCE FORMS

1 All resonance forms must be proper Lewis structures; that is, they must have the proper number of bonds to each atom, or else that atom must have a formal charge.

2 The positions of the nuclei must be the same in all resonance structures.

3 All atoms taking part in the resonance must lie in the same plane.

4 Saturated carbon atoms do not participate in resonance.

The resonance forms of naphthalene and phenanthrene are shown in figure 3-16.

Draw all resonance structures of the following compounds:

PROBLEM 3-18

(a) (b)

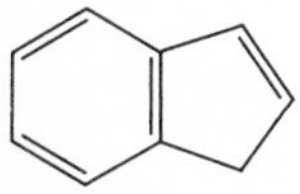

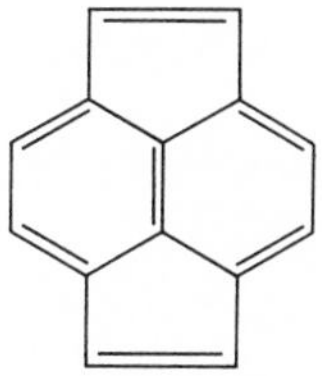

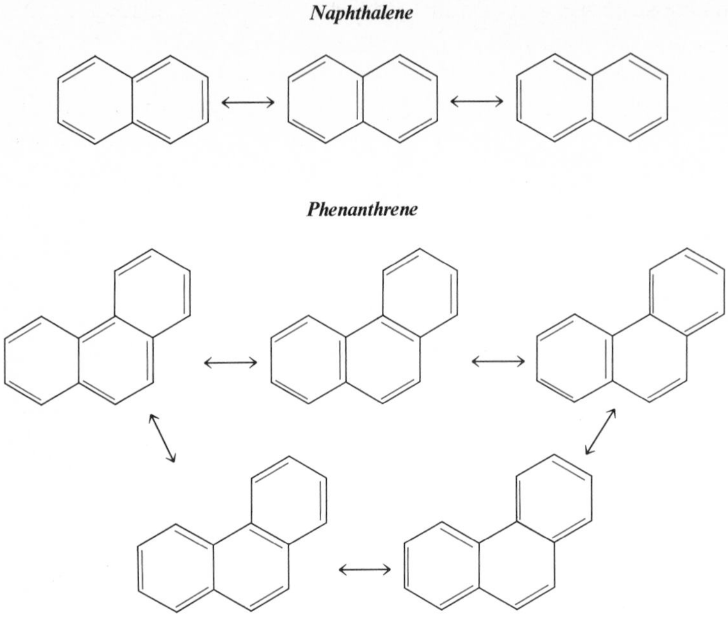

62

Figure 3-16
Resonance structures of naphthalene and phenanthrene.

Naphthalene

Phenanthrene

PROBLEM 3-19   The use of circles to indicate resonance sometimes leads to incorrect conclusions. Show that the following structure does not represent a real compound:

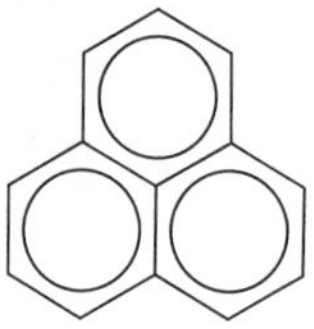

PROBLEM 3-20   Draw structures of naphthalene and phenanthrene and indicate the positions of all hydrogen atoms.

PROBLEM 3-21   How many compounds called methylnaphthalene are there?

MOLECULAR ORBITALS IN BENZENE   The inadequacy of a single structure for describing benzene suggests that the π bonding in benzene is not like that in simple olefins. This is indeed true. Benzene is built up of *delocalized* π orbitals (π orbitals which extend over more than two atoms) rather than localized orbitals between pairs of carbon atoms. Three different π orbitals are used to accommodate the six π electrons in benzene.

In order to understand the nature of the π bonding in benzene, we must first consider the framework on which that bonding is built. The σ bonds in benzene are built up from sp^2 hybrid orbitals of carbon and s orbitals of hydrogen. This leaves one p orbital on each carbon atom free to participate in the π system. The π molecular orbitals are built up from combinations of *all six p* orbitals. The three π orbitals which are occupied in benzene are shown in figure 3-17.

AROMATIC HYDROCARBONS AND CANCER

Aromatic hydrocarbons differ in many important ways from aliphatic hydrocarbons—in physical properties, in chemical reactivity, and in their effects on living things. Aromatic compounds occur commonly in our environment. They are present in coal and petroleum, they are widely used in the chemical industry, and in recent years they have become common and dangerous pollutants.

In the eighteenth century it was observed that chimney soot caused cancer in chimney sweeps. Later, workers in the dye and petroleum industries were found to have a high incidence of cancer. More recently it has been shown that both cigarette smoke and pollutants arising from the incomplete combustion of coal cause cancer. In all these cases, the high incidence of cancer is due to the same type of compounds, the polycyclic aromatic hydrocarbons. Although neither benzene nor naphthalene causes cancer, most aromatic compounds containing four or more aromatic rings do.

A number of known cancer-causing hydrocarbons are shown in figure 3-18. Radioactive-labeling studies have shown that these compounds bind covalently to many important materials in the body, including DNA, RNA, and proteins. However, exactly how these compounds cause cancer is still unknown.

3-8 REACTIONS OF ALKANES

The chemical reactions of alkanes are singularly uninteresting. Alkanes are refractory to most reagents. Of all the parts of organic molecules, the saturated hydrocarbon parts are the least reactive. The two reactions of alkanes which are well known are hard to control and are of little use in organic chemistry.

COMBUSTION

Nearly all organic compounds are capable of undergoing combustion, or burning:

$$C_mH_n + \left(m + \frac{n}{4}\right)O_2 \longrightarrow mCO_2 + \frac{n}{2}H_2O + \text{energy}$$

This reaction is of tremendous commercial importance; about 500,000

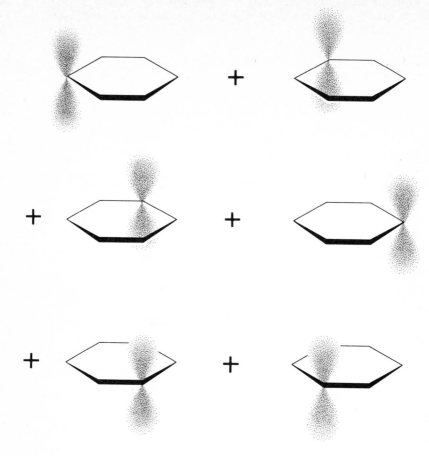

p atomic orbitals

Figure 3-17
Benzene.

The π molecular orbitals are built up by combining a p orbital from each of the six carbon atoms into a set of π molecular orbitals which extend over the entire six-carbon framework of the molecule.

tons of crude oil and 560 billion tons of coal are produced in the United States each year. Most of this material is burned to produce energy.

HALOGENATION Saturated hydrocarbons undergo free-radical halogenation in the presence of chlorine or bromine and light or heat

$$CH_4 + Cl_2 \xrightarrow{\text{light}} CH_3Cl + CH_2Cl_2 + CHCl_3 + CCl_4 + HCl$$

A variety of products containing different numbers of halogen atoms is obtained. This reaction is a very high-energy reaction and is of little use in organic chemistry because of the complex product mixtures obtained.

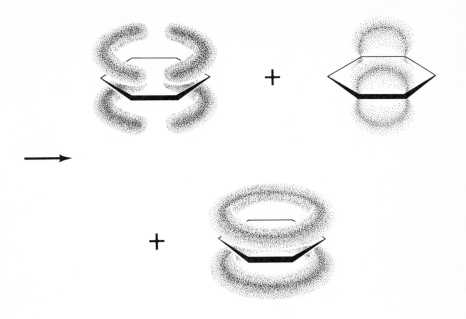

π molecular orbitals

3-9 REACTIONS OF ALKENES

Whereas carbon-carbon σ bonds and carbon-hydrogen bonds are very unreactive under most circumstances, carbon-carbon π bonds react readily with a wide variety of *electrophilic,* or electron-seeking, reagents. This difference between σ and π bonds is a result of the very different electron distributions in the two types of bonds (section 2-3). In the carbon-carbon σ bond, the electron density is highest in the region immediately between the two nuclei; this location effectively protects the electrons from being approached by electrophilic reagents. On the other hand, π bonds are located much farther from the nuclei and are thus not so effectively protected from attack. In addition, the *overlap* of atomic orbitals involved in forming a π bond is not so favorable as that in a σ bond, so that a π bond is weaker than a σ bond and is more easily broken.

Figure 3-18
Some cancer-causing
hydrocarbons.

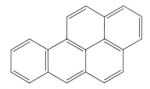

1,2,5,6-dibenzanthracene
occurs in cigarette smoke, coal smoke,
mineral oils, smoked meat, oranges, and
many other places

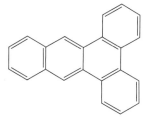

3,4-benzpyrene
the most ubiquitous and potent cancer-
causing hydrocarbon in polluted air;
produced by the incomplete combustion
of coal; also found in tobacco smoke
and many foods

1,2,3,4-dibenzanthracene
does not cause cancer, in spite of
its structural similarity to the
compounds above

ORGANIC
REACTION
MECHANISMS

Most organic reactions do not proceed all the way from starting materials to products in a single step but instead proceed through a sequence of *reaction intermediates:* unstable molecules which have lifetimes ranging from microseconds or less to perhaps a few seconds. The reaction steps which interconvert starting materials, intermediates, and products are usually simple steps involving changes in a small number of bonds. If these simple steps can be understood, the reaction can be understood. For this reason, we will concentrate on understanding the individual steps in chemical reactions. We frequently give not only the structures of the starting material, intermediates, and products in a reaction but also diagrams which show the bond changes occurring in each step as well. To distinguish them from other structures, these *step diagrams* will be shown on a shaded background. Reaction intermediates will be identified by enclosing them in brackets. Thus, starting materials and products will be clearly distinguished from other types of structures.

Structural changes will be shown in step diagrams by means of curved arrows; each curved arrow represents the movement of one pair of electrons in that particular reaction step. We will see that most reaction steps involve changes in one, two, or at most three pairs of electrons. It must be emphasized at the outset that these step diagrams are bookkeeping devices to help us understand the structural changes occurring in organic reactions; they do not represent stable molecules which can be isolated and studied.

ADDITION OF HBr TO OLEFINS

Alkenes are more reactive than alkanes because of the carbon-carbon π bond. In most reactions of alkenes, the π bond is broken and is replaced by two new σ bonds. For example, when an olefin reacts with hydrogen bromide, the carbon-carbon π bond is broken and a new carbon-hydrogen σ bond and a carbon-bromine σ bond are formed

$$\underset{H}{\overset{CH_3}{}}C=C\underset{CH_3}{\overset{H}{}} + HBr \longrightarrow CH_3-\underset{Br}{\overset{H}{C}}-\underset{H}{\overset{H}{C}}-CH_3$$

This addition occurs because of the ease with which the electron-deficient hydrogen ion, H^+, attacks the electron-rich π bond. The reaction is shown in detail in mechanism 3-1.

Hydrogen chloride and hydrogen iodide react with olefins in the same way that hydrogen bromide does. The reaction of cyclopentene with hydrogen chloride proceeds by way of a carbonium-ion intermediate

cyclopentene   chlorocyclopentane

Write out the mechanism for the last reaction.   *PROBLEM 3-22*

$$\underset{CH_3}{\overset{H}{}}C=C\underset{H}{\overset{CH_3}{}} + H^+ + Br^-$$

In solution, HBr ionizes to a mixture of H^+ and Br^-. The hydrogen ion is electron-poor, and the double bond is electron-rich. The reaction begins because of this complementarity.

$$H^+$$
$$\underset{CH_3}{\overset{H}{}}C=C\underset{H}{\overset{CH_3}{}}$$

electrophilic
attack

In the first step, the proton attacks the carbon-carbon π bond, pulling the electrons away from one carbon and toward itself. The proton thus forms a bond to one of the carbon atoms, and the other carbon atom becomes electron-deficient.

$$\left[CH_3-\underset{\underset{H}{+}}{\overset{H}{C}}-\overset{H}{\underset{}{C}}-CH_3 \right]$$

carbonium-ion
intermediate

*This electron-deficient intermediate is called a **carbonium ion.** One carbon atom of the carbonium ion has only three bonds and has a formal positive charge. Since such an electron-deficient carbon atom is very reactive, this intermediate exists for only a fraction of a second. As soon as it encounters an electron-rich ion or molecule with which it can react, it does so.*

$$CH_3-\underset{\underset{H}{+}}{\overset{H}{C}}-\overset{H}{C}-CH_3$$
$$Br^-$$

bromide attack

Reaction of the electron-deficient carbon with a Br^- ion forms a new carbon-bromine bond by using an electron pair from the bromine. The only change which occurs in this step is the formation of this carbon-bromine σ bond.

$$CH_3-\underset{Br}{\overset{H}{C}}-\underset{H}{\overset{H}{C}}-CH_3$$

The final product of the reaction is a saturated alkyl halide.

*MARKOVNIKOV'S
RULE*
The two carbons of a double bond suffer dissimilar fates during addition reactions; one becomes bonded to a hydrogen and the other to a halogen. Sometimes different products might be obtained according to which atom suffers which fate. For example, two different compounds might theoretically be formed by reaction of 1-butene with HBr

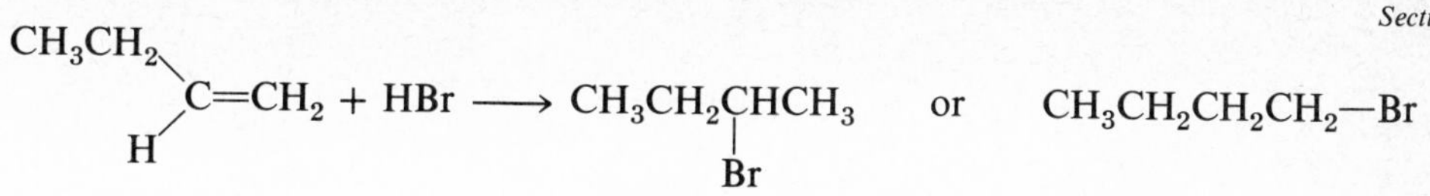

$$CH_3CH_2 \diagdown C=CH_2 + HBr \longrightarrow CH_3CH_2\underset{\underset{Br}{|}}{C}HCH_3 \quad \text{or} \quad CH_3CH_2CH_2CH_2-Br$$

1-butene   observed product   not formed

The rule which explains which of the two products is formed bears the name of Vladimir W. Markovnikov, a Russian chemist who derived the rule in 1869, long before the reason for the preference was known. *Markovnikov's rule* states that in the addition of H^+ X^- to carbon-carbon double bonds (where $X^- = Cl^-$, Br^-, I^-, or H_2O), *the positive ion becomes attached to the unsaturated carbon carrying the smaller number of alkyl groups.*

The preference embodied in Markovnikov's rule occurs because of differences in the structures of the carbonium-ion intermediates which would be formed in the two cases:

$$[CH_3CH_2\overset{+}{C}HCH_3] \quad \text{and} \quad [CH_3CH_2CH_2\overset{+}{C}H_2]$$

more favorable   less favorable

These two structures differ in a significant way. The positively charged carbon atom in the right structure has only one other carbon atom attached to it, whereas the positively charged carbon in the structure on the left has two other carbon atoms attached to it. The number of such carbons bonded to the positively charged carbon is the determining factor in these reactions. The more carbons bonded, the more easily that carbonium ion can be formed and the more favorable that particular reaction pathway. Thus, it is actually the ease of formation of the carbonium-ion intermediate which governs the course of this reaction.

This preference is also illustrated by the reaction of 1-methylcyclopentene with HCl

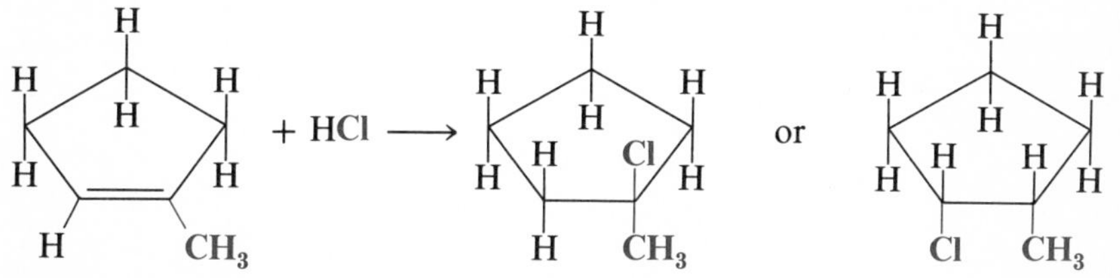

1-methylcyclopentene   observed product   not formed

The terpene *limonene,* $C_{10}H_{16}$, is a common constituent of a number of oils, including oil of lemon, orange, caraway, dill, and bergamot. Reaction of this compound with HCl was an important step in the

proof of its structure

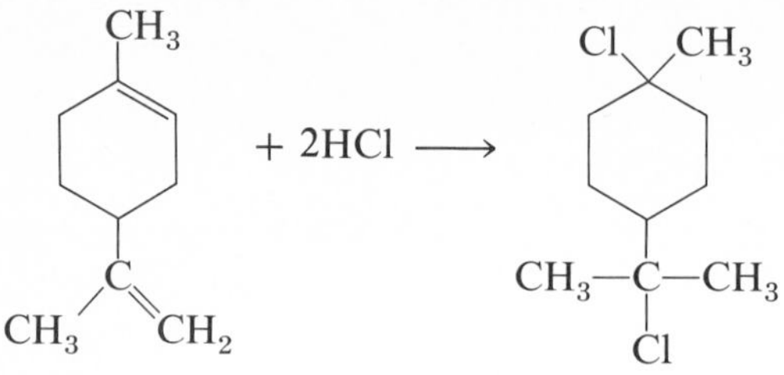

limonene

$+ 2HCl \longrightarrow$

Since limonene contains two carbon-carbon double bonds, it sequentially adds two equivalents of HCl. In a case like this, addition of the first HCl and of the second HCl occurs separately rather than at the same time; no double carbonium ion having two positive charges is formed as an intermediate. Both additions follow Markovnikov's rule.

ADDITION OF WATER   In the presence of a small amount of acid, water can be added to a carbon-carbon double bond, forming an alcohol. Sulfuric acid is most commonly used for this purpose

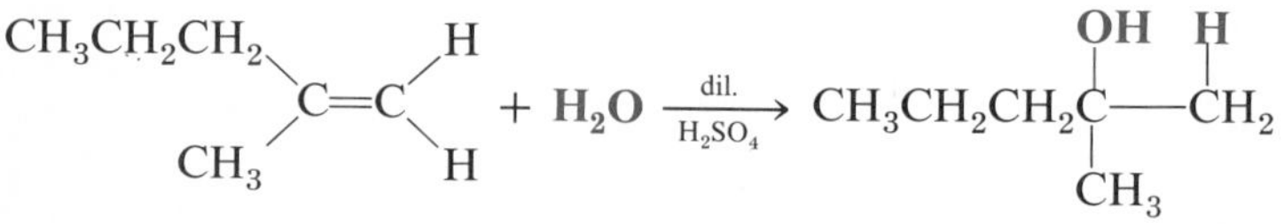

The mechanism is similar to the mechanism of addition of HBr to olefins. In the first step a proton attacks the π bond, breaking it and forming a carbonium ion. In the second step, the carbonium ion reacts with water, forming an alcohol and releasing a proton from the water. Markovnikov's rule is followed.

The reaction of olefins with dilute sulfuric acid provides a useful way to prepare alcohols from olefins. Alcohols can be converted back into olefins by what is in effect the reverse of the above reaction (chapter 5).

PROBLEM 3-23   Give the mechanism and product of the reaction

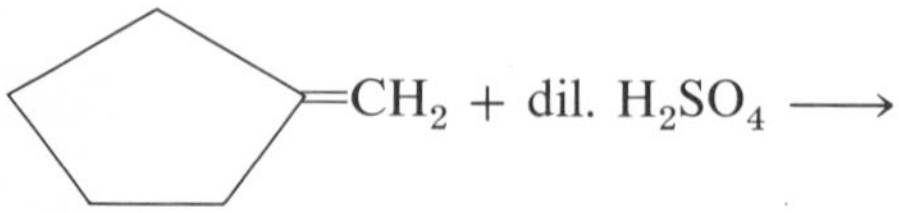

ADDITION OF BROMINE TO CARBON-CARBON DOUBLE BONDS

Bromine, Br_2, chlorine, Cl_2, and iodine, I_2, add readily to carbon-carbon double bonds

$$CH_2{=}CH_2 + Br_2 \longrightarrow Br{-}CH_2{-}CH_2{-}Br$$

ethylene — 1,2-dibromoethane

The mechanism of this addition (mechanism 3-2) is similar to that of the addition of acids to double bonds in that it occurs in two steps. However, the intermediate in this case is a positively charged bromonium ion, in which the positive charge resides primarily on the bromine.

Chlorine also reacts readily with carbon-carbon double bonds

$$CH_3CH{=}CH_2 + Cl_2 \longrightarrow CH_3\overset{\displaystyle Cl}{\overset{|}{C}}HCH_2{-}Cl$$

STEREOCHEMISTRY OF ADDITION

The reaction of bromine with cyclopentene might conceivably form either the trans isomer or the cis isomer

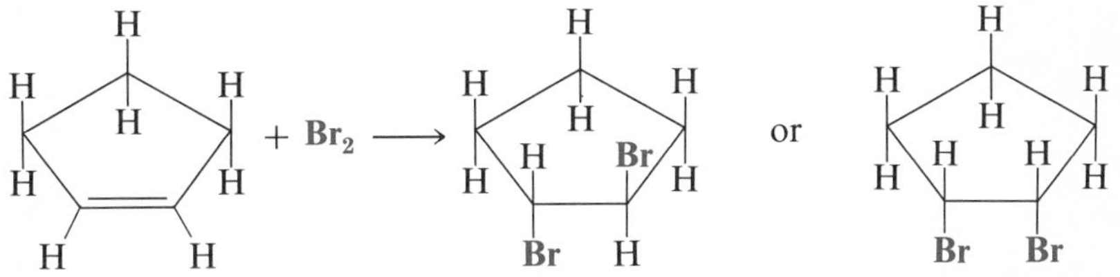

trans isomer, observed product — cis isomer, not formed

Only the first of the two possible products is observed. This preference occurs because of the geometry of the bromonium-ion intermediate. The bromine in this intermediate prevents the bromide ion from approaching the side of the ring to which the first bromine has become attached. Approach from the opposite side results in formation of the trans isomer. Such addition is called *anti addition.*

PROBLEM 3-24

Build a model of the bromonium-ion intermediate formed in the above case. Use tape to help construct the three-membered ring if necessary. Confirm that addition of the second bromine from the side opposite the first one results in formation of the trans isomer.

Mechanism 3-2
Addition of
Bromine to Olefins

$$CH_3 \quad H$$
$$\diagdown C=C \diagup \quad + \; Br_2$$
$$CH_3 \diagup \quad \diagdown H$$

The electrophile in this reaction is Br_2.

:Br—Br

$$CH_3 \quad H$$
$$\diagdown C=C \diagup$$
$$CH_3 \diagup \quad \diagdown H$$

electrophilic
attack

In the first step, the electrophile attacks the carbon-carbon double bond. However, the attack is more complex than the attack of a proton on a double bond. In this case the bromine contributes a pair of electrons and is thereby able to form bonds to both carbon atoms simultaneously.

$$\left[\begin{array}{c} CH_3 \quad Br^+ \quad H \\ \diagdown \overset{|}{C}-\overset{|}{C} \diagup \\ CH_3 \diagup \qquad \diagdown H \end{array} \right] + \; Br^-$$

bromonium-ion
intermediate

*The intermediate formed by this attack is a **bromonium ion,** a three-membered ring structure with a positive charge on the bromine. This intermediate is very unstable and reacts rapidly with any electron-rich species present.*

$$CH_3 \quad Br^+ \quad H$$
$$\diagdown \overset{|}{C}-\overset{|}{C} \diagup$$
$$CH_3 \diagup \qquad \diagdown H$$
$$Br^-$$

bromide attack

The second step is the reaction of the bromonium ion with bromide ion. In this step a new carbon-bromine bond is formed, and one carbon-bromine bond in the bromonium ion is broken.

$$\begin{array}{ccc} & CH_3 & Br \\ & | & | \\ CH_3- & C - & C -H \\ & | & | \\ & Br & H \end{array}$$

The product of this reaction is a dibromide with bromines on adjacent carbon atoms.

CATALYTIC HYDROGENATION

In the presence of a platinum or palladium catalyst, alkenes react with hydrogen to form alkanes

$$CH_3 \quad H \qquad\qquad\qquad CH_3$$
$$\diagdown C=C \diagup \quad + \; H_2 \; \xrightarrow{\text{Pt or}\atop\text{Pd}} \; CH_3-\overset{|}{\underset{|}{C}}-CH_3$$
$$CH_3 \diagup \quad \diagdown H \qquad\qquad\qquad H$$

an alkene   an alkane

One equivalent of hydrogen is consumed for each double bond present.

Thus, a compound with two double bonds will consume two equivalents of hydrogen

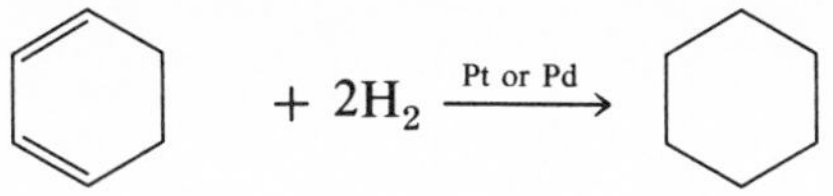

1,3-cyclohexadiene   cyclohexane

The addition of hydrogen to the π bond occurs on the surface of the metal catalyst. Unlike the addition of Br_2 to olefins, the addition of H_2 occurs entirely on one side of the double bond. This is called *syn* (same side) addition. For example

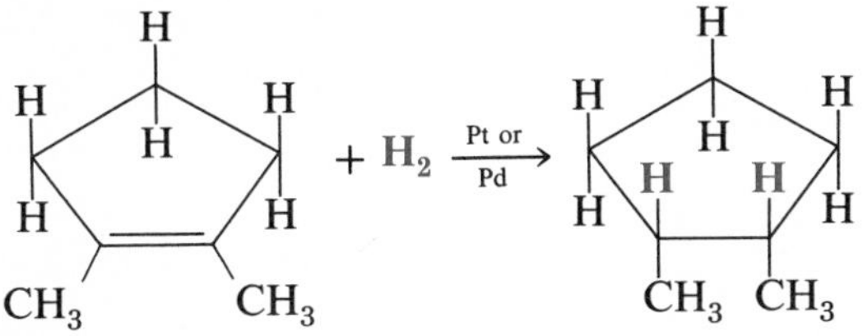

Aromatic compounds are not reduced by this procedure.*

Another reaction which is useful in structure determination is ozonolysis: reaction with ozone. Olefins react with ozone to form compounds which react with a mixture of acid and zinc to form aldehydes and ketones. In the course of this reaction, the carbon-carbon double bond is completely cleaved and is replaced by two carbon-oxygen double bonds

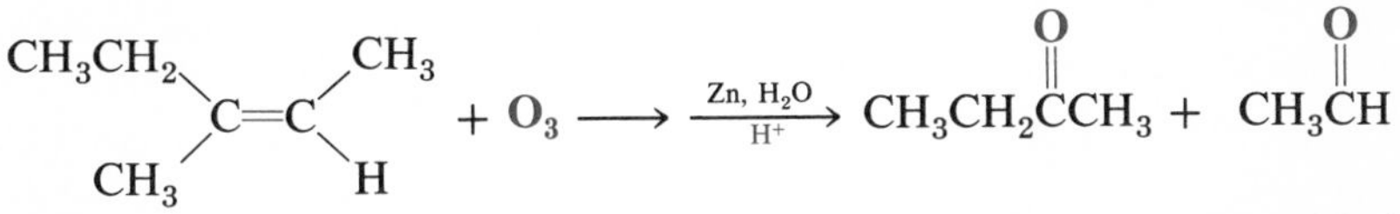

OZONOLYSIS

Catalytic hydrogenation and ozonolysis are often useful in determining the structures of unknown molecules. Catalytic hydrogenation can be used to establish the number of double bonds, and ozonolysis can be used to determine their locations. For example, an unknown compound C_6H_{10} reacts with hydrogen in the presence of a catalyst to form C_6H_{12}.

STRUCTURE DETERMINATION

---

*However, at high temperature and a high pressure of hydrogen, reduction of aromatic compounds can be achieved.

The original compound undergoes ozonolysis to form the two compounds

$$CH_3\!-\!\overset{\displaystyle O}{\overset{\|}{C}}H \quad \text{and} \quad$$

From this information, the structure of the unknown must be

$$CH_3CH=$$

PROBLEM 3-25 Write the structure of the olefin which on ozonolysis gives

$$CH_3\!-\!\overset{\displaystyle O}{\overset{\|}{C}}\!-\!CH_2CH_2CH_2\!-\!\overset{\displaystyle O}{\overset{\|}{C}}\!-\!CH_3$$

3-10 REACTIONS OF ALKYNES

Like alkenes, alkynes are electron-rich and are quite susceptible to attack by a variety of electrophilic reagents. Alkynes undergo two successive reactions with electrophiles, leading first to a substituted alkene and then to a substituted alkane

$$H\!-\!C\!\equiv\!C\!-\!H + 2Br_2 \longrightarrow \underset{\substack{Br \quad\quad H}}{\overset{\substack{H \quad\quad Br}}{C=C}} \overset{Br_2}{\longrightarrow} H\!-\!\underset{\substack{Br \quad Br}}{\overset{\substack{Br \quad Br}}{C\!-\!C}}\!-\!H$$

first product   final product

HBr, HCl, and other acids also add to carbon-carbon triple bonds. In such cases, Markovnikov's rule is followed, and the product which is formed has two halogen atoms on the same carbon

$$CH_3C\!\equiv\!CCH_3 + HBr \longrightarrow \underset{\substack{Br \quad\quad CH_3}}{\overset{\substack{CH_3 \quad\quad H}}{C=C}} \overset{HBr}{\longrightarrow} CH_3\!-\!\underset{\substack{Br}}{\overset{\substack{Br}}{C}}\!-\!CH_2CH_3$$

first product   final product

The mechanisms of these reactions are similar to those of additions to carbon-carbon double bonds (mechanisms 3-1 and 3-2).

Alkynes also react with hydrogen in the presence of a metal catalyst

$$CH_3C\!\equiv\!CH + 2H_2 \xrightarrow{\substack{Pt\ or \\ Pd}} CH_3CH_2CH_3$$

Although benzene has π bonds, they are very different from the π bonds in simple alkenes. Benzene does not ordinarily undergo any of the reactions we have considered in this chapter except ozonolysis. Discussion of the reactions of benzene will be deferred to chapter 11.

3-26 Give the structures of all compounds with empirical formula *PROBLEMS* (*a*) C_3H_6BrCl; (*b*) C_5H_8; (*c*) C_3H_5I.

3-27 Give the structure of a compound which is (*a*) an alkane; (*b*) an alkene; (*c*) an alkyne; (*d*) unsaturated; (*e*) saturated; (*f*) aromatic.

3-28 Give the structures of (*a*) a pair of saturated compounds which are geometrical isomers; (*b*) a pair of acyclic compounds which are geometrical isomers.

3-29 Give the mechanism and product of each of the following reactions: (*a*) 1-hexene + HCl; (*b*) 1-hexene + Cl_2; (*c*) 1-hexene + H_2 (platinum catalyst).

3-30 The halogenation of saturated hydrocarbons in the presence of light (section 3-8) often forms a mixture of every possible halogenated hydrocarbon having the same carbon skeleton as the starting hydrocarbon. Give the structures of all products which might be formed by reaction of ethane with bromine in the presence of light.

3-31 Give the structures of all structural and geometrical isomers of heptene.

3-32 Draw all structural and geometrical isomers of dichlorocycloheptane.

3-33 Draw a Newman projection of the most stable conformation of 1-bromopropane.

3-34 Why doesn't cyclopentyne exist?

3-35 Give the structures of all cyclic compounds having the empirical formula C_6H_{12}.

3-36 Draw the structures of all isomers of methyladamantane (see figure 3-14 for the structure of adamantane). Use molecular models to confirm your answer.

3-37 Show that retinal (figure 3-7) follows the isoprene rule.

3-38 Give the structure of (*a*) 3-methyl-4-*t*-butylheptane; (*b*) 2,2,3,3-tetramethylbutane; (*c*) *trans*-4-ethyl-3-octene; (*d*) *cis,cis*-2,6-decadiene; (*e*) 3-*n*-propyl-2-hexene; (*f*) 2-methyl-3-heptyne.

3-39 Give the structures of two different olefins that give the same product on reaction with HBr.

3-40 Would you expect cycloheptane to be planar? Why or why not? Build a model of cycloheptane to confirm your answer.

3-41 Draw two geometrical isomers of 1,3-dichlorocyclohexane. For each geometrical isomer, draw two conformational isomers and indicate which is of lower energy.

3-42 Use molecular models to help you draw the structures of all isomers of dimethylspiropentane. The structure of spiropentane is

3-43 Draw a Newman projection of the product formed when bromine reacts with *cis*-2-butene. Do the same for the trans isomer. Are these two compounds the same or different? Use molecular models to confirm your answer.

3-44 Give all resonance structures for each of the following compounds:

3-45 Give the product of the reaction of 1-methylcyclopentene with (*a*) Br_2; (*b*) Cl_2; (*c*) H_2 (palladium catalyst); (*d*) HCl; (*e*) dil. H_2SO_4; (*f*) O_3, then Zn, H_3O^+; (*g*) HI.

3-46 The terpene α-terpinene has empirical formula $C_{10}H_{16}$. Reaction with hydrogen in the presence of a platinum catalyst forms $C_{10}H_{20}$. Reaction of α-terpinene with ozone, followed by reaction with Zn + H_3O^+, forms

$$\underset{\text{HC}-\text{CH}}{\overset{\text{O O}}{\|\ \|}} \quad \text{and} \quad \underset{\text{CH}_3-\text{C}-\text{CH}_2\text{CH}_2-\text{C}-\text{CHCH}_3}{\overset{\text{O}\qquad\quad\text{O CH}_3}{\|\qquad\quad\|\ |}}$$

What is the structure of α-terpinene?

3-47 Give the structure of the product or products of each of the following reactions:

(*a*) $CH_3CH=C\overset{\displaystyle CH_3}{\underset{\displaystyle CH_3}{\big<}}$ + HCl $\longrightarrow$

(*b*) [benzene ring with] $C\equiv CCH_3$ + Br_2 $\longrightarrow$

(*c*) $CH_3C\equiv CCH_3 + H_2 \xrightarrow{\text{Pt}}$

(*d*) [cyclopentadiene] + $Cl_2 \longrightarrow$

(e) $CH_3CH_2CH_2CH{=}CH_2 + H_2SO_4 \longrightarrow$

(f) *trans*-3-Methyl-3-hexene + HBr $\longrightarrow$

(g) Cyclohexene + HI $\longrightarrow$

(h) Cyclooctene + dil. $H_2SO_4 \longrightarrow$

(i)

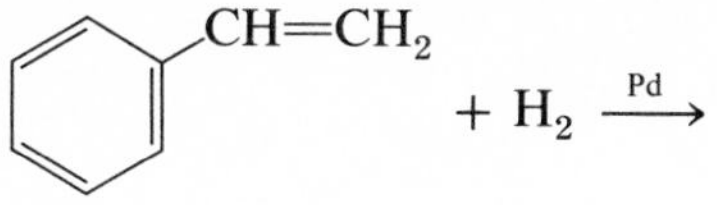

$+ H_2 \xrightarrow{Pd}$

(j)

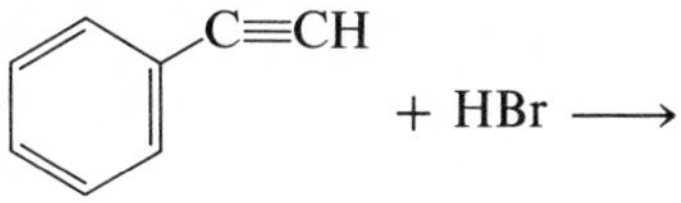

$+ HBr \longrightarrow$

Banks, J. E.: *Naming Organic Compounds,* Saunders, Philadelphia, 1967. A programmed introduction to nomenclature.

Breslow, R.: *Organic Reaction Mechanisms,* 2d ed., Benjamin, Menlo Park, Calif., 1969. A useful introduction to organic reaction mechanisms. Includes a number of biological examples.

Hubbard, R., and A. Kropf: Molecular Isomers in Vision, p. 405 in *Organic Chemistry of Life: Readings from Scientific American,* Freeman, San Francisco, 1973. The chemistry of the visual process.

Lambert, J. B.: The Shapes of Organic Molecules, p. 21 in *Organic Chemistry of Life: Readings from Scientific American,* Freeman, San Francisco, 1973. Well-illustrated introduction to stereochemistry.

Natta, G., and M. Farina: *Stereochemistry,* Harper & Row, New York, 1972. A good introduction to the subject, especially the stereochemistry of proteins and other biological molecules.

Price, C. C.: *Geometry of Molecules,* McGraw-Hill, New York, 1971. Covers many details of stereochemistry and bonding.

Traynham, J. G.: *Organic Nomenclature: A Programmed Introduction,* Prentice-Hall, Englewood Cliffs, N.J., 1966. A comprehensive treatment of nomenclature.

SUGGESTED READINGS

**Optical
Isomerism**

Microorganisms are amazingly adaptable; they can grow under a variety of conditions using many different chemicals as their source of nourishment. If a typical yeast or bacterium is grown in an environment containing chemically synthesized alanine

$$CH_3-\underset{\underset{\textstyle NH_2}{|}}{CH}-\overset{\overset{\textstyle O}{||}}{C}-OH$$

alanine

(one of the twenty amino acids required for growth) as the only carbon-containing nutrient, it will thrive because it can produce other essential compounds from this amino acid. However, when half of the alanine has been used up, growth will stop. If this experiment is conducted with alanine isolated from nature, instead of synthetic alanine, growth will continue until all the alanine is consumed.

This difference between a synthetic and a natural compound occurs because the synthetic compound is actually a 50 : 50 mixture of two different isomers of alanine, only one of which supports growth. This isomerism is such an important and subtle phenomenon that it is the subject of this entire chapter.

4-1 CHIRALITY

Although reading a newspaper is simple, trying to read one in a mirror is a difficult task indeed. The appearance of the newspaper in the mirror (called the *mirror image*) is different from the appearance of the newspaper itself, and there is no way we can turn the newspaper to eliminate this difference.

Not all objects present this dichotomy. A chair, for example, looks the same in a mirror as when viewed directly; that is, a chair is identical with its mirror image. Some examples of nonidentical mirror images are given in figure 4-1. An object which is not identical with its mirror image is said to be *chiral,* from the Greek word for hand.

PROBLEM 4-1 Which of the following objects are chiral? (*a*) A baseball; (*b*) a screw; (*c*) a book; (*d*) a book with no printing; (*e*) a coffee cup; (*f*) a pair of scissors.

Figure 4-1
Chiral objects. *A left hand and a right hand are mirror images, as are a left-handed batter and a right-handed batter.*

4-2 CHIRAL COMPOUNDS WITH A SINGLE ASYMMETRIC CARBON ATOM*

A carbon atom which has four different substituent groups attached to it is called an *asymmetric carbon atom*. If a model of a compound containing an asymmetric carbon atom is viewed in a mirror, the mirror image is similar to the original compound, but not identical to it (figure 4-2). That is, molecular models indicate that a compound containing an asymmetric carbon atom is chiral and may exist as two stereoisomers; these two nonidentical mirror images are called *enantiomers*.

Some other chiral molecules having an asymmetric carbon atom are shown in figure 4-3. Although most chiral compounds have at least one asymmetric carbon atom, the presence or absence of asymmetric carbons does not guarantee the presence or absence of chirality. More general criteria for chirality will be considered later in this chapter.

PROBLEM 4-2   Build models of two enantiomers of $CH_3CH_2CHClCH_3$ and show that they are different.

*The chirality of organic molecules can be appreciated only if one has a clear picture of the three-dimensional structures of molecules. Molecular models are essential for this purpose.

Figure 4-2
The two enantiomers of lactic acid.

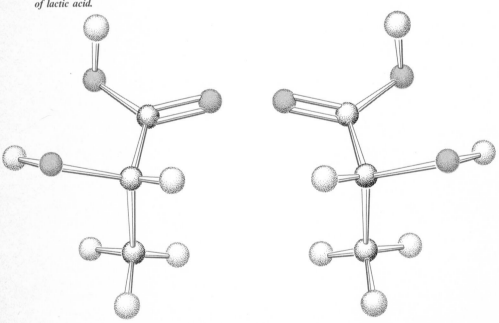

L-Lactic acid   D-Lactic acid

Which of the following compounds are chiral? Use molecular models **PROBLEM 4-3**
to confirm your answers: (*a*) 2-methylhexane; (*b*) 3-methylhexane;
(*c*) 3-methylpentane; (*d*) 3-methyl-1-pentene; (*e*) *cis,trans*-4-methyl-
2,5-heptadiene.

4-3 POLARIZED LIGHT

Chirality is intimately related to *optical activity,* the ability of a compound to rotate the plane of polarized light passing through it. In
this section we will consider polarized light and the nature and use
of this effect.

According to classical physics, light travels in the form of waves. The *LIGHT WAVES*
rise and fall of the waves occurs perpendicular to the direction of
motion of the light. If light could be viewed on a microscopic scale,
it might appear as a collection of waves of different orientations, a
few of which are shown in figure 4-4.

When a *polarizing filter* is introduced into the light beam, it removes
all light except for waves moving parallel to the axis of the filter. The
collection of waves of different orientations is thus reduced to a thin
slice of waves, all vibrating in the same direction. Filtered light of this
kind is called *plane-polarized light* because the motion of the light
waves is now polarized (restricted to vibration in one plane).

If a second filter is inserted between the first filter and the observer,
the amount of light reaching the observer depends on the relative
orientations of the two polarizing filters. If the axes of the two filters
are parallel, the light passes through the second filter unchanged, but
if the two filters are arranged in any other orientation, the second

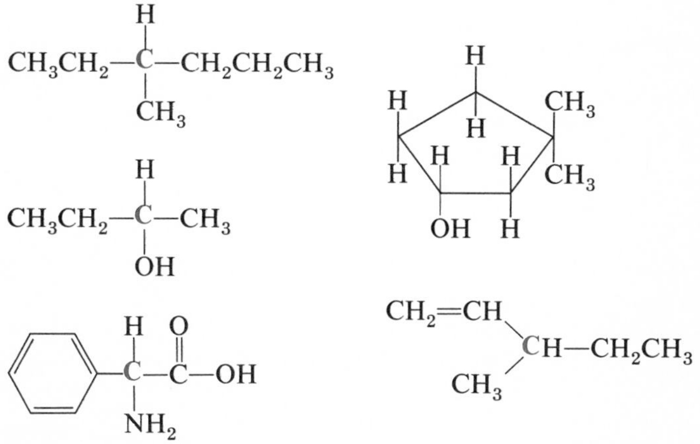

Figure 4-3
Some chiral
compounds having a
single asymmetric
carbon atom.

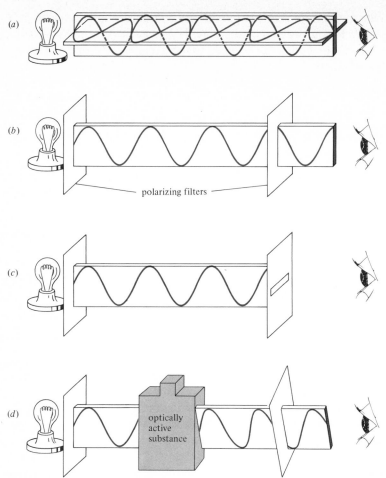

polarizing filters

Figure 4-4
Polarized light.
(a) Polarized light can be viewed (on a microscopic scale) as a collection of waves vibrating in different directions perpendicular to the direction of motion. (b) If a polarizing filter is inserted in the light beam, all waves are filtered out except those which vibrate in the same direction as the axis of the filter. If a second polarizing filter is inserted in the beam with its axis parallel to the first, the light passes through it unchanged. (c) If the second polarizing filter has its axis perpendicular to that of the first, no light passes through it. (d) An optically active substance rotates the plane of the polarized light. In order to maximize light transmission, the second filter must be placed at an angle to the first. This angle is the **optical rotation.**

filter stops at least part of the light. The greater the angle between the two filters, the less light transmitted. When the filters are arranged at 90° angles, no light is transmitted at all.

If a solution of one enantiomer of a chiral compound is placed in the light beam between the two polarizing filters in figure 4-4, the compound will rotate the plane of the polarized light. That is, the maximum transmission of light through the two filters will no longer occur when their axes are parallel; instead, the two filters must be rotated with respect to each other by some characteristic angle, called the *optical rotation*. In fact, it is this particular property, the ability to rotate the plane of polarized light, that most clearly characterizes chiral substances. A substance which rotates the plane of polarized light passing through it is said to be *optically active*.* Compounds which are identical except for different arrangements of groups about one or more saturated carbon atoms are called *optical isomers*. Thus, optical isomers are a particular class of stereoisomers (chapter 3).

EFFECT OF A CHIRAL SUBSTANCE

The greatest amount of optical rotation is observed for a particular compound when only one of the two enantiomers of the compound is present. Opposite enantiomers of a compound rotate polarized light equal amounts but *in opposite directions* (figure 4-5). A *racemic mixture* is a mixture of equal amounts of both enantiomers; it does not rotate the plane of polarized light at all because the effects of the two enantiomers cancel each other out.

The configurations of most asymmetric carbon atoms are stable and have no tendency to change with time. In some cases, however, a simple chemical reaction is capable of converting a particular asymmetric carbon atom into its mirror image. Such a conversion is called *inversion*. Likewise, *racemization* is the conversion of an optically active substance into an equal mixture of two enantiomers (a racemic mixture).

The optical rotation caused by a given amount of a compound is characteristic of that particular compound; the greater the amount of material, the greater the rotation. However, if a sample of a compound contains some of both enantiomers, rather than just one, the rotation observed will depend on the excess of one enantiomer over the other; the greater the excess, the greater the rotation.

---

* Note in this discussion the subtle but important difference between *chiral* and *optically active:* a chiral molecule is one which is not superimposable on its mirror image; an optically active material is one which rotates the plane of polarized light. A particular chemical compound is always either chiral or not, but individual samples of the compound may or may not be optically active, depending on the proportions of the two enantiomers present.

Figure 4-5
Enantiomers.
Enantiomers have
equal and opposite
optical rotations. A
racemic mixture, a
mixture of equal
amounts of the two
enantiomers, has no
optical rotation.

Enantiomers

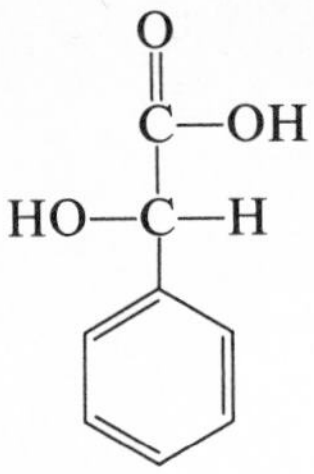

(S)-mandelic acid
optical rotation of a
1% solution in a 10-cm
tube +155°

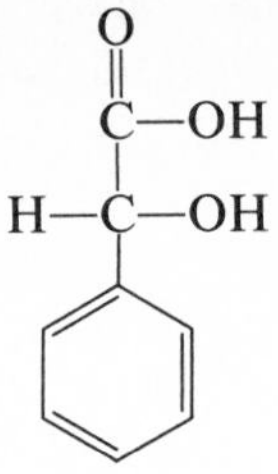

(R)-mandelic acid
optical rotation of a
1% solution in a 10-cm
tube −155°

Racemic Mixture

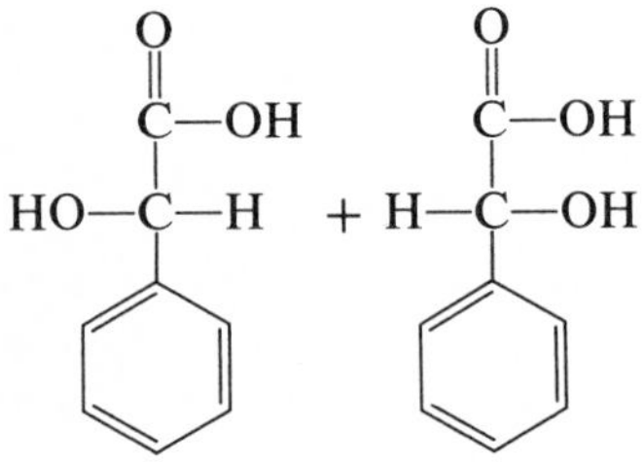

racemic mandelic acid
optical rotation zero

CHEMICAL AND
BIOLOGICAL
SPECIFICITY

Chemical reactions which produce chiral products from nonchiral starting compounds always produce racemic mixtures. In order to obtain an optically active product, it is necessary to use an optically active starting material or an optically active catalyst. The chemical reactions which occur in living systems, on the other hand, usually produce (and consume) only one enantiomer because the catalysts which control reactions in living systems (enzymes) are themselves optically active.

This specificity explains the experiment described in the introduction to this chapter. Alanine is chiral; the chemically synthesized alanine was a racemic mixture, only one enantiomer of which could be utilized by the organism for growth. The natural alanine was exclusively the enantiomer found commonly in nature and could be consumed completely.

PROBLEM 4-4

The synthetic alanine used in the experiment described in the introduction to this chapter showed no optical rotation. However, the alanine remaining after the microorganism had grown showed a very strong negative optical rotation. Explain. Is the optical rotation of naturally occurring alanine positive or negative?

Almost all organic compounds which exist in nature are optically active. Geochemists searching for evidence for the existence of life on the moon and elsewhere in the universe have made use of this fact by looking for optically active organic compounds in soil and rock samples from the moon and in meteorites. The presence of optical activity in such materials would be evidence for the existence of life.

SEARCHING FOR LIFE

4-4   *REPRESENTATION AND SPECIFICATION OF CONFIGURATION*

A chemist who wants to describe a particular compound to another chemist must either use a molecular model of the compound or name or describe the compound in an unambiguous way. Of course, organic compounds are not adequately characterized by their empirical formulas. Structural isomerism can easily be handled by means of the nomenclature rules given in chapter 3 and elsewhere. Stereoisomerism involving double bonds and rings can also be handled easily. However, the description of the structure of an optically active compound is more difficult because it is necessary to distinguish between the compound in question and its enantiomer. Several methods used to distinguish between enantiomers are described in the following sections.

The simplest way to distinguish an optically active compound from its enantiomer is by the sign of the optical rotation it produces. This was, in fact, the earliest method for describing structures, since it can be used even when the structure of the compound is not known. A compound is called $(+)$, or d (for *dextrorotatory*), when it rotates the plane of polarized light to the right. Conversely, a compound is called $(-)$, or l (for *levorotatory*), when it rotates polarized light to the left.

$(+), (-)$ *AND d,l*

Unfortunately, there is no simple correlation between the sign of the optical rotation produced by a compound and the configuration of the asymmetric carbon atom. For example, when $(+)$-lactic acid is dissolved in water, the solution has a positive optical rotation; when $(+)$-lactic acid is dissolved in aqueous sodium hydroxide, converting the acid into its sodium salt, the solution has a negative rotation, even though the configuration of the asymmetric carbon is unchanged.

The *absolute configuration* of a carbon atom is the arrangement of the substituents in space around that carbon atom. A pair of enantiomers have opposite absolute configurations at all asymmetric carbon atoms.

FISCHER PROJECTIONS AND THE D,L *SYSTEM*

A useful method for representing the absolute configurations of asymmetric carbon atoms in many molecules was defined by the famous nineteenth-century German chemist Emil Fischer (figure 4-6). The carbon chain of the compound is arranged vertically, with the most oxidized carbon above the asymmetric carbon.* The groups above and below the asymmetric carbon are considered to project back from the plane of the paper, and the groups on the sides project forward.

When the structure of the compound is pictured in this way, the compound is D (dextro) if the functional group is on the right, or L (levo) if the functional group is on the left. This rule can often be used for compounds containing several asymmetric centers in the same carbon chain. It works well for amino acids (chapter 12) and for carbohydrates (chapter 13), but it is difficult to apply to some types of compounds.† As a result, a more comprehensive system has been devised.

---

*If the chain contains no oxidized carbons, the lower-numbered carbons are placed above the asymmetric carbon atom.

†A complete definition of the stereochemistry of a molecule requires that a connection be made between the sign of the optical rotation [(+) or (−)] of an enantiomer and its absolute configuration (D or L). When Fischer established the D,L system, this connection was unknown and he was forced to make a guess. Sixty years later, his choice was proved correct.

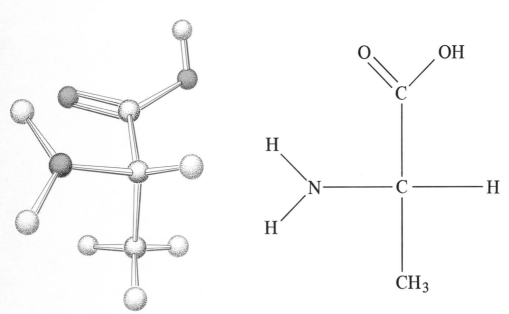

*Figure 4-6
Two views of
L-alanine in the
same orientation.*

On the left is a molecular model. The Fischer projection on the right is assumed to be in the same orientation; the two horizontal substituents project toward the reader, and the two vertical substituents project away from the reader.

Draw Fischer projections of the D isomers of the acyclic compounds **PROBLEM 4-5**
shown in figure 4-3.

It is sometimes necessary to compare two Fischer projections which *MANIPULATING*
are not drawn in the same orientation. In that case, one or both *FISCHER*
molecules must be redrawn so their orientations are identical. The *PROJECTIONS*
following rules govern redrawing of Fischer projections:

1 Interchanging any pair of groups in a Fischer projection inverts the
   configuration of the asymmetric carbon.
2 Interchanging two pairs of groups retains the original configuration.
3 Exchanging the positions of any three groups retains the original
   configuration.

With the help of molecular models, prove the three rules above. **PROBLEM 4-6**

The R and S system is the best method for specifying the absolute *THE R AND S*
configuration of an asymmetric carbon atom. Such a carbon atom *SYSTEM*
has four different substituent groups attached. In the R and S system,
priorities are assigned to the four groups, and the carbon atom is
designated as either R or S according to the arrangement of groups
around the asymmetric carbon atom.

Three rules are used to establish the priority of groups:

1 Atoms of higher atomic number take precedence over those of lower
   atomic number.
2 If two identical atoms are attached to the asymmetric carbon, the
   priority is derived from the second atoms away from the asymmetric
   carbon. When several different second atoms are attached to the
   same first atom, the second atom of highest priority is taken. If no
   choice can be made on this basis, the second atoms of next lower
   priority are compared, etc. If all second atoms are identical, then
   the third atoms are considered, etc. Rule 1 is used to establish
   priorities throughout.
3 If a substituent atom has a double or triple bond, the atom attached
   by that bond is considered to be doubled or tripled before rule 1
   is applied.

According to rule 1, the order of priority of attached atoms is

$I > Br > Cl > F > O > N > C > H$

Thus, in the compound

$$CH_3CH_2\underset{\underset{H}{|}}{\overset{\overset{Cl}{|}}{C}}CH_3$$

the Cl has the highest priority and H the lowest.

According to rule 2, the substituent —CH$_2$OH has priority over —CH$_2$NH$_2$, which in turn has priority over —CH$_2$CH$_3$. Among hydrocarbon groups, —C(CH$_3$)$_3$ has priority over —CH(CH$_3$)$_2$, which has priority over —CH$_2$CH$_3$, which has priority over —CH$_3$.

Rule 3 requires that substituents with double or triple bonds be altered before the priorities are established. For example,

—CH=CH$_2$ is considered to be   —CH—CH$_2$
$|$
C

—CH=O   is considered to be   —CH—O
$|$
O

C
$|$
—C≡CH   is considered to be   —C—CH
$|$
C

Thus —C≡CH has priority over —CH=CH$_2$, which has priority over —CH$_2$—CH$_3$.

After the priority of groups attached to the asymmetric carbon has been established, the assignment of the configuration as R or S is simple. The compound is arranged in space so that the substituent of lowest priority is farthest from the viewer, that is, behind the asymmetric carbon. The three remaining substituents are then seen to be arranged with the priority decreasing clockwise, in which case the configuration is R (for *rectus,* right), or with the priority decreasing counterclockwise, in which case the configuration is S (for *sinister,* left). The operation of the sequence rules is shown in figure 4-7.

WHICH SYSTEM TO USE?   The R and S system is more versatile and less ambiguous than the D,L system, but the D,L terminology has been in such common use in chemistry for so long that it is unlikely to disappear. Thus, for the present at least we should be familiar with both systems. The D,L system continues to be used for carbohydrates and amino acids.

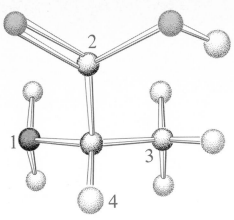

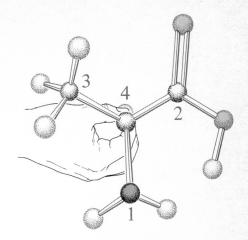

assigning priorities		viewing the model in the proper orientation

(a) *Priorities are first assigned to the four groups attached to the asymmetric carbon atom according to the rules in section 4-4.* (b) *The molecule is then* — *viewed with the lowest priority group at the rear.* (c) *Priorities are seen to decrease in a counterclockwise direction, so the compound is S.*

Figure 4-7 Assigning the S configuration to L-alanine.

Draw Fischer or Haworth projections of (a) 3(R)-methylhexane; (b) 3(S)-methyl-1-pentene; (c) 2(S)-chloro-1,1-dimethylcyclopentane; (d) 3(R)-methylcyclobutene. Use molecular models to check your configurations.

PROBLEM 4-7

4-5 COMPOUNDS WITH MORE THAN ONE ASYMMETRIC CARBON ATOM

If a compound has more than one asymmetric carbon, it has more than two optical isomers. In general, each asymmetric carbon may be either R or S, and a compound with n asymmetric carbons has 2^n different optical isomers, which can be divided into $2^n/2$ pairs of mirror images. However, in some cases fewer isomers exist because some optical isomers are identical with their mirror images. In the following discussion we explore the chirality of compounds with more than one asymmetric center.

When studying compounds with more than two optical isomers, it is useful to consider two structures at a time before trying to understand all structures together. A particular pair of structures may be either enantiomers or diastereomers. The structures are *enantiomers* if they are *mirror images;* that is, every carbon which has the R configuration in one isomer has the S configuration in the other, and vice versa. Stereoisomers which are not enantiomers are called *diastereomers;* at least one asymmetric carbon atom in each structure has the opposite configuration, and at least one has the same configuration.*

A compound with two asymmetric carbons should have $2^2 = 4$ optical isomers, that is, two pairs of enantiomers. Figure 4-8 shows the four optical isomers of 3-chloro-2-butanol. R and S configurations are assigned to asymmetric carbon atoms in molecules like these just as for molecules with a single asymmetric carbon.

PROBLEM 4-8 Build models of the four optical isomers of 3-chloro-2-butanol and confirm that they are all different. Assign R or S configurations to all asymmetric carbon atoms. Draw Fischer projections of the four isomers and indicate the configurations of all asymmetric carbon atoms.

MESO ISOMERS If a compound has two asymmetric centers but both centers have the same four substituents attached, the total number of isomers is three rather than four (figure 4-9). One isomer of such a compound is not chiral because it is identical with its mirror image. This is an example of a *meso compound,* a compound that has asymmetric carbon atoms but is superimposable on its mirror image.

*Diastereomers are defined as any stereoisomers which are not enantiomers. Thus, geometrical isomers are also diastereomers.

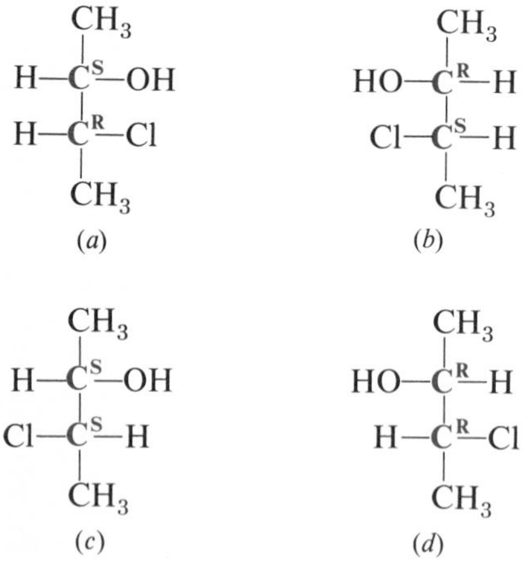

(a)

(b)

(c)

(d)

Figure 4-8
Fischer projections
of the four optical
isomers of
3-chloro-2-butanol.

All asymmetric carbons are labelled as R or S. Structures (a) and (b) are enantiomers, as are (c) and (d). All other pairs are diastereomers. Structures

(a) and (b) are also called erythro (same side) isomers, and (c) and (d) are threo (opposite side) isomers.

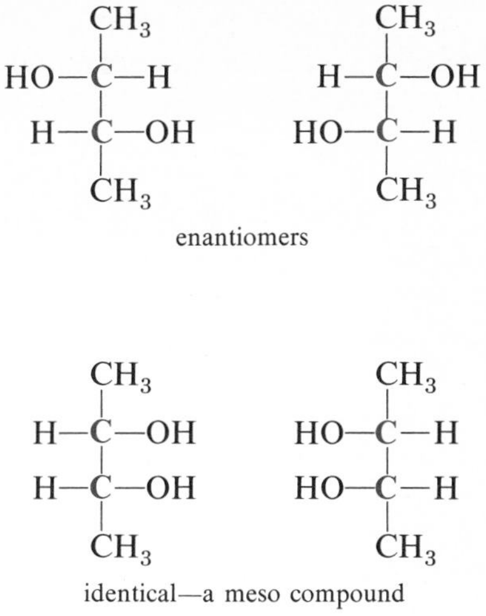

CH₃
|
HO—C—H
|
H—C—OH
|
CH₃

CH₃
|
H—C—OH
|
HO—C—H
|
CH₃

enantiomers

Figure 4-9
Fischer projections
of the three optical
isomers of
2,3-butanediol.
The isomer in which
the two hydroxyl
groups are on the
same side in the
Fischer projection is
identical with its
mirror image and is
therefore a meso
isomer.

CH₃
|
H—C—OH
|
H—C—OH
|
CH₃

CH₃
|
HO—C—H
|
HO—C—H
|
CH₃

identical—a meso compound

Build models of the three optical isomers of 2,3-butanediol. Derive a rule for predicting the absolute configurations of the asymmetric carbon atoms in a meso isomer.

PROBLEM 4-9

Draw Fischer projections of all isomers of each of the following compounds. Assign configurations to all asymmetric carbon atoms, and identify all meso isomers. Use molecular models to check your answers. (*a*) 2,4-dichlorohexane; (*b*) 2-chloro-5-methylhexane; (*c*) 2,6-heptanediol; (*d*) 1,2,4-tribromopentane.

PROBLEM 4-10

When the asymmetric carbon atoms in a chiral compound are part of a ring, the isomerism is more complex than in acyclic compounds. The chirality of some cyclopentanes is illustrated in figure 4-10. A cyclic compound which has two different asymmetric carbons, that is, two asymmetric carbons with different sets of substituent groups attached, has a total of $2^2 = 4$ optical isomers, an enantiomeric pair of cis isomers and an enantiomeric pair of trans isomers. However, when the two asymmetric centers have the same set of substituent groups attached, the cis isomer is a meso isomer and only the trans isomer is chiral.

CYCLIC
COMPOUNDS

Draw Haworth projections of all optical isomers of all compounds in figure 4-10. Use molecular models to help you. Identify pairs of enantiomers.

PROBLEM 4-11

PROBLEM 4-12   Draw Haworth projections of all isomers of dimethylcyclohexane. Indicate which isomers are chiral, and identify pairs of enantiomers. Use molecular models to check your answers.

COMPOUNDS WITH MORE THAN TWO ASYMMETRIC CARBONS

It is a general principle of organic chemistry that the possibilities for isomerism increase with increasing molecular size. Thus, the number of isomers of the saturated acyclic hydrocarbons increases rapidly with the number of carbon atoms (section 3-1). Similarly, the number of optical isomers increases with the number of asymmetric carbons.

Most naturally occurring compounds have at least one asymmetric carbon, and many have a large number. The enzyme *lysozyme,* which degrades the carbohydrate in bacterial cell walls, has 128 asymmetric centers. Thus, there are $2^{128} = 3.4 \times 10^{38}$ optical isomers of lysozyme, but only one of these exists in nature. This very high natural selection for a particular optical isomer is nearly always the rule in living systems.

RECOGNIZING CHIRAL COMPOUNDS

How do you know whether or not a given structure represents a chiral compound? As we have seen, chiral compounds usually contain one or more asymmetric carbon atoms, but we have also seen compounds containing asymmetric carbon atoms which are not chiral. To make matters worse, a few compounds exist which are chiral but contain no asymmetric carbon atoms. Thus, a definitive criterion for chirality is needed.

As we said earlier, a compound is chiral if it is not superimposable on its mirror image. The best way to use this criterion is to build a

Figure 4-10 Chirality of some substituted cyclopentanes.

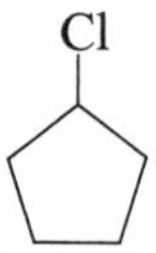

chlorocyclopentane	3-chlorocyclopentene	4-chlorocyclopentene
not chiral	chiral	not chiral

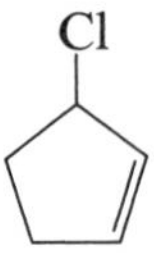

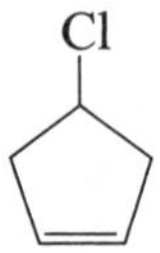

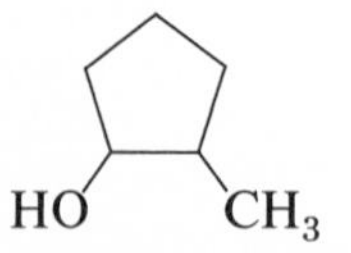

2-methylcyclopentanol
has an enantiomeric pair
of cis isomers and an
enantiomeric pair of
trans isomers

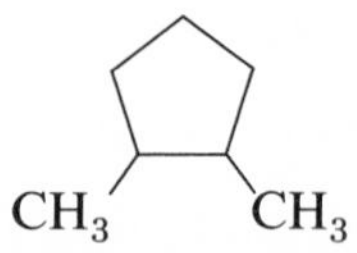

1,2-dimethylcyclopentane
has an enantiomeric pair
of trans isomers; the
cis isomer is a meso
isomer

molecular model of the compound and its mirror image and see whether or not they are identical. Frequently, however, this is not practical. A second method which is often useful is to make careful drawings of the compound and its mirror image.

A useful criterion for chirality is that *a chiral compound does not have a plane of symmetry*. That is, if a compound is not chiral, it is possible to visualize a plane passing through the molecule which divides the molecule into two equal halves which are the mirror images of each other.*

4-6 RESOLUTION

The separation of a racemic mixture into a pair of enantiomers is called *resolution*. This process is frequently of concern in organic chemistry because a single enantiomer, rather than a racemic mixture, may be necessary for studies of reaction mechanism or biological activity. Resolution can be achieved by chemical or biological means. Although these two methods are superficially different, they both make use of the same principle: enantiomers interact differently with an optically active compound.

PROPERTIES OF ENANTIOMERS

A pair of enantiomers have identical physical properties such as appearance, melting point, solubility, boiling point, and light absorption.† Because of this identity, it is not possible to separate enantiomers by any of the standard techniques for separating organic compounds;‡ instead, more subtle methods are needed.

The chemical properties of enantiomers are identical also, except when the reagent with which they are reacting is itself optically active. For example, the R and S enantiomers of 3-methyl-1-pentene

$$CH_3CH_2\overset{\overset{\displaystyle CH_3}{|}}{C}HCH{=}CH_2$$

3-methyl-1-pentene

---

*There are a handful of exceptions to this rule.

†Enantiomers may differ in taste and smell, but this occurs because our taste and smell receptors are themselves optically active.

‡There are occasional exceptions to this generalization. For a few compounds, crystallization from a supersaturated solution results in the formation of some crystals which are pure R and some which are pure S, rather than racemic. It is one of the fortunate accidents of science that the first compound to be resolved, sodium ammonium tartrate (resolved by Pasteur in 1848), is such a compound. Pasteur's demonstration of the production of an optically active compound from optically inactive starting material predated the concept of the tetrahedral carbon atom by nearly twenty years. Even so, it is clear from Pasteur's writings that he conceived of optical activity in terms of some molecular asymmetry, even though he did not know the nature of the asymmetry.

react at identical rates with H_2SO_4, and the product of reaction of the R isomer is the mirror image of the product of reaction of the S isomer. If a racemic mixture is reacted, the product alcohol has no optical activity.

PROBLEM 4-13   Use molecular models to show that the 2-bromobutane formed by reaction of HBr with 1-butene is racemic.

CHEMICAL RESOLUTION   Although enantiomers have identical reactivity toward reagents which are not optically active, they have different reactivities toward reagents which are optically active. This difference is the basis of the most common method of chemical resolution.

When a racemic mixture reacts with an optically active reagent, the two enantiomers may react at different rates and the products formed from the two enantiomers are not enantiomers but diastereomers. Unlike enantiomers, diastereomers have different physical properties and can therefore be separated. This forms the basis for the most important method of resolution,* illustrated in figure 4-11.

Thus, chemical resolution depends on the availability of an optically active compound to use as a resolving agent. Fortunately, many such compounds are available from natural sources.

---

*This method of resolution is exactly analogous to the separation of a sackful of gloves (a racemic mixture) into a pile of left gloves and a pile of right gloves. The gloves weigh and feel exactly the same, but if the person who is separating them tries to fit each glove on his left hand and puts those which fit into one pile and those which do not into another, the racemic mixture will be resolved. Note that the resolving agent (the person's left hand) is optically active.

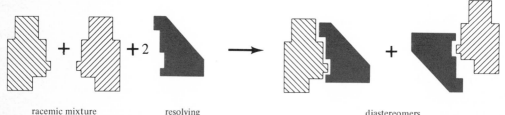

racemic mixture   resolving   diastereomers
                          agent

Figure 4-11
Resolution of a
chiral molecule by
reaction with an
optically active
reagent.

The products of reaction of the two enantiomers with the optically active resolving agent are diastereomers, rather *than enantiomers, and can therefore be separated.*

Enzymes, the catalysts which accelerate and control nearly all chemical reactions in living systems, are optically active, being made exclusively from L-amino acids. Most enzymes will catalyze the reaction of only one enantiomer of a substance; the other enantiomer may be inert, or it may actually inhibit the action of the enzyme on the reactive enantiomer.

The ability of enzymes to distinguish between enantiomers arises from the presence on the enzyme of a *chiral binding site,* a hole in the enzyme with a shape that is complementary to the shape of the enantiomer whose reaction it catalyzes. The reactive enantiomer binds to this site because of specific attractions between particular parts of the reactive enantiomer and corresponding areas of the binding site. Only after this recognition and binding have occurred can the enzyme proceed with the chemical reaction.

This enantiomeric specificity of enzymes forms the basis of a very useful method of resolution. If an enzyme is allowed to react with a racemic mixture, only one enantiomer will be consumed and the other will be left behind. An example of this method of resolution is shown in figure 4-12. Unfortunately, the isomer consumed by this process is frequently the isomer wanted for further studies. In fortunate cases, two different enzymes are available with activities toward opposite enantiomers.

It is sometimes possible to accomplish this same resolution by growing a microorganism and using a racemic mixture of the desired compound as the principal nutrient. If the microorganism utilizes only one enantiomer efficiently for growth, the other enantiomer will be left behind and can be recovered after growth is complete (recall problem 4-4). In addition to his separation of tartrate salts by crystallization, Pasteur separated them by this method.

4-7 OPTICAL ACTIVITY IN LIVING SYSTEMS

Nature is optically active; almost all compounds of importance in living systems have one or more asymmetric carbon atoms. The configurations of optically active compounds in nature form a consistent and interlocking pattern throughout all living things. The majority of naturally occurring compounds exist in nature in the form of only one of the many possible optical isomers. This is particularly true of such important and complex molecules as DNA, RNA, proteins, and carbohydrates.

The enzymes which catalyze the reactions occurring in living things are large molecules called *proteins* (chapter 12). These proteins are composed of many amino acids held together by covalent bonds. All but one of the twenty important amino acids are chiral, and when they occur in proteins, they are always in the L configuration—whether that

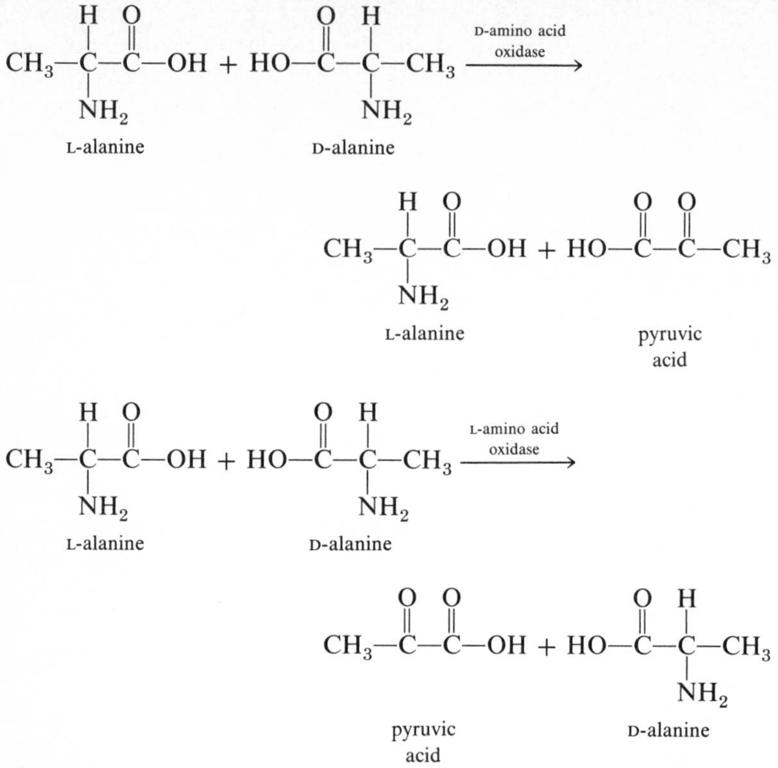

$$CH_3-\overset{\overset{\displaystyle H}{|}}{\underset{\underset{\displaystyle NH_2}{|}}{C}}-\overset{\overset{\displaystyle O}{\|}}{C}-OH + HO-\overset{\overset{\displaystyle O}{\|}}{C}-\overset{\overset{\displaystyle H}{|}}{\underset{\underset{\displaystyle NH_2}{|}}{C}}-CH_3 \xrightarrow{\text{D-amino acid oxidase}}$$

L-alanine   D-alanine

$$CH_3-\overset{\overset{\displaystyle H}{|}}{\underset{\underset{\displaystyle NH_2}{|}}{C}}-\overset{\overset{\displaystyle O}{\|}}{C}-OH + HO-\overset{\overset{\displaystyle O}{\|}}{C}-\overset{\overset{\displaystyle O}{\|}}{C}-CH_3$$

L-alanine   pyruvic acid

$$CH_3-\overset{\overset{\displaystyle H}{|}}{\underset{\underset{\displaystyle NH_2}{|}}{C}}-\overset{\overset{\displaystyle O}{\|}}{C}-OH + HO-\overset{\overset{\displaystyle O}{\|}}{C}-\overset{\overset{\displaystyle H}{|}}{\underset{\underset{\displaystyle NH_2}{|}}{C}}-CH_3 \xrightarrow{\text{L-amino acid oxidase}}$$

L-alanine   D-alanine

$$CH_3-\overset{\overset{\displaystyle O}{\|}}{C}-\overset{\overset{\displaystyle O}{\|}}{C}-OH + HO-\overset{\overset{\displaystyle O}{\|}}{C}-\overset{\overset{\displaystyle H}{|}}{\underset{\underset{\displaystyle NH_2}{|}}{C}}-CH_3$$

pyruvic acid   D-alanine

Figure 4-12 Resolution of alanine by enzymes.   *The enzyme L-amino acid oxidase can bind L-alanine but not D-alanine. As a result, if the enzyme is given a mixture of the two enantiomers, the L isomer is converted to pyruvic acid, whereas the D isomer remains unchanged. This provides a convenient preparation of D-alanine. The opposite result is obtained with D-amino acid oxidase.*

protein is from man, from another animal, from a plant, or from any other living thing. Why all configurations are L rather than D has been the subject of much speculation, but no compelling theory has been advanced. It is possible that the selection of L over D was simply a matter of chance at some point in the early evolution of life. If that is so, there is a chance that elsewhere in the universe a world exists that is exclusively D. If we met intelligent beings from such a world, we might be able to communicate with them but we could not share their food.*

*The possibility of such a meeting seems to have been discussed first by Lewis Carroll. In *Through the Looking-Glass,* Alice speculates on the possibility that her kitten would find that looking-glass milk is not good to drink.

Write a two- to three-page science-fiction story about space travelers   *PROBLEM 4-14*
from earth landing on a planet populated by D people.

Although D-amino acids are not widespread in the environment, they   *D-AMINO ACIDS*
occur to a small extent, both through synthesis by microorganisms and
insects and through racemization of L-amino acids. The enzymes
involved in amino acid metabolism are for the most part not reactive
toward D-amino acids. A special enzyme present in mammals, called
D-*amino acid oxidase,* oxidizes amino acids to ketones, thus destroying
the optical activity (figure 4-12). The presence of this enzyme ensures
that D-amino acids do not accumulate in potentially harmful amounts.

The cell walls of many bacteria are constructed from carbohydrate
(chapter 13) and a special structural protein which contains two amino
acids of the D configuration. The destruction of cell walls usually
involves degradation of the protein by enzymes; since the protein here
is partially made up of D-amino acids, it is resistant to degradation
by enzymes and the cell is safer from destruction than it would have
been if only L-amino acids had been used.

Drugs exert their effects by specifically altering biological structure or   *OPTICAL*
reactivity. Because of the pervasiveness of optical activity in living   *ACTIVITY OF*
systems, it is not surprising that only one optical isomer of most drugs   *DRUGS*
has biological activity.

Morphine, probably the most used analgesic of all time, occurs in
nature exclusively as one enantiomer

morphine

The mirror image of this compound has been synthesized (it does not
occur in nature) and it lacks analgesic properties.

L-Dopa, used in the treatment of Parkinson's disease (section 6-7), is biologically active (it occurs naturally in the central nervous system), but its mirror image, D-dopa, is without activity.

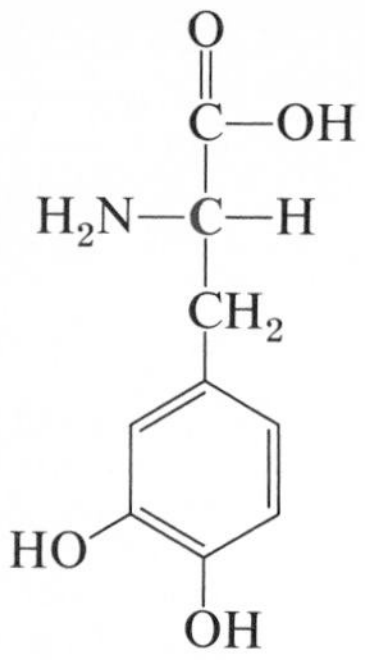

$$
\begin{array}{c}
O \\
\| \\
C-OH \\
| \\
H_2N-C-H \\
| \\
CH_2
\end{array}
$$

HO

OH

L-dopa

Some drugs which affect the central nervous system show a much lower optical specificity. *Ephedrine*, which is used in the treatment of asthma and other conditions, occurs naturally in a number of plants and has been used in Chinese medicine for over 5000 years.

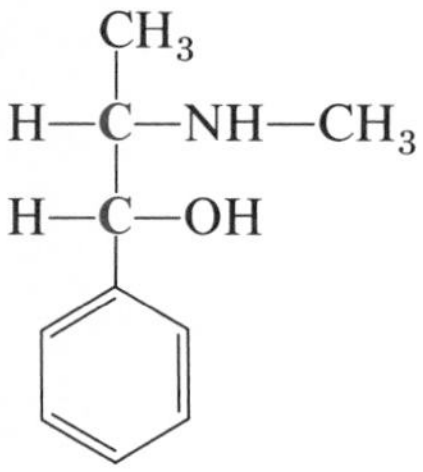

$$
\begin{array}{cc}
CH_3 & CH_3 \\
| & | \\
H-C-NH-CH_3 & H-C-NH-CH_3 \\
| & | \\
H-C-OH & HO-C-H
\end{array}
$$

natural
ephedrine

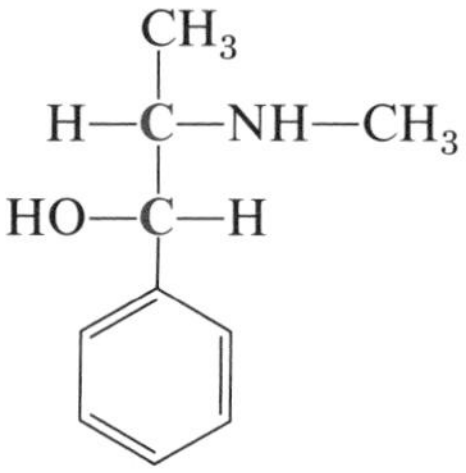

ψ-ephedrine
(chemically synthesized),
a diastereomer of
natural ephedrine

The natural material is exclusively of one configuration, but its enantiomer (which was synthesized chemically) is of nearly equal potency. Ephedrine has two asymmetric carbon atoms, and the two diastereomers of natural ephedrine (called ψ-ephedrine) are of much lower biological activity.

The chemical synthesis of drugs leads to the formation of racemic mixtures unless the starting materials are themselves optically active. Although only one enantiomer of the racemic mixture may be biologically active, it is often the practice to administer the racemic mixture to avoid the expensive and time-consuming process of resolving the compound. Of course, this is possible only when the unwanted enantiomer has no harmful effects.

4-15 Give the structure or structures of (*a*) a meso compound; (*b*) a **PROBLEMS** pair of enantiomers; (*c*) a pair of diastereomers; (*d*) a meso compound with four asymmetric carbon atoms.

4-16 Draw the structures of all isomers of methyloctane and indicate which are chiral. Do the same for methylnonane.

4-17 If 0.2 gram of a pure enantiomer of a chiral compound rotates polarized light 2.8°, how much rotation will be obtained from 0.1 gram of its mirror image under the same conditions? How much rotation will be obtained from 0.5 gram of a mixture of equal parts of the two enantiomers?

4-18 The barrier to rotation of the carbon-carbon single bond in ethane is very small, and the ends rotate nearly freely with respect to each other. If the rotation barrier were so large that the carbon-carbon single bond were frozen in a particular conformation, which of the following compounds would be chiral: (*a*) ethane (eclipsed conformation); (*b*) ethane (staggered conformation); (*c*) propane (staggered conformation); (*d*) 1,2-dichloroethane (gauche conformation); (*e*) 1,2-dichloroethane (trans conformation)?

4-19 Many hydrocarbons are chiral. Draw as many chiral hydrocarbons as you can which have six or fewer carbon atoms (there are at least twenty such compounds).

4-20 Inositol has nine optical isomers, seven meso isomers and one pair of enantiomers. Draw them.

inositol

4-21 Assign R and S configurations to all asymmetric carbons in the following compounds. Assume that structures (*a*), (*b*), and (*d*) are Fischer projections.

$$CH$$
$$\|\|$$
$$C$$

(d) $CH_3-\overset{\displaystyle |}{\underset{\displaystyle |}{C}}-H$

$CH=CH_2$

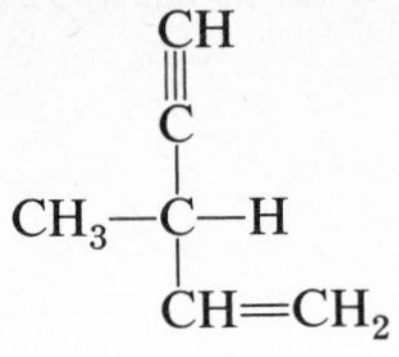

(e)

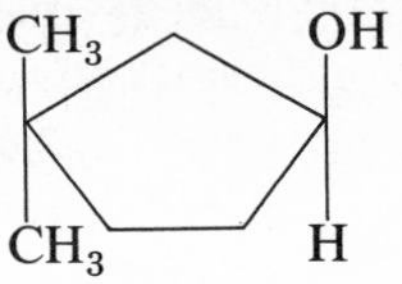

4-22 Which of the following compounds are chiral? Confirm your answers by use of molecular models.

(a)

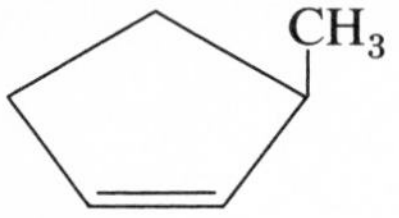

CH$_3$

(b)

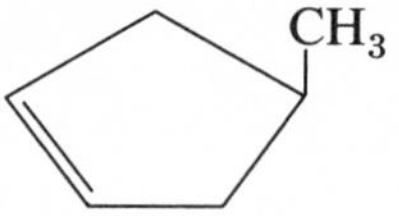

CH$_3$

(c)

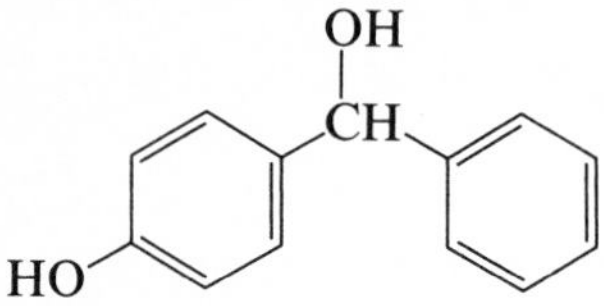

OH

(d)

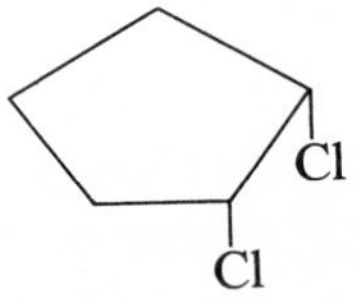

Cl

Cl

(e) CH$_3$

Br

CH$_3$

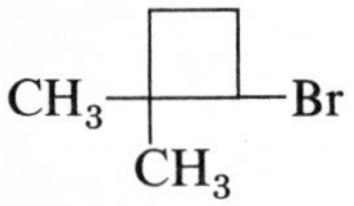

SUGGESTED READINGS   See references on stereochemistry at the end of chapter 3.

5

**Alcohols,
Alkyl Halides,
and Related
Compounds**

In chapter 3 we discussed the structures and chemistry of hydrocarbons at all oxidation levels. In this chapter and the next we discuss the structures and chemistry of organic compounds at oxidation level 1. The compounds to be considered are shown in figure 5-1. With the exception of olefins, all compounds at this oxidation level contain at least one atom other than carbon and hydrogen.

Most of the chemical reactions to be considered in this chapter are of two types:

1 *Nucleophilic substitution,* in which one group on a carbon atom is substituted for another

2 *Elimination,* in which a double bond is formed by loss of a hydrogen and some other group from adjacent carbon atoms

5-1 STRUCTURE AND BONDING

ALKYL HALIDES   Carbon forms strong σ bonds to fluorine, chlorine, bromine, and iodine. The structure of a typical alkyl halide, chloromethane, is shown in figure 5-2.

With the exception of iodine, all the halogens are more electronegative than carbon, and carbon-halogen σ bonds are therefore

Figure 5-1
The types of compounds which have a carbon atom at oxidation level 1.

Hydrocarbons
(chapter 3)

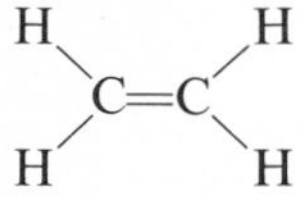

$$H \quad\quad H$$
$$\diagdown\quad\quad\diagup$$
$$C{=}C$$
$$\diagup\quad\quad\diagdown$$
$$H \quad\quad H$$
olefins

Oxygen Compounds

$CH_3{-}OH$
alcohols

$CH_3{-}O^-$ Na^+
alkoxide salts

$CH_3{-}O{-}CH_3$
ethers

Alkyl Halides

$CH_3{-}F$
alkyl fluorides

$CH_3{-}Cl$
alkyl chlorides

$CH_3{-}Br$
alkyl bromides

$CH_3{-}I$
alkyl iodides

Nitrogen Compounds
(chapter 6)

$CH_3{-}NH_2$
amines

$CH_3{-}NH_3{}^+$ Cl^-
ammonium salts

Sulfur Compounds

$CH_3{-}SH$
mercaptans

$CH_3{-}S{-}CH_3$
sulfides

$CH_3{-}S{-}S{-}CH_3$
disulfides

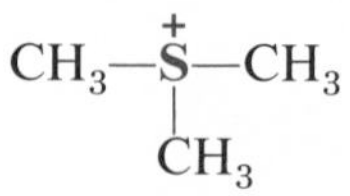

$CH_3{-}\overset{+}{S}{-}CH_3$
$|$
CH_3
sulfonium salts

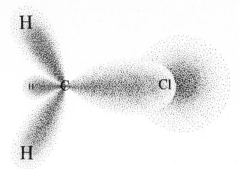

The carbon atom is sp³-hybridized and forms four σ bonds to the other atoms. The chlorine atom has six unshared outer-shell electrons (*shown here in brown*). Because of the greater electronegativity of chlorine, the carbon-chlorine bond is polar.

**Figure 5-2
Chloromethane.**

polar; the electron density is greater near the halogen atom and lower near the carbon. The carbon atom is thus rendered slightly *electron-deficient* and is subject to attack by electron-rich reagents, called *nucleophiles* (section 5-6).

ALCOHOLS

The structure of methanol, the simplest alcohol, is shown in figure 5-3. The oxygen atom has two unshared pairs of electrons occupying sp^3 hybrid orbitals. These electrons may become involved in hydrogen bonding (section 5-5). One electron pair (but not both) can become bonded to a proton when methanol acts as a base (section 5-3).

Since oxygen is more electronegative than carbon, the carbon atom in alcohols is slightly electron-deficient and the oxygen is slightly

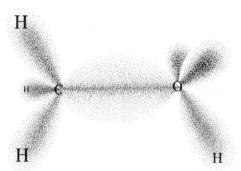

The carbon atom is sp³-hybridized and forms four σ bonds to other atoms. The oxygen atom is also sp³-hybridized; two of the hybrid orbitals are occupied by unshared pairs of electrons (*shown here in brown*). Because oxygen is more electronegative than carbon, the carbon-oxygen bond is polar.

**Figure 5-3
Methanol.**

electron-rich. This polarity has an important influence on the properties of alcohols.

PROBLEM 5-1 Considering that the oxygen in methanol is sp^3-hybridized, what would you expect the C—O—H bond angle to be?

PROBLEM 5-2 When methanol reacts with strong acids, a small amount of the conjugate acid of methanol, CH_3—OH_2^+, is formed (section 5-3). The oxygen atom in the conjugate acid is still sp^3-hybridized. What is its geometry?

5-2 NOMENCLATURE

A *primary* alcohol or halide is one which has *one carbon atom* attached to the carbon bearing the hydroxyl group or the halogen; a *secondary* compound has *two* carbons attached, and a *tertiary*, three:

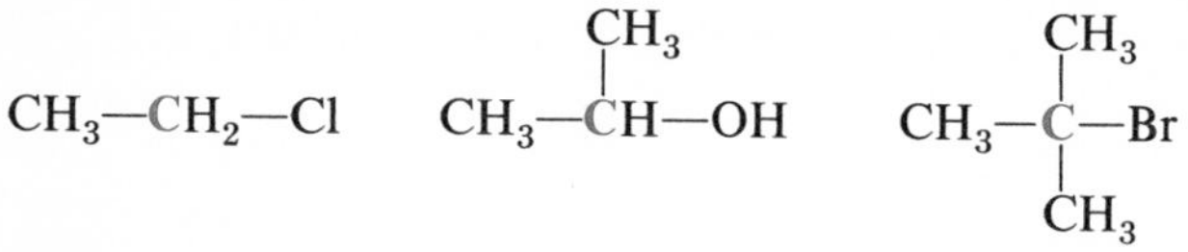

$$CH_3-CH_2-Cl \qquad CH_3-\overset{\displaystyle CH_3}{\underset{\displaystyle }{CH}}-OH \qquad CH_3-\overset{\displaystyle CH_3}{\underset{\displaystyle CH_3}{C}}-Br$$

a primary chloride   a secondary alcohol   a tertiary bromide

FUNCTIONAL CLASS NAMES

Two systems for naming alcohols and alkyl halides are in common use. Simple compounds containing up to about five carbon atoms are frequently named by *functional class names* using the group names in table 3-2. For more complex compounds the more systematic IUPAC names should be used.

Functional class names are constructed simply by adding the appropriate word *alcohol, chloride, bromide*, etc., to the group name from table 3-2. Examples are

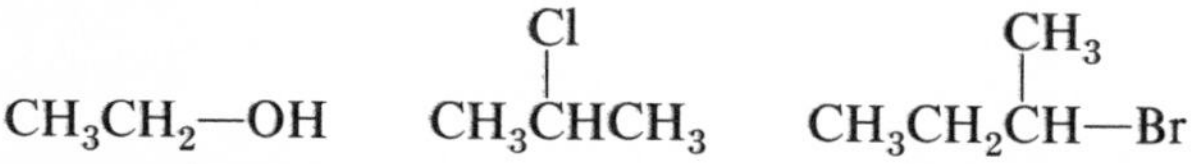

$$CH_3CH_2-OH \qquad CH_3\overset{\displaystyle Cl}{\underset{\displaystyle }{CH}}CH_3 \qquad CH_3CH_2\overset{\displaystyle CH_3}{\underset{\displaystyle }{CH}}-Br$$

ethyl   isopropyl   s-butyl
alcohol   chloride   bromide

IUPAC NAMES OF HALIDES

The IUPAC names of alkyl halides and of alcohols are based on the same system as the names of hydrocarbons. Alkyl halides are named by using the following rules:

1 Find the longest continuous carbon chain containing the halogen atom. The final part of the name is the name of that hydrocarbon (table 3-1).

2 Number the chain starting from the end which gives the halogen atom the smaller number.

3 Indicate the nature and position of the halogen by number and by the prefix *fluoro-, chloro-, bromo-,* or *iodo-.*

4 Name and number carbon substituents on the chain using the group names from table 3-2.

5 The presence of two, three, or more identical groups is indicated by the prefix *di-, tri-,* etc., and a corresponding set of position numbers.

The names of some halides are shown in figure 5-4.

The systematic names of alcohols are derived like those of other compounds. In this case, the ending *-ol* is added to the parent hydrocarbon name to indicate the presence of the hydroxyl group.

IUPAC NAMES OF ALCOHOLS

1 Find the longest continuous chain of carbon atoms containing the hydroxyl group. The final part of the name of the compound is the name of that hydrocarbon, with the alcohol ending *-ol* substituted for the final *-e* of the hydrocarbon name.

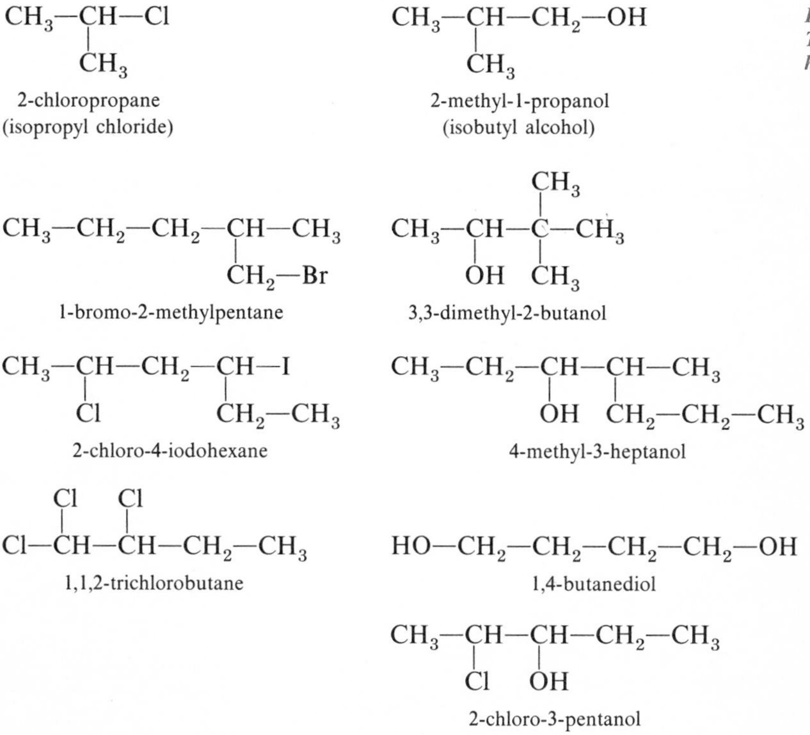

CH_3—CH—Cl
    |
    CH_3

2-chloropropane
(isopropyl chloride)

CH_3—CH—CH_2—OH
    |
    CH_3

2-methyl-1-propanol
(isobutyl alcohol)

Figure 5-4
The names of some halides and alcohols.

CH_3—CH_2—CH_2—CH—CH_3
              |
              CH_2—Br

1-bromo-2-methylpentane

CH_3—CH—C—CH_3
    | |
    OH CH_3

(with CH_3 above the C)

3,3-dimethyl-2-butanol

CH_3—CH—CH_2—CH—I
    | |
    Cl CH_2—CH_3

2-chloro-4-iodohexane

CH_3—CH_2—CH—CH—CH_3
              | |
              OH CH_2—CH_2—CH_3

4-methyl-3-heptanol

Cl   Cl
| |
Cl—CH—CH—CH_2—CH_3

1,1,2-trichlorobutane

HO—CH_2—CH_2—CH_2—CH_2—OH

1,4-butanediol

CH_3—CH—CH—CH_2—CH_3
    | |
    Cl   OH

2-chloro-3-pentanol

2 Number the carbon atoms of the main chain sequentially starting from the end nearer the hydroxyl group.

3 Locate and name substituents on the main chain as usual.

4 Compounds with two, three, or more hydroxyl groups are called *-diol, -triol,* etc., and the numbers of all of the hydroxyl groups are given.

The names of some alcohols are shown in figure 5-4.

5-3 ACID-BASE PROPERTIES OF ALCOHOLS

The acid-base properties of alcohols are very similar to those of water. Both water and alcohols can be protonated by strong proton donors, both can give up a proton to strong proton acceptors, and both can form salts. An understanding of the acid-base chemistry of water will make the acid-base properties of alcohols clear.

ACIDS AND BASES
According to the Brønsted-Lowry definition of acids and bases, an acid is a *proton donor* and a base is a *proton acceptor.* Thus, for example, when dry HCl dissolves in water, an acid-base reaction ensues:

$$HCl + H_2O \rightleftharpoons H_3O^+ + Cl^-$$

acid   base   conjugate   conjugate
                   acid   base

The acid, HCl, transfers a proton to the base, H_2O, forming the acid H_3O^+, which is the *conjugate acid* of H_2O, and the base Cl^-, which is the *conjugate base* of HCl. Like all acid-base reactions, this one is an equilibrium. In this case the equilibrium lies almost entirely on the right because HCl is a much stronger acid than H_3O^+.

The proton, or hydrogen ion, occupies a unique place in chemistry. Hydrogen forms strong bonds to many nuclei, including oxygen, carbon, chlorine, and nitrogen, but in a number of cases these bonds can be made and broken very rapidly—much more rapidly than bonds involving any other nucleus. Chemical reactions in which a proton is transferred between oxygen, nitrogen, or sulfur atoms are among the fastest chemical reactions known. Proton transfers involving carbon, however, are much slower.

ACID STRENGTH
The strength of an acid is indicated by the *acid dissociation constant.* For ionization of the acid HA according to the equilibrium

$$HA \rightleftharpoons H^+ + A^-$$

the acid dissociation constant K_a is defined by

$$K_a = \frac{[H^+][A^-]}{[HA]}$$

where the brackets represent concentrations. It is more usual to report pK_a than K_a:

$$pK_a = -\log K_a$$

Thus, the stronger an acid, the *smaller* its pK_a. For example, ammonium ion, NH_4^+, pK_a 9.2, is a much weaker acid than HCl, pK_a about -7.

When an acid HA and a base B are mixed, the equilibrium

$$HA + B \rightleftharpoons A + HB$$

(in which charges have been omitted) will lie on the right side if the pK_a of HA is less than that of HB; it will lie on the left if the pK_a of HB is less than that of HA. If the pK_a values for the two acids differ by three units or less, there will be appreciable amounts of all species present at equilibrium.

By this means we can calculate what happens when NH_3 and HCl are mixed. The pK_a of HCl is less than that of NH_4^+; as a result, an equimolar mixture of NH_3 and HCl will exist as NH_4^+ and Cl^-.

An alternative, completely equivalent approach to the relative strengths of acids is to consider instead the relative strengths of the conjugate bases. In the above example if the equilibrium lies on the right, we can say either that HA is a stronger acid than HB or that B is a stronger base than A. The conjugate base of a strong acid is a weak base; the conjugate base of a weak acid is a strong base. For example, HCl is a strong acid, and its conjugate base, Cl^-, is a very weak base; NH_4^+, on the other hand, is a weak acid, and its conjugate base, NH_3, is a moderately strong base.

BASE STRENGTH

Many inorganic compounds are strong acids; for example, HCl and H_2SO_4. By contrast, few organic compounds are strong acids, and many organic compounds are not acidic at all. The pK_a values for some organic and inorganic compounds are given in table 5-1.

Alcohols have acid-base properties much like those of water. In the presence of a strong acid, such as HCl, an alcohol can be converted into its conjugate acid

THE STRENGTHS OF ORGANIC ACIDS

$$CH_3-O-H + HCl \rightleftharpoons CH_3-\overset{\overset{\displaystyle H}{|}}{O}{}^+-H + \quad Cl^-$$

base   acid   conjugate   conjugate
                                   acid   base

Table 5-1
Acid Dissociation
*Constants**

Acid	Conjugate Base	pK_a
HCl	Cl$^-$	-7
$\overset{\displaystyle \text{H}}{\underset{\displaystyle \quad}{\text{CH}_3\text{—O}^+\text{—CH}_3}}$	CH$_3$—O—CH$_3$	-3.8
CH$_3$—OH$_2$$^+$	CH$_3$—OH	-2.5
H$_3$O$^+$	H$_2$O	-1.7
H$_3$PO$_4$	H$_2$PO$_4$$^-$	2.1
$\overset{\displaystyle \text{O}}{\overset{\displaystyle \|}{\text{CH}_3\text{—C—OH}}}$	$\overset{\displaystyle \text{O}}{\overset{\displaystyle \|}{\text{CH}_3\text{—C—O}^-}}$	4.8
NH$_4$$^+$	NH$_3$	9.2
CH$_3$CH$_2$—SH	CH$_3$CH$_2$—S$^-$	10.5
CH$_3$—OH	CH$_3$—O$^-$	15.5
H$_2$O	OH$^-$	15.7
CH$_3$CH$_2$—OH	CH$_3$CH$_2$—O$^-$	16
$\underset{\displaystyle \text{CH}_3}{\overset{\displaystyle \text{CH}_3}{\text{CH}_3\text{—C—OH}}}$	$\underset{\displaystyle \text{CH}_3}{\overset{\displaystyle \text{CH}_3}{\text{CH}_3\text{—C—O}^-}}$	19

*Data from H. A. Sober, *Handbook of Biochemistry,* Chemical Rubber Co., Cleveland, Ohio, 1968.

just as water is converted into its conjugate acid, H$_3$O$^+$. The pK_a values for H$_3$O$^+$ and CH$_3$OH$_2$$^+$ are similar, indicating that these two species are acids of similar strength (table 5-1). Consequently, H$_2$O and CH$_3$OH are bases of similar strength.

Alcohols, like water, also act as acids. In so doing, they are converted into their conjugate bases, alkoxide ions

$$\text{CH}_3\text{—O—H} + \text{base} \rightleftharpoons \text{CH}_3\text{—O}^- + {}^+\text{H—base}$$

$$\underset{\substack{\text{alcohol} \\ \text{(acid)}}}{\qquad} \qquad \underset{\substack{\text{alkoxide} \\ \text{(conjugate base)}}}{\qquad}$$

The pK_a values of alcohols are similar to that of water (table 5-1).

ALKOXIDE SALTS   The conjugate base of water, hydroxide ion, forms stable, isolable salts with many metal ions; for example, Na$^+$ $^-$OH and Ca^{2+} ($^-$OH)$_2$. The conjugate bases of alcohols, alkoxide ions, also form such salts; for example, Na$^+$ $^-$O—CH$_3$ and K$^+$ $^-$O—CH$_2$CH$_2$CH$_3$. Such salts are frequently prepared by the reaction of an alcohol with an alkali metal:

$$CH_3CH_2{-}O{-}H + Na \longrightarrow CH_3CH_2{-}O^- \quad Na^+ + \tfrac{1}{2}H_2$$

$\qquad$ ethanol $\qquad\qquad\qquad\qquad\qquad$ sodium ethoxide

Although sodium ethoxide is an ionic compound, being composed of equal numbers of Na^+ ions and $CH_3CH_2{-}O^-$ ions, its structure is frequently represented as $CH_3CH_2{-}ONa$, just as the structure of sodium hydroxide is frequently represented as NaOH.

Alkoxides are strong bases and readily remove a proton from a variety of acids:

$$HCl + Na^+ \;\; {}^-O{-}CH_2CH_3 \longrightarrow Na^+ \;\; Cl^- + H{-}O{-}CH_2CH_3$$
$$HPO_4{}^{2-} + CH_3{-}O^- \longrightarrow PO_4{}^{3-} + CH_3{-}O{-}H$$

Complete each of the following acid-base reactions and use the pK_a values given in table 5-1 to predict the positions of the equilibria: (*a*) $HCl + CH_3CH_2O^-$; (*b*) $CH_3OH_2{}^+ + NH_3$; (*c*) $CH_3OH + CH_3OCH_3$; (*d*) $H_2O + (CH_3)_3CO^-$. $\qquad$ **PROBLEM 5-3**

5-4 HALIDES AND ALCOHOLS IN BIOCHEMISTRY

Very few halogen compounds are important in the metabolism of higher animals. Chloride ion is required for health because of the essential role of chloride as an electrolyte in all cells. However, it does not appear that this chloride is ever converted into chlorine-containing organic compounds in the course of normal metabolism.

Iodide ion is also an essential nutrient. Ingested iodide is concentrated in the thyroid gland, where a sequence of chemical reactions results in the formation of *thyroxine,* a hormone which controls the rates of many reactions in metabolism.

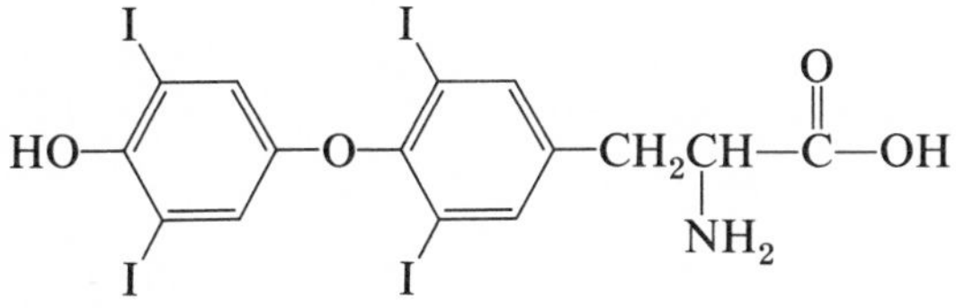

thyroxine

Although few halogen-containing compounds are natural components of animal metabolism, a number of such compounds have found wide use as drugs. Several examples are shown in figure 5-5. Some of these compounds were first isolated from natural sources, but most were first produced by chemical synthesis. *HALOGEN-CONTAINING DRUGS*

Figure 5-5 Some halogen-containing drugs.

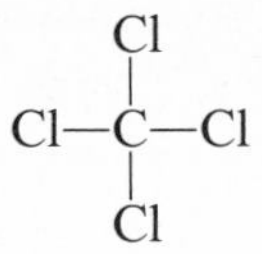

carbon tetrachloride
once used as an anesthetic; highly toxic to the liver

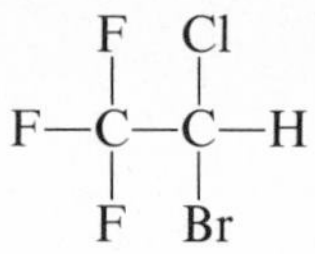

halothane
widely used anesthetic; nontoxic and nonflammable

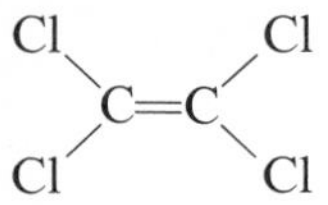

tetrachloroethylene
effective treatment for hookworm

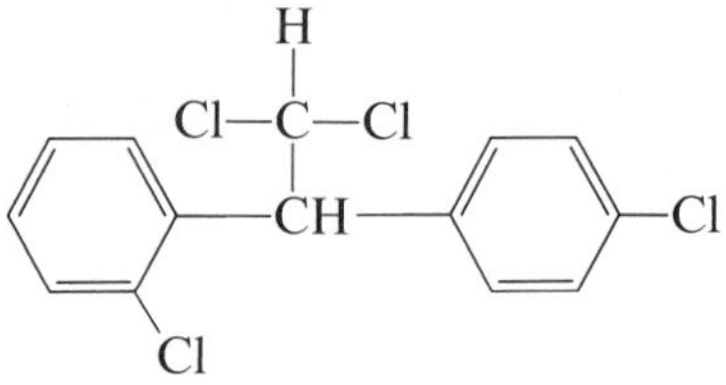

mitotane
used in treatment of brain cancer; close structural relative of DDT

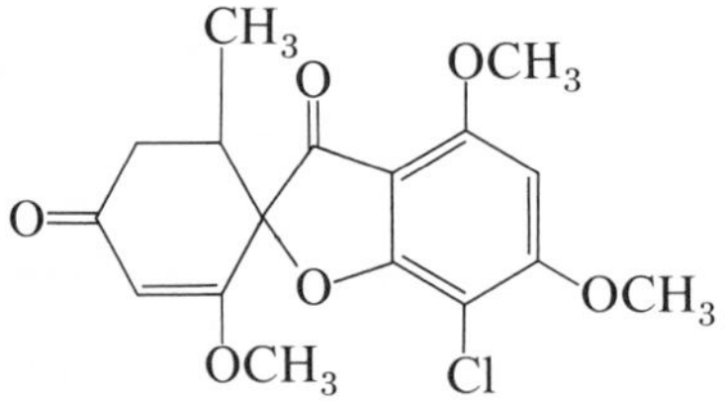

griseofulvin
used in treatment of fungal infections; effective because it interferes with the synthesis of the cell wall of the fungus

ALCOHOLS IN METABOLISM

Unlike halides, alcohols are integral components of the chemistry of every plant and animal. The naturally occurring alcohols range from simple alcohols such as ethanol to complex multifunctional compounds such as DNA, proteins, and carbohydrates. A number of important naturally occurring alcohols are shown in figure 5-6.

STEROIDS

The metabolism of an animal is a closely regulated process. The rates at which chemical reactions occur in metabolism are often controlled by special compounds called *hormones,* which can increase and decrease the rates of metabolic processes according to the needs of the organism.

An important class of hormones is the *steroids,* a number of which

$$CH_3-CH_2-OH$$

ethanol
produced by fermentation

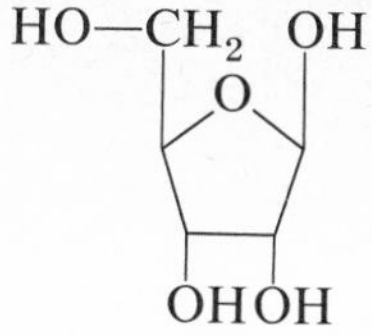

ribose
a sugar

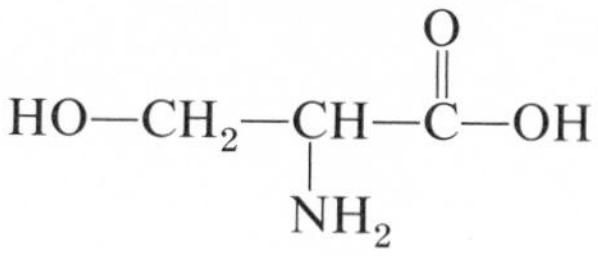

$$HO-CH_2-\underset{\underset{NH_2}{|}}{CH}-\overset{\overset{O}{\|}}{C}-OH$$

serine
an amino acid

$$CH_3-\underset{\underset{}{\overset{\overset{OH}{|}}{CH}}}-\overset{\overset{O}{\|}}{C}-OH$$

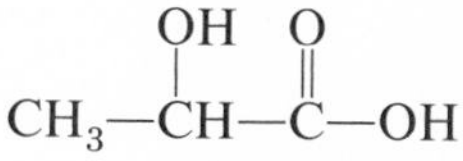

lactic acid
important intermediate
in metabolism

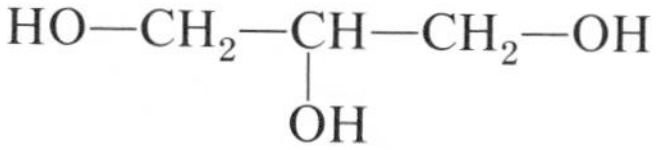

$$HO-CH_2-\underset{\underset{OH}{|}}{CH}-CH_2-OH$$

glycerol
important component
of lipids

Figure 5-6
Some important
naturally occurring
alcohols.

control the functioning of the reproductive organs. All steroids have a common structural feature, a rigidly connected set of five- and six-membered rings:

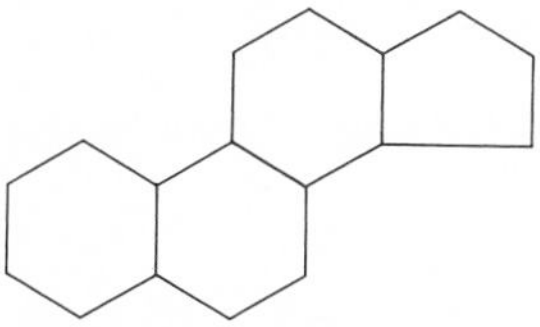

the steroid ring system

Many of these compounds are alcohols. A number of important steroids are shown in figure 5-7.

All the compounds in figure 5-7 except the last one are synthesized by animals as a natural part of their metabolism (though not all individuals synthesize all possible steroids). Following the structure elucidation of a number of naturally occurring steroids in the 1930s and 1940s, chemists began searching for artificial sex hormones which could be used in fertility control. As a result of a massive synthesis and screening program in the 1940s and 1950s, a small number of candidates were identified, and since that time some of them have reached the market in the form of birth-control pills. Norlutin is one such compound.

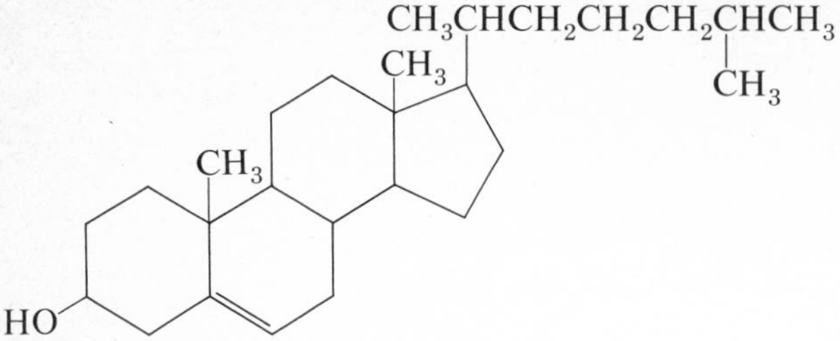

cholesterol
the principal steroid
of mammals

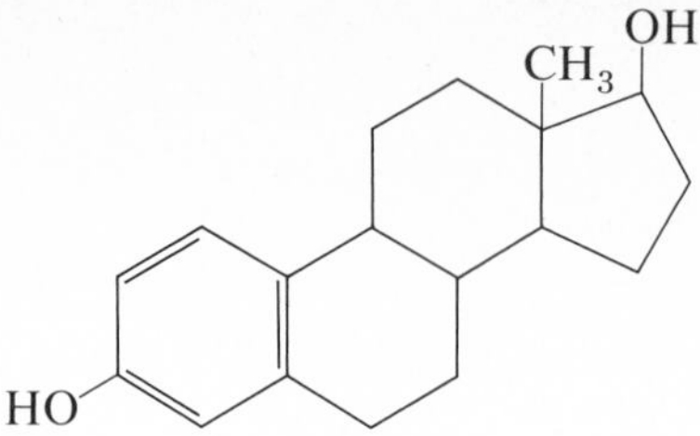

estradiol
an estrogen

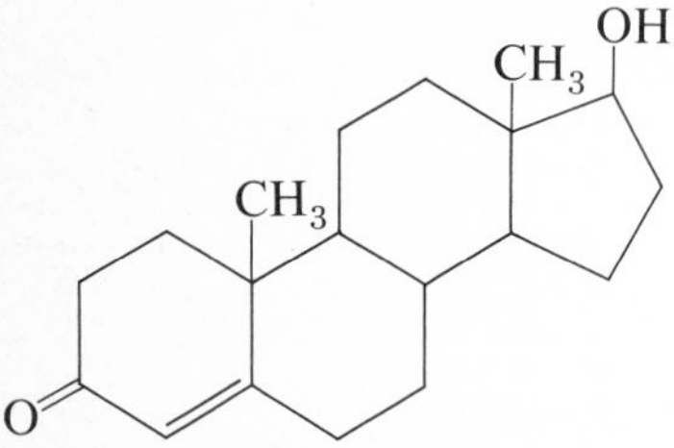

testosterone
an androgen

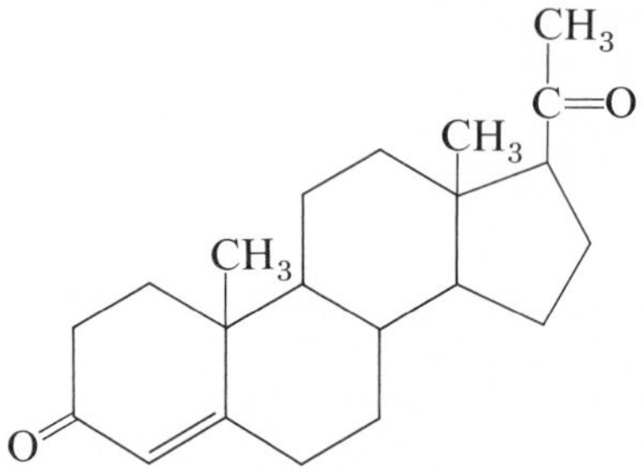

progesterone
an estrogen
secreted during pregnancy

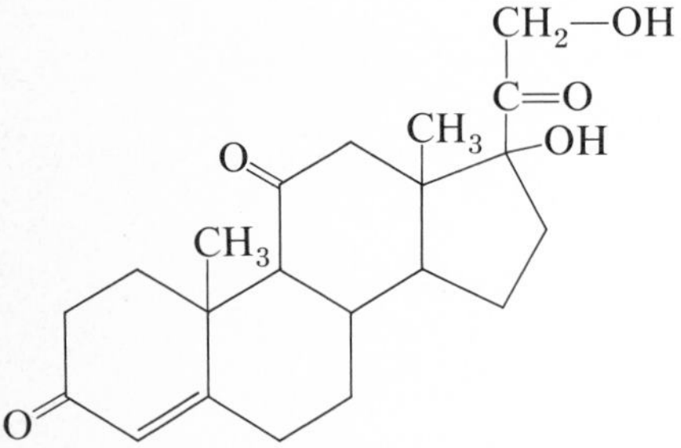

cortisone
an adrenal hormone

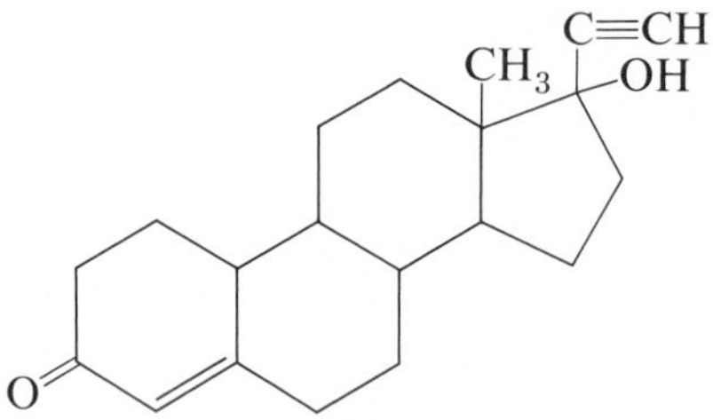

Norlutin
used in birth control pills

Figure 5-7
Some important
steroids. Estrogens
are female sex
hormones, and
androgens are male
sex hormones.

5-5 THE POLARITY OF ORGANIC COMPOUNDS

The lower-molecular-weight alcohols are very much like water in a number of respects. The most important (in addition to the similarity of their acid-base properties) is that, like water, they are polar liquids. Polarity and its implications must now be considered briefly.

In section 5-1 we mentioned that bonds between unlike atoms are frequently polar; that is, the electron distribution in the bond is asymmetric, one atom getting more than its share of the electrons and the other getting correspondingly less. Examples of polar bonds include carbon-oxygen and carbon-chlorine bonds, in both of which carbon is the electron-poor nucleus, and the hydrogen-oxygen bond, in which hydrogen is the electron-poor nucleus. Neither carbon-carbon bonds nor carbon-hydrogen bonds are polar.

A molecule is polar if it contains polar bonds. Whether the molecule is very polar or only slightly polar depends on the proportion of polar and nonpolar bonds in the molecule; the greater the proportion of polar bonds, the more polar the molecule, as illustrated in figure 5-8.

The properties of water and alcohols are affected not only by the polarity of their C—O and O—H bonds but also by the presence of extensive hydrogen bonding (figure 5-9). A hydrogen bond is a weak, mobile bond between a hydrogen atom already bound to oxygen or nitrogen and an unshared electron pair of another oxygen or nitrogen. Other atoms are seldom involved in hydrogen bonding. Carbon, in particular, is almost never involved in hydrogen bonding.

HYDROGEN BONDING

Hydrogen bonds have bond strengths of only about 5 kilocalories/mole and thus are weak by comparison with covalent bonds (bond strengths 80 to 150 kilocalories/mole). They are generally formed and broken very rapidly. Hydrogen bonds to a particular electron pair in

polar part	nonpolar part	
H—O—H		**water** very polar
H—O—CH_2—CH_2—CH_3		**1-propanol** moderately polar
H—O—CH_2—CH_2—CH_2—CH_2—CH_2—CH_2—CH_2—CH_3		**1-octanol** very slightly polar
CH_3—CH_2—CH_2—CH_2—CH_2—CH_2—CH_2—CH_3		**octane** nonpolar

The polarity of a molecule is governed by the number and nature of the polar groups in the molecule and by the relative numbers of polar and nonpolar bonds. Water, with two polar bonds and no nonpolar bonds, is one of the most polar substances known. Octane, with twenty-five nonpolar bonds and no polar bonds, is nonpolar.

Figure 5-8
Polar and nonpolar molecules.

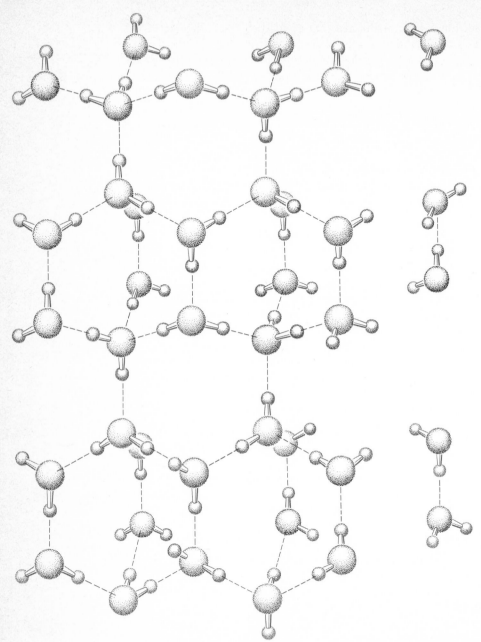

Figure 5-9
Hydrogen bonds in
ice.

In addition to covalent bonds, ice is held together by hydrogen bonds (shown here in color) between oxygens and hydrogens of adjacent molecules in the ice crystal. Similar hydrogen bonds exist in liquid water, but there are fewer of them.

(Reprinted from Linus Pauling, The Nature of the Chemical Bond, 3d ed. Copyright © 1939 and 1940, 3d edition copyright © 1960, by Cornell University. Used by permission of Cornell University Press.)

water, for example, may form and break up to 10^{12} times in a second.

Although hydrogen bonds are weak and fluctuate rapidly, they serve to make molecules more polar. Because of hydrogen bonding, CH_3—OH is more polar than CH_3—Cl.

Polarity is important in biochemistry. Water, the medium which envelops all living cells, is a very polar solvent and interacts favorably with polar molecules. Molecules or parts of molecules which are themselves polar and can interact favorably with water are called *hydrophilic,* or water-loving. Nonpolar molecules are called *hydrophobic.*

The hydrophobic molecules of nature tend to associate with themselves, thereby limiting their association with water. This hydrophobic aggregation is important in determining the structures of cells and subcellular particles. Membranes, the natural barriers which surround cells and sections of cells, are hydrophobic barriers. Polar substances cannot pass through membranes except when special transport mechanisms exist. This membrane selectivity serves an important function in controlling the access of polar molecules to the cell or organ beyond the membrane.

Which compound in each of the following pairs is more polar? (*a*) n-$C_{10}H_{22}$ and $CH_3(CH_2)_9OH$; (*b*) CH_3CH_2Cl and CH_3CH_2OH; (*c*) CH_3CH_2OH and $CH_3CH_2CH_2CH_2OH$; (*d*) $CH_3CH_2OCH_2CH_3$ and CH_3CH_2OH; (*e*) $CH_3CH_2OCH_2CH_3$ and $CH_3CH_2CH_2CH_2CH_3$.

5-6 SUBSTITUTION REACTIONS OF ALKYL HALIDES*

Nucleophilic substitution, a reaction by which one group attached to a saturated carbon atom is replaced by another, is an important reaction of alkyl halides. There are two mechanisms for this reaction, a displacement mechanism and a mechanism involving a carbonium-ion intermediate.

Nucleophilic displacement reactions involve the direct displacement of a halogen or some other similar atom or group by some atom or group which has an unshared pair of electrons. The displaced atom or group is called the *leaving group,* and the displacing group is called

---

*The discussion in this chapter applies equally well to alkyl chlorides, bromides, and iodides. Alkyl fluorides, however, seldom undergo substitution or elimination reactions, and they will not be considered.

the *nucleophile:*

$$CH_3-I + OH^- \longrightarrow CH_3-OH + I^-$$

nucleo-
phile

leaving
group

Mechanism 5-1 shows the details of this important reaction. The nucleophile carries with it a pair of electrons; in the course of the reaction this pair becomes bonded to the carbon atom which originally held the leaving group. When the leaving group departs, it carries with it the pair of electrons it originally shared with carbon.

Many nucleophiles and many types of compounds are capable of taking part in this reaction. Cyanide ion, for example, is an excellent nucleophile:

benzyl chloride   cyanide   phenylacetonitrile   chloride

PROBLEM 5-5   Identify the nucleophile and the leaving group in the last reaction and write the mechanism.

STEREOCHEMISTRY OF DISPLACEMENT   During the course of the displacement, both the nucleophile and the leaving group are electron-rich. Thus, it is natural for these two groups to stay as far apart as possible during the reaction. As a consequence, the nucleophile approaches carbon from the side *opposite* the leaving group. In the course of the displacement step, the configuration of the carbon is inverted, rather like an umbrella blown inside out. This configuration change is called *Walden inversion.*

If displacement is conducted with an optically active halide, Walden inversion is clearly visible because the configuration of the chiral center is inverted, as shown in figure 5-10. Walden inversion can also be seen in many reactions of cyclic compounds. By this process, a cis isomer will be converted into a trans isomer, and vice versa:

trans
isomer

cis
isomer

$$OH^- + CH_3{-}I$$

Nucleophilic displacements at saturated carbon atoms occur in a single step. The nucleophile has an unshared pair of electrons which is readily available for bond formation. The carbon to which the iodine is attached is susceptible to attack by the electron-rich nucleophile.

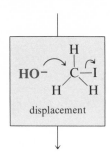

displacement

Two simultaneous changes occur in the displacement step. The new carbon-oxygen bond is formed using an unshared electron pair from the hydroxide; simultaneously, the carbon-iodine bond is broken. Iodine leaves in the form of iodide ion, taking with it the electron pair which formerly constituted the carbon-iodine bond. The nucleophile approaches the carbon from the more accessible side, the side opposite the leaving group. As displacement occurs, the carbon is temporarily sp²-hybridized.

$$HO{-}CH_3 + I^-$$

The reaction is complete when the iodide is free of its association with carbon and the carbon-oxygen bond is completely formed.

Use molecular models to demonstrate that Walden inversion in the cyclopentane case above converts a trans isomer into a cis isomer.

Water and alcohols are very poor nucleophiles. Displacement reactions with them require heating for an extended time

$$CH_3CH_2CH_2{-}Br + H_2O \xrightarrow{heat} CH_3CH_2CH_2{-}OH + HBr$$

Note that in this case, the displacement reaction involves *three* bond

Nucleophilic displacement causes the reactive carbon atom to invert its configuration and form a new bond to the nucleophile on the side opposite where *the bond to the leaving group was. This reaction therefore causes the inversion of a chiral center.*

Figure 5-10
Walden inversion.

changes. One of the hydrogens which is part of water in the starting material is transferred to the bromine in the course of the reaction and forms HBr.

Hydroxide and alkoxide ions are excellent nucleophiles and readily displace halide ions from alkyl halides

$$CH_3CH_2CH_2{-}Br + OH^- \longrightarrow CH_3CH_2CH_2{-}OH + Br^-$$

The structure of the organic compound formed is the same whether water or hydroxide is used as the nucleophile; the reaction occurs more readily when hydroxide is used. Alkoxides are also excellent nucleophiles

$$CH_3CH_2{-}I + \underset{\underset{CH_3}{|}}{CH_3CH}{-}O^- \longrightarrow CH_3CH_2{-}O{-}\underset{\underset{CH_3}{|}}{CHCH_3} + I^-$$

alkyl halide   alkoxide   an ether

This reaction is frequently used to make ethers, but it is useful only with primary alkyl halides because secondary and tertiary halides usually undergo elimination (section 5-7).

REACTIVITIES OF ALKYL HALIDES   The structure of the alkyl halide affects its reactivity. Displacements do not occur on carbon atoms which are multiply bonded to other carbon atoms because the π electrons prevent approach of the nucleophile. Displacements at saturated carbon atoms follow a predictable pattern. Methyl compounds are most reactive, followed by primary compounds, then secondary compounds; tertiary compounds do not undergo nucleophilic displacement. Thus, the reactivity of iodides decreases in the order

$$CH_3{-}I > CH_3CH_2{-}I > \underset{\underset{CH_3}{|}}{CH_3CH}{-}I > CH_3{-}\underset{\underset{CH_3}{|}}{\overset{\overset{CH_3}{|}}{C}}{-}I$$

most reactive   not reactive

Substitution in secondary and tertiary compounds may occur by a different mechanism, involving a carbonium-ion intermediate, which will be considered in a later section.

NUCLEOPHILES AND LEAVING GROUPS   Many nucleophiles and many leaving groups can participate in displacement reactions with some facility. The reactivity of nucleophiles and of leaving groups correlates in a general way with basicity: very basic species are good nucleophiles but poor leaving groups; very

	NUCLEOPHILES	LEAVING GROUPS
greatest	$^-$SH, $^-$SCH$_3$	I$^-$
	$^-$C≡N	Br$^-$
	enolates (chapter 7)	
	I$^-$	H$_2$O, CH$_3$OH
REACTIVITY	NH$_3$	
	$^-$OH, $^-$OCH$_3$	Cl$^-$
	Br$^-$	
	$^-$O—$\overset{\overset{\textstyle O}{\|\|}}{C}CH_3$	CH$_3$SCH$_3$
	Cl$^-$	
least	H$_2$O, CH$_3$OH	F$^-$

Figure 5-11
Relative reactivities
of some nucleophiles
and some leaving
groups. (*From J.*
March, Advanced
Organic Chemistry,
McGraw-Hill Book
Company, 1968.)

nonbasic species are poor nucleophiles but good leaving groups. Thus, water is a poor nucleophile but a good leaving group; hydroxide is a good nucleophile but a poor leaving group. The relative reactivities of a number of nucleophiles and leaving groups are shown in figure 5-11. Some additional displacement reactions are the following:

$$CH_3CH_2-SH + I-CH_2-\overset{\overset{\textstyle O}{\|\|}}{C}-OH \longrightarrow CH_3CH_2-S-CH_2-\overset{\overset{\textstyle O}{\|\|}}{C}-OH + HI$$

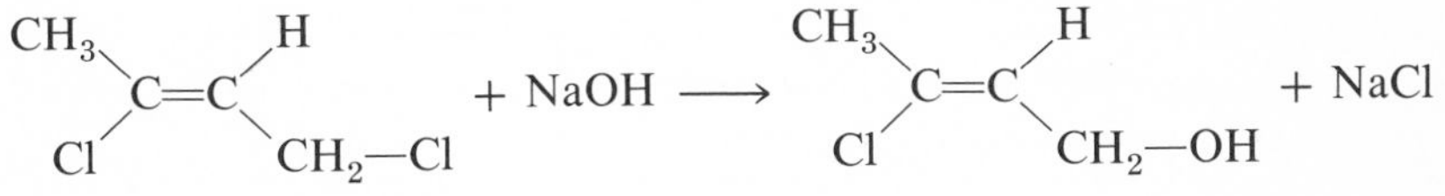

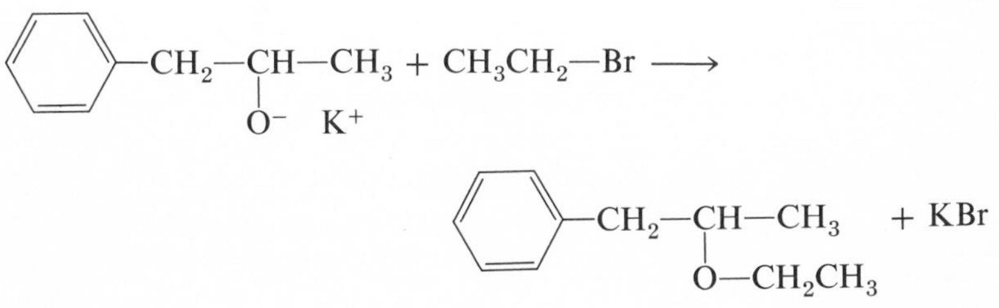

Nucleophilic substitution can occur by two mechanisms: (1) the displacement mechanism just discussed and (2) a two-step mechanism involving a carbonium-ion intermediate. This mechanism occurs commonly with tertiary and secondary halides.

If 2-bromo-2-methylpropane is dissolved in water containing no nucleophiles other than water itself, substitution occurs

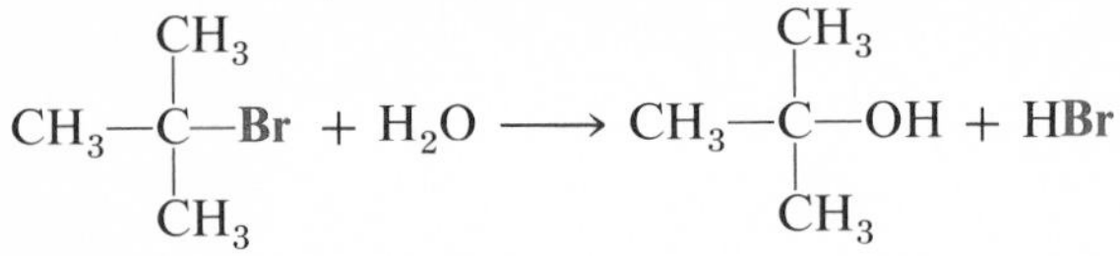

$$CH_3-\underset{\underset{\displaystyle CH_3}{|}}{\overset{\overset{\displaystyle CH_3}{|}}{C}}-Br + H_2O \longrightarrow CH_3-\underset{\underset{\displaystyle CH_3}{|}}{\overset{\overset{\displaystyle CH_3}{|}}{C}}-OH + HBr$$

This reaction follows a two-step sequence, as shown in mechanism 5-2.

Carbonium ions like that formed in this reaction are very unstable intermediates, and this reaction occurs only if the solvent is polar (water and alcohols are often used) and therefore able to stabilize the carbonium ion. Tertiary alkyl halides often react by the carbonium-ion mechanism and never react by a nucleophilic displacement mechanism. Secondary halides often react by a carbonium-ion mechanism in the absence of a good nucleophile. Primary halides seldom react by a carbonium-ion mechanism.

If a carbonium-ion intermediate is formed in the reaction of an optically active halide, the product will be racemic because the carbonium-ion intermediate is not chiral and therefore yields a racemic product

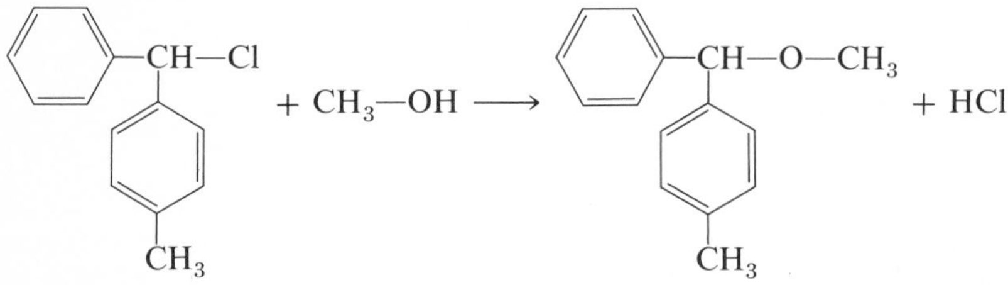

optically active   racemic

PROBLEM 5-7 Write the mechanism of the above reaction and use molecular models to show why a racemic product is formed.

5-7 ELIMINATION REACTIONS OF ALKYL HALIDES

Hydroxide ion and alkoxide ions are nucleophiles. In section 5-6 we considered displacement reactions involving these ions. However, hydroxide ion and alkoxide ions are also bases. We must now consider the reactions of secondary and tertiary alkyl halides with hydroxide ion and with alkoxide ions, where the latter act as bases rather than nucleophiles.

$$CH_3-\underset{\underset{CH_3}{|}}{\overset{\overset{CH_3}{|}}{C}}-Br \ + \ H_2O$$

This reaction occurs most often with tertiary halides, although secondary halides occasionally react by this mechanism.

$$CH_3-\underset{\underset{CH_3}{|}}{\overset{\overset{CH_3}{|}}{C}}\!\frown\!Br$$

bond cleavage

In a polar solvent, an unassisted breaking of the carbon-halogen bond may occur. The pair of electrons which originally constituted that bond goes with the halogen atom, leaving the carbon atom electron-deficient.

$$\left[CH_3-\underset{\underset{CH_3}{|}}{\overset{\overset{CH_3}{|}}{C^+}} \right] Br^-$$

carbonium-ion
intermediate

The intermediate which is formed is a carbonium ion. The three carbons lie in the same plane because the center carbon is sp^2-hybridized. The carbonium ion is very unstable and will react with whatever electron-rich ion or molecule comes along.

$$CH_3-\underset{\underset{CH_3}{|}}{\overset{\overset{CH_3}{|}}{C^+}} \quad \underset{H}{\overset{..}{O}}-H$$

attack on
water

The unshared electrons of the solvent water are the most common site of attack by the electron-deficient carbon atom. A new carbon-oxygen bond is formed using two electrons from oxygen.

$$\left[CH_3-\underset{\underset{CH_3}{|}}{\overset{\overset{CH_3}{|}}{C}}-\overset{+}{O}-H \right]$$
$$CH_3 \ H$$

The product of this step is the conjugate acid of an alcohol.

proton
transfer
to solvent

$$CH_3-\underset{\underset{CH_3}{|}}{\overset{\overset{CH_3}{|}}{C}}-OH \ + \ H^+ \ + \ Br^-$$

Loss of a proton results in formation of an alcohol and HBr.

Secondary and tertiary halides usually react with alkoxides and with hydroxide to form olefins, rather than ethers or alcohols

$$H-\underset{\underset{H}{|}}{\overset{\overset{H}{|}}{C}}-\underset{\underset{CH_3}{|}}{\overset{\overset{CH_3}{|}}{C}}-Cl + {}^-OH \longrightarrow \underset{H}{\overset{H}{>}}C=C\underset{CH_3}{\overset{CH_3}{<}} + HOH + Cl^-$$

alkyl   base   olefin
halide

This reaction is called *elimination* because the reaction results in the elimination of a proton and a halide ion from adjacent carbon atoms. The entire change occurs in a single step, as shown in mechanism 5-3.

Alkoxides are often used in this reaction instead of hydroxide. Sodium ethoxide is a frequent choice:

$$\text{(cyclohexyl bromide with } CH_3) + Na^+ \; {}^-O-CH_2CH_3 \longrightarrow \text{(methylenecyclohexane)} + \text{(1-methylcyclohexene)} + HO-CH_2CH_3 + NaBr$$

PROBLEM 5-8 Write a mechanism for the last reaction above. Show two different elimination steps to illustrate how the two products are formed.

STEREO-CHEMISTRY OF ELIMINATION Elimination reactions convert two tetrahedral sp^3-hybridized carbon atoms into planar sp^2-hybridized carbons. As the elimination step occurs, the hydrogen and the halogen which are being eliminated are situated on opposite sides of the forming carbon-carbon double bond in an anti configuration (figure 5-12). Elimination in this fashion, called *anti elimination,* is analogous to the anti addition of Br_2 to olefins (section 3-9).

EFFECT OF HALIDE STRUCTURE The reactions of alkyl halides with hydroxide and alkoxides are profoundly affected by the structure of the alkyl halide. Methyl halides give substitution exclusively:

$$CH_3-Br + Na^+ \; {}^-O-CH_2CH_3 \longrightarrow CH_3-O-CH_2CH_3 + NaBr$$

methyl   substitution
halide   product

$$\text{H—CH}_2\text{—}\underset{\underset{\text{CH}_3}{|}}{\overset{\overset{\text{CH}_3}{|}}{\text{C}}}\text{—Cl} + \text{OH}^-$$

If a secondary or tertiary alkyl halide is dissolved in a solvent containing sodium hydroxide or sodium ethoxide, a one-step elimination reaction results.

$$\text{HO}^\frown\text{H—CH}_2\text{—}\underset{\underset{\text{CH}_3}{|}}{\overset{\overset{\text{CH}_3}{|}}{\text{C}}}\text{—Cl}$$

elimination

A total of three bond changes occur in a single step. Rather than acting as a nucleophile, hydroxide acts as a base and pulls off a proton from a carbon adjacent to the carbon to which the halogen atom is bound. Simultaneously, the carbon-halogen bond breaks.

$$\underset{\text{H}}{\overset{\text{H}}{}}\text{C=C}\underset{\text{CH}_3}{\overset{\text{CH}_3}{}} + \text{HOH} + \text{Cl}^-$$

As a result of these changes, a new π bond is formed, and the product is an olefin. The base (in this case hydroxide ion) is converted into its conjugate acid (water).

Primary halides give mostly substitution, with only a very small amount of elimination:*

$$\text{CH}_3\text{CH}_2\text{—Br} + \text{Na}^+ \ ^-\text{O—CH}_2\text{CH}_3 \longrightarrow$$

primary
halide

$$\text{CH}_3\text{CH}_2\text{—O—CH}_2\text{CH}_3 + \underset{\text{H}}{\overset{\text{H}}{}}\text{C=C}\underset{\text{H}}{\overset{\text{H}}{}} + \text{NaBr}$$

substitution
product, 99%

elimination
product, 1%

Secondary halides give more elimination than substitution:

$$\text{CH}_3\underset{\underset{\text{CH}_3}{|}}{\text{CH}}\text{—Br} + \text{Na}^+ \ ^-\text{O—CH}_2\text{CH}_3 \longrightarrow$$

secondary
halide

$$\text{CH}_3\underset{\underset{\text{CH}_3}{|}}{\text{CH}}\text{—O—CH}_2\text{CH}_3 + \underset{\text{H}}{\overset{\text{H}}{}}\text{C=C}\underset{\text{CH}_3}{\overset{\text{H}}{}} + \text{NaBr} + \text{HO—CH}_2\text{CH}_3$$

substitution
product, 21%

elimination
product, 79%

*The percentages of substitution and elimination products given in this section are typical, but the actual percentages vary according to reaction conditions and compound structure.

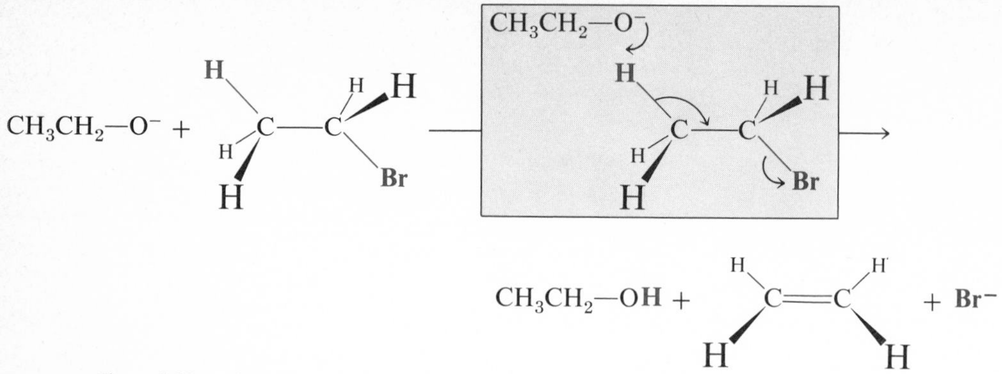

$$CH_3CH_2{-}O^- +$$ [structure] $\longrightarrow$

[Figure: reaction mechanism box with $CH_3CH_2{-}O^-$ attacking]

$$CH_3CH_2{-}OH + \quad [olefin] \quad + Br^-$$

Figure 5-12 *Stereochemistry of elimination.* *The hydrogen and the bromine which will be eliminated are arranged in an anti conformation at the beginning of the reaction, and the elimination occurs from this conformation.*

Tertiary halides give elimination exclusively:

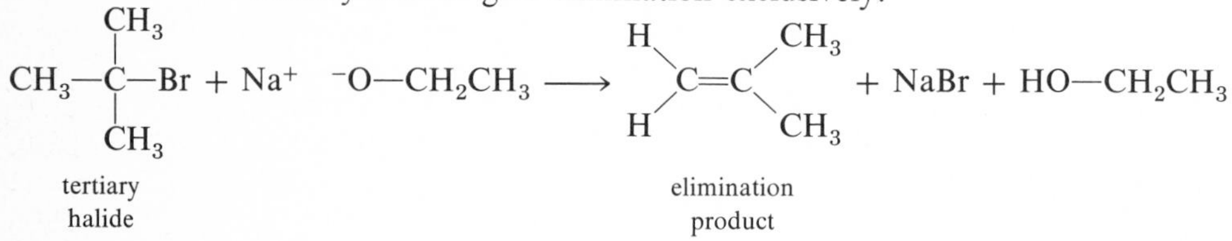

$$CH_3{-}\overset{\overset{\displaystyle CH_3}{|}}{\underset{\underset{\displaystyle CH_3}{|}}{C}}{-}Br + Na^+ \quad {}^-O{-}CH_2CH_3 \longrightarrow \quad [olefin] \quad + NaBr + HO{-}CH_2CH_3$$

tertiary halide

elimination product

DIRECTION OF ELIMINATION
Although only one elimination product is possible in the reactions above, in many cases two or even three different elimination products are possible, depending on which hydrogen is lost. Elimination reactions of alkyl halides usually give a higher yield of the more highly substituted olefin, that is, the compound with the larger number of carbon groups on the double bond.* However, some of the less substituted olefin is usually formed as well. The following examples illustrate this preference:

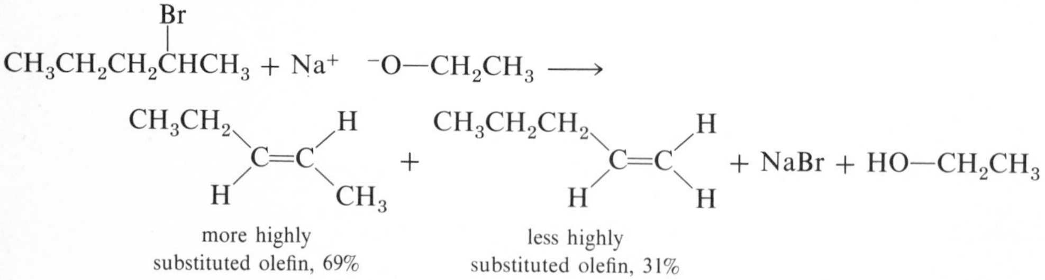

$$CH_3CH_2CH_2\overset{\overset{\displaystyle Br}{|}}{C}HCH_3 + Na^+ \quad {}^-O{-}CH_2CH_3 \longrightarrow$$

[olefin] + [olefin] $+ NaBr + HO{-}CH_2CH_3$

more highly substituted olefin, 69% | less highly substituted olefin, 31%

*This generalization is called the *Saytzeff rule,* after the Russian chemist credited with the first description of this preference. It has been maintained that the rule is much older and is, in fact, the oldest known general rule of physical organic chemistry, being clearly stated in Matthew 25:29: "Unto every one that hath shall be given . . . : but from him that hath not shall be taken away even that which he hath." The foundation of the Markownikoff rule is also contained in this quotation.

In the above example, some of the nucleophilic displacement product is also formed. The percentages given are percentages of the total olefin formed. No substitution product is formed in the following case:

$$CH_3CH_2\overset{\overset{\displaystyle CH_3}{|}}{\underset{\underset{\displaystyle CH_3}{|}}{C}}-Br + Na^+ \ {}^-O-CH_2CH_3 \longrightarrow$$

$$\underset{\substack{\text{more highly} \\ \text{substituted olefin, 70\%}}}{\overset{CH_3}{\underset{H}{}}\diagdown C=C \diagup \overset{CH_3}{\underset{CH_3}{}}} \quad + \quad \underset{\substack{\text{less highly} \\ \text{substituted olefin, 30\%}}}{\overset{CH_3CH_2}{\underset{CH_3}{}}\diagdown C=C \diagup \overset{H}{\underset{H}{}}} \quad + \ NaBr + HO-CH_2CH_3$$

5-8 SUBSTITUTION AND ELIMINATION REACTIONS OF ALCOHOLS

In the preceding section we discussed the reactions of alkyl halides with alcohols and with their conjugate bases, the alkoxide ions. In this section we will examine reactions in which the oxygen of the alcohol acts as a leaving group in a substitution or elimination reaction.

Hydroxide ion and alkoxide ions do not usually function as leaving groups, and reactions such as

$$I^- + CH_3-OH \overset{}{\longrightarrow}\!\!\!/\!\!\!\longrightarrow CH_3-I + OH^-$$

(where the slashed arrow means "does not occur") are rare. However, if this same reaction is conducted in *acidic solution,* displacement does occur

$$H^+ \ \ I^- + CH_3-OH \longrightarrow CH_3-I + HOH$$

In this case hydroxide is not the leaving group; instead, water is the leaving group, as shown in mechanism 5-4.

Thus, displacement reactions of alcohols can occur only in acidic solution; protonation of the alcohol oxygen makes it possible for water to act as the leaving group. Elimination reactions similar to those discussed in section 5-7 do not occur with alcohols.

Primary alcohols also undergo nucleophilic displacement

$$CH_3CH_2CH_2-OH + HI \longrightarrow CH_3CH_2CH_2-I + H_2O$$

Secondary and tertiary alcohols usually undergo elimination reactions under these conditions.

Write the mechanism of the last reaction above.   **PROBLEM 5-9**

128

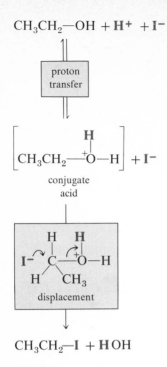

Mechanism 5-4
Nucleophilic
Displacement on an
Alcohol in Acid
Solution

$CH_3CH_2{-}OH + H^+ + I^-$

proton transfer

$$\left[CH_3CH_2{-}\overset{H}{\underset{}{\overset{|}{\overset{+}{O}}}}{-}H \right] + I^-$$

conjugate
acid

H   H
I⁻ C—O—H
H   CH₃

displacement

$CH_3CH_2{-}I + HOH$

The oxygen atom of an alcohol is slightly basic, and if the alcohol is dissolved in strong acid, the oxygen can be protonated.

The concentration of this protonated alcohol is low, but it is quite reactive because it contains a good leaving group, water. If a good nucleophile is present, displacement will occur.

The displacement step for this reaction is similar to those for other nucleophilic displacements we have considered. The iodide uses its pair of electrons to form a bond to carbon, and simultaneously water departs, carrying with it the electron pair that was formerly the carbon-oxygen bond.

The result of this reaction is formation of an alkyl halide and a molecule of water.

CARBONIUM IONS FROM ALCOHOLS

When tertiary alcohols are dissolved in concentrated sulfuric or phosphoric acid, they undergo a sequence of steps leading ultimately to the elimination of water and formation of an olefin

$$CH_3{-}\overset{CH_3}{\underset{CH_3}{\overset{|}{\underset{|}{C}}}}{-}OH \xrightarrow{\text{conc. } H_2SO_4} \overset{H}{\underset{H}{}}C{=}C\overset{CH_3}{\underset{CH_3}{}} + H_2O$$

The mechanism of this reaction (mechanism 5-5) is precisely the reverse of the mechanism by which alcohols are formed from olefins in dilute sulfuric acid (section 3-9). The difference between the two reactions lies in the concentration of acid used.

The ease with which this reaction occurs is directly related to the stability of the carbonium ion formed; the more stable the carbonium ion, the more rapid the reaction. When primary alcohols are exposed to strong acid in the presence of a nucleophile, they undergo nucleophilic displacement. However, in the absence of a nucleophile, elimination occurs

$$CH_3CH_2{-}OH \xrightarrow[\text{heat}]{\text{conc. } H_2SO_4} \overset{H}{\underset{H}{}}C{=}C\overset{H}{\underset{H}{}} + H_2O$$

$$H-CH_2-\underset{\underset{CH_3}{|}}{\overset{\overset{CH_3}{|}}{C}}-OH + H^+$$

If a tertiary alcohol is dissolved in concentrated acid solution, the alcohol is converted to a small extent into its conjugate acid.

Mechanism 5-5
Carbonium-ion Formation from Alcohols

proton transfer

$$\left[H-CH_2-\underset{\underset{CH_3}{|}}{\overset{\overset{CH_3}{|}}{C}}\overset{+}{\underset{}{O}}\underset{}{\overset{H}{-}}H \right]$$

conjugate acid

With primary and secondary alcohols, this conjugate acid would be subject to nucleophilic displacement if nucleophiles were present. With tertiary alcohols, however, displacement is not possible because too many groups are crowded around the center carbon atom.

$$H-CH_2-\underset{\underset{CH_3}{|}}{\overset{\overset{CH_3}{|}}{C}}\overset{+}{O}-H$$

water loss

Instead, the carbon-oxygen bond is broken and water is lost. When the water molecule leaves, it carries with it the pair of electrons which formerly constituted the carbon-oxygen bond.

$$\left[H-CH_2-\underset{\underset{CH_3}{|}}{\overset{\overset{CH_3}{|}}{C^+}} \right] + HOH$$

carbonium-ion intermediate

The intermediate which is formed by the loss of water is a tertiary carbonium ion.

$$H-\ddot{O}H-CH_2-\overset{+}{\underset{\underset{CH_3}{|}}{\overset{\overset{CH_3}{|}}{C}}}$$

proton removal by solvent

Because it is electron-deficient, this intermediate tries to gain a pair of electrons. In some cases this is achieved by reacting with an electron-rich species, but in this case it is achieved by removal of a proton by the solvent. A pair of electrons which formerly constituted a carbon-hydrogen bond becomes a π-electron pair in a carbon-carbon double bond.

$$\underset{H}{\overset{H}{>}}C=C\underset{CH_3}{\overset{CH_3}{<}} + H^+ + H_2O$$

The products of the reaction are an olefin and water.

Secondary alcohols also undergo elimination

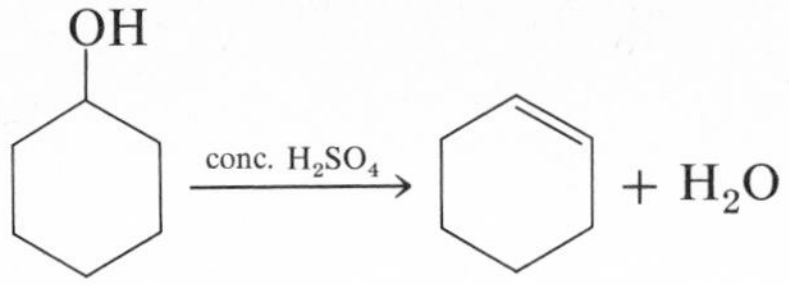

$$\text{(cyclohexanol with OH)} \xrightarrow{\text{conc. } H_2SO_4} \text{(cyclohexene)} + H_2O$$

PROBLEM 5-10   Give the mechanism of the above reaction.

Comparison of the reactions in this section with those of sulfuric acid in section 3-9 reveals that the elimination of water from alcohols to form olefins is a *reversible reaction*. Treatment of an alcohol with concentrated acid results in formation of an olefin. Treatment of an olefin with dilute acid results in formation of an alcohol

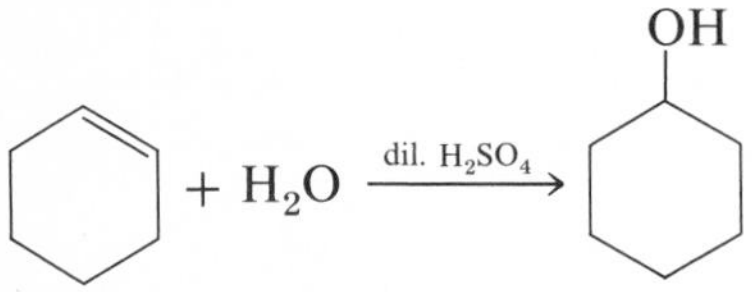

$$\text{(cyclohexene)} + H_2O \xrightarrow{\text{dil. } H_2SO_4} \text{(cyclohexanol with OH)}$$

ELIMINATION REACTIONS OF ALCOHOLS IN BIOCHEMISTRY   Many of the hydroxy compounds which occur in living systems are synthesized by the addition of water to olefins by a mechanism similar to that discussed in chapter 3. Likewise, many olefins are synthesized in living systems by elimination mechanisms similar to those discussed in this chapter.

A particularly well-known and important enzyme is *fumarase*, which catalyzes the reaction

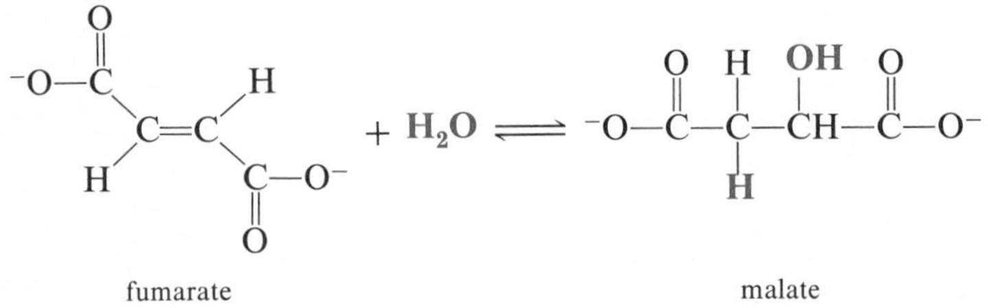

fumarate   malate

This reaction forms a part of the citric acid cycle, a central energy-producing cycle in the metabolism of all living things.

Fumarase catalyzes both the hydration of fumaric acid and the dehydration of malic acid—any catalyst must catalyze its reaction in both directions. Thus, if we understand the mechanism of hydration of fumarate, shown in figure 5-13, we will also understand the mechanism of dehydration of malate. Like elimination reactions of alcohols in organic chemistry, this elimination reaction involves a carbonium-ion intermediate.

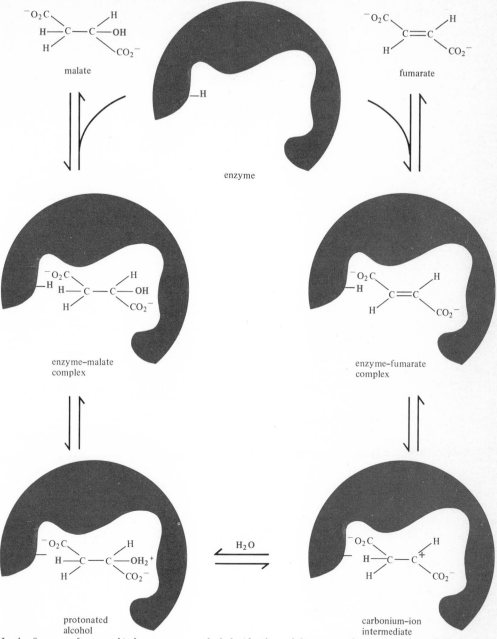

In the first step, fumarate binds to a specific site on the enzyme whose shape is complementary to the shape of fumarate. In the second step, the enzyme transfers a proton to the carbon-carbon double bond, forming a carbonium ion. In the third step, the carbonium ion reacts with water, forming a protonated alcohol. After loss of the proton, the product malate dissociates from the enzyme, and the enzyme is ready to catalyze reaction of another molecule of fumarate. The reverse reaction, the dehydration of malate, occurs by precisely the same mechanism.

Figure 5-13
Mechanism of action of fumarase.

Ordinarily, the formation of a carbonium ion from an alcohol requires the presence of fairly concentrated acid. Fumarase, however, is able to catalyze formation of the carbonium ion in neutral solution at body temperature without a strong acid catalyst. Its ability to do this is a reflection of the importance of enzyme geometry in such reactions. The enzyme has a strategically placed acidic group which can donate a proton to the hydroxyl group and aid in the formation of the carbonium ion. This acidic group makes up in position what it lacks in strength.

NUCLEOPHILIC DISPLACEMENT REACTIONS OF ALCOHOLS IN BIOCHEMISTRY Nucleophilic displacement is not among the most common reactions in biochemistry. Alkyl halides are not common in nature and are therefore unavailable to the living organism for use as substrates in displacement reactions. Alcohols, although abundant in nature, are not good candidates for displacement because such reactions occur only in strongly acidic solution.

As usual, nature has evolved a solution to this dilemma. In order to effect nucleophilic displacement, an alcohol is first converted into a new compound with a better leaving group. An example is shown in figure 5-14. A hydroxyl group of the starting sugar phosphate, ribose 5-phosphate, is converted by reaction with adenosine triphosphate (ATP) into a reactive pyrophosphate ester. Displacement of this pyrophosphate by ammonia occurs easily, and the product of the reaction is used in further transformations.

5-9 OXIDATION OF ALCOHOLS

The reactions discussed so far in this chapter convert compounds having one carbon skeleton and a particular functional group into compounds having the same carbon skeleton but a different functional group at the same oxidation level. We will now consider one of the most important methods for converting alcohols into compounds at higher oxidation states.

The course of alcohol oxidation varies according to whether the alcohol is primary, secondary, or tertiary. Primary alcohols are readily oxidized to aldehydes by chromic acid, $H_2Cr_2O_7$, or potassium permanganate, $KMnO_4$,

primary
alcohol

aldehyde

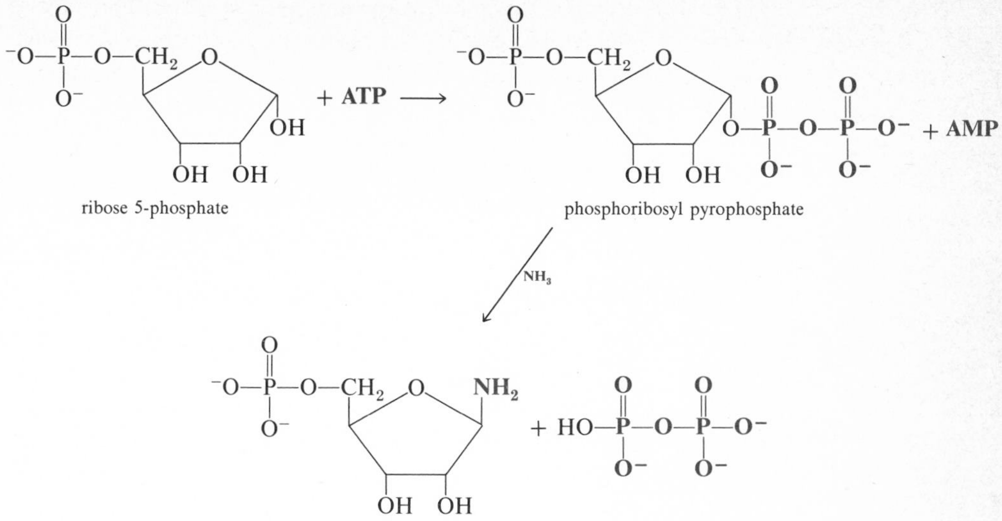

ribose 5-phosphate

+ ATP ⟶

phosphoribosyl pyrophosphate + AMP

NH₃

5-phosphoribosyl-1-amine   pyrophosphate

The hydroxyl group is first converted into a pyrophosphate ester by reaction with adenosine triphosphate (ATP).

Pyrophosphate is a good leaving group and can be displaced by ammonia.

Figure 5-14
Nucleophilic
displacement of a
hydroxyl group in
biochemistry.

These reactions are frequently conducted at elevated temperature so that the aldehyde is distilled out of the reaction mixture as soon as it is formed. If the aldehyde is not removed, it will be oxidized to a carboxylic acid

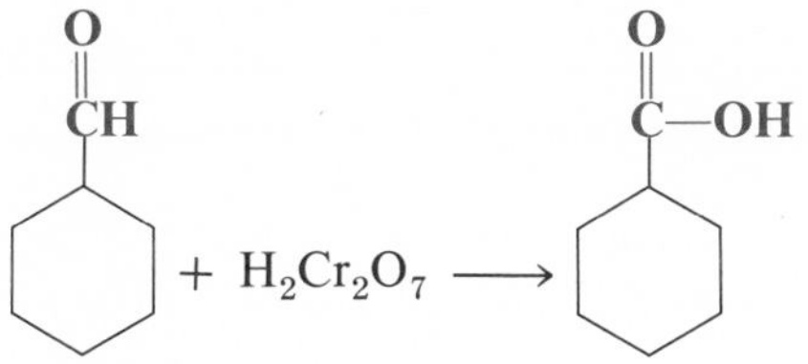

aldehyde   + $H_2Cr_2O_7$ ⟶   carboxylic acid

The oxidation of secondary alcohols is simpler than that of primary alcohols. The first product formed is a ketone, which is not susceptible to further oxidation under the reaction conditions

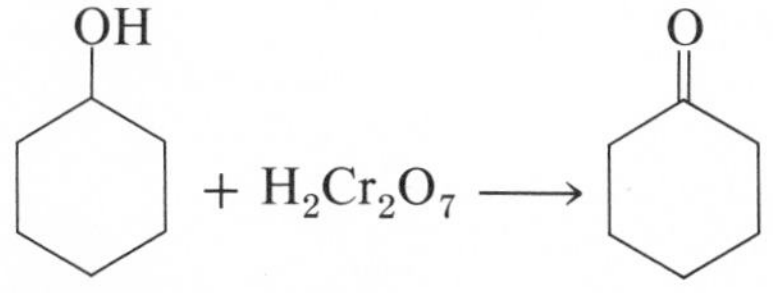

secondary alcohol   + $H_2Cr_2O_7$ ⟶   ketone

Tertiary alcohols do not react with ordinary oxidizing agents.

The mechanism of chromic acid oxidation is shown in mechanism 5-6. The key step in the mechanism is an elimination reaction which forms a carbon-oxygen double bond.

Molecules which contain more than one functional group frequently react as if the functional groups were in separate molecules. For example, the chromic acid oxidation of 2-cyclopentene-1,4-diol proceeds without interaction of the two functional groups

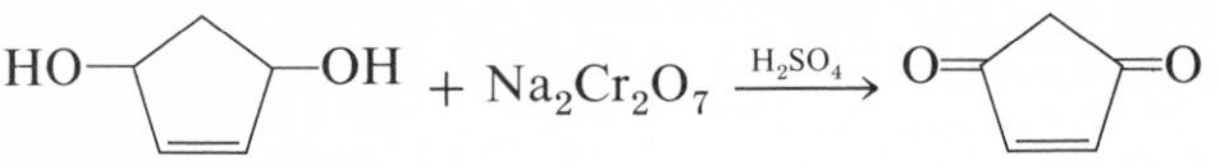

$$HO-\langle\rangle-OH + Na_2Cr_2O_7 \xrightarrow{H_2SO_4} O=\langle\rangle=O$$

PROBLEM 5-11 Give the mechanism of chromic acid oxidation of 1-pentanol to the corresponding aldehyde.

5-10 ETHERS

Divalent oxygen compounds in which both substituents on the oxygen atom are organic are called *ethers*. The structures of a number of important ethers are given in figure 5-15. The substituents on the oxygen atom in ethers may be either aliphatic or aromatic.

PROBLEM 5-12 Considering the geometry of methanol (section 5-1), what do you think the geometry of methyl ether would be?

Figure 5-15
Some important ethers.

CH_3-O-CH_3
methyl ether

$CH_3-O-CH_2CH_3$
methyl ethyl ether

$CH_3CH_2-O-CH_2CH_3$
ethyl ether
("ether")
anesthetic

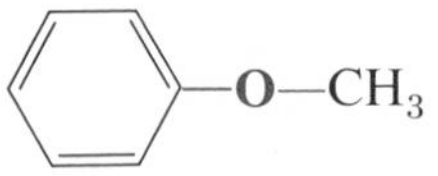

$\langle\rangle-O-CH_3$
anisole

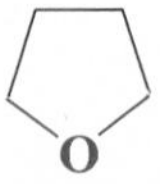

tetrahydrofuran
solvent

Mechanism 5-6
Chromic Acid
Oxidation

OH
|
CH_3CHCH_3 +

$$H^+ + HO-\overset{\overset{O}{\|}}{\underset{\underset{O}{\|}}{Cr}}-O^-$$

The oxidizing agent exists in acidic solution as an equilibrium mixture of at least two forms. The active species is believed to be $HCrO_4^-$.

oxygen attack, then loss of water

In the first step of the reaction, the oxygen atom of the alcohol acts as a nucleophile and attacks the chromium atom. A proton transfer from this oxygen occurs, and a water molecule is expelled from chromium.

$$H-\overset{\overset{CH_3}{|}}{\underset{\underset{CH_3}{|}}{C}}-O-\overset{\overset{O}{\|}}{\underset{\underset{O}{\|}}{Cr}}-OH + H_2O$$

chromate
ester

The chromate ester which is formed by the loss of water is the key intermediate in the oxidation sequence.

$$H_2\overset{..}{O} \curvearrowright H-\overset{\overset{CH_3}{|}}{\underset{\underset{CH_3}{|}}{C}}-O-\overset{\overset{O}{\|}}{\underset{\underset{O}{\|}}{Cr}}-OH$$

elimination

In the key step, an elimination reaction takes place. Water acts as a base and removes a proton from carbon. Simultaneously, a chromium-oxygen bond breaks. The analogy of this step to the elimination reactions of alkyl halides is obvious.

$$CH_3-\overset{\overset{O}{\|}}{C}-CH_3 + HOH_2^+ + HCrO_3^- + H_2O$$

The products of the reaction are a ketone and a reduced chromium species. This chromium species reacts further and is eventually reduced to the Cr(III) state.

Ethers are usually synthesized by nucleophilic displacement reactions. A number of syntheses of ethers have already been given. In addition, the following examples are typical:

MAKING ETHERS BY NUCLEOPHILIC DISPLACEMENT

These reactions follow mechanism 5-1.

The reaction of ethers with nucleophiles in hot acid solution follows a course exactly parallel to that of alcohols (mechanism 5-4). Examples of this reaction are the following:

$$CH_3CH_2CH{-}O{-}CH_2CH_2CH_2CH_3 + HBr \longrightarrow$$
$$\underset{\displaystyle CH_3}{|}$$

$$CH_3CH_2CH{-}OH + CH_3CH_2CH_2CH_2{-}Br$$
$$\underset{\displaystyle CH_3}{|}$$

PROBLEM 5-13 Give the mechanism of the last reaction above, and explain why the product is 2-butanol rather than 1-butanol.

AN ETHER AND A GYPSY MOTH The gypsy moth has become a major insect pest in the northeastern United States since it was accidentally introduced into Massachusetts in 1869. For a number of years its population was controlled by the use of DDT, but after DDT was banned, gypsy moth populations began to rise again. The need for a method which would specifically control this one species led to an investigation of the sex attractant emitted by egg-bearing females.

Abdominal tips were collected from 78,000 female gypsy moths, and from them was obtained less than a milligram of *disparlure,* the active attractant. Disparlure is a nineteen-carbon cyclic ether with no other functional groups

$$\underset{\displaystyle CH_3}{\overset{\displaystyle CH_3}{|}}$$
$$CH_3CHCH_2CH_2CH_2CH_2CH{-}CHCH_2CH_2CH_2CH_2CH_2CH_2CH_2CH_2CH_2CH_3$$
disparlure

Knowledge of this structure makes it possible to control gypsy moths by use of traps containing disparlure and a poison. However, male gypsy moths are very particular and are not attracted by analogs of disparlure which differ only slightly from the natural material. Changing chain length, position of methyl group, or position of oxygen greatly diminishes the biological activity of these compounds.

PROBLEM 5-14 Disparlure reacts with HI to give a 50:50 mixture of two products. What are they?

5-11 THE SULFUR ANALOGS OF ALCOHOLS AND ETHERS

Because sulfur and oxygen are in the same column of the periodic table, they have similar (though not identical) chemistry in organic compounds. The difference arises principally because sulfur can expand its valence shell beyond eight electrons, an ability not shared by oxygen, and because divalent sulfur can form stable sulfur-sulfur bonds.

The sulfur analogs of alcohols are called *mercaptans* (when named by *MERCAPTANS* functional class names) or *-thiols* (when named by IUPAC names).

$$CH_3—SH \qquad CH_3CH_2CH_2CH_2—SH$$

methyl mercaptan   *n*-butyl mercaptan
(methanethiol)   (1-butanethiol)

Like alcohols, mercaptans react with strong bases to form salts

$$CH_3CH_2—SH + NaOH \rightleftharpoons CH_3CH_2—S^-\ Na^+ + HOH$$

    mercaptan   strong   salt
                  base

Mercaptans are more acidic than alcohols. The pK_a of ethanethiol is 10.5, whereas that of ethanol is 16. As a result, the equilibrium in the above reaction favors salt formation, whereas the analogous equilibrium for an alcohol does not.

Mercaptans are more nucleophilic than alcohols. It is possible to make sulfides by reaction of mercaptans with alkyl halides

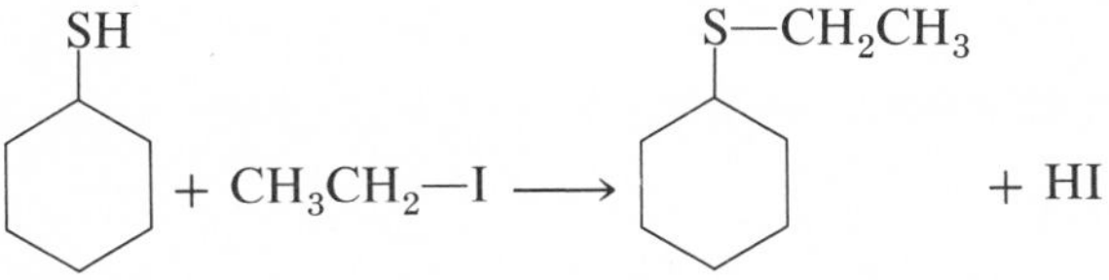

This reaction occurs by a direct displacement mechanism (mechanism 5-1).

Mercaptide ions, the conjugate bases of mercaptans, are more nucleophilic than alkoxide ions. The following reaction involves two nucleophilic displacements

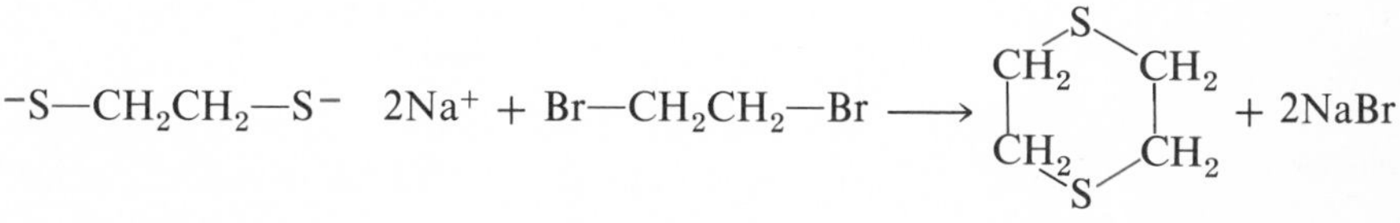

Sulfides are the sulfur analogs of ethers. They are more reactive than *SULFIDES* ethers, just as mercaptans are more reactive than alcohols. Sulfides

are sufficiently nucleophilic to react under certain circumstances with alkyl halides to form sulfonium salts

$$CH_3CH_2-S-CH_2CH_3 + CH_3CH_2-I \longrightarrow CH_3CH_2-\overset{+}{\underset{\underset{CH_2CH_3}{|}}{S}}-CH_2CH_3 \quad I^-$$

sulfide   alkyl halide   sulfonium salt

Although sulfonium salts are frequently formed by nucleophilic displacement reactions, they are themselves reactive toward nucleophiles

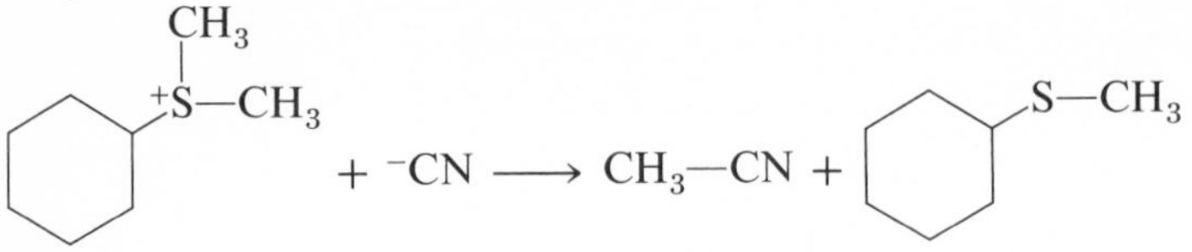

This reaction is very important in biochemistry, being the primary means by which methyl groups become attached to oxygen, nitrogen, and sulfur (section 5-12).

PROBLEM 5-15   Give the mechanism of the last reaction above.

DISULFIDES   Sulfur forms stable bonds to itself more readily than oxygen does.* Sulfur-sulfur bonds are usually formed by oxidation of a mercaptan with a mild oxidizing agent. The amino acid cysteine is readily oxidized to the disulfide form, called *cystine,*

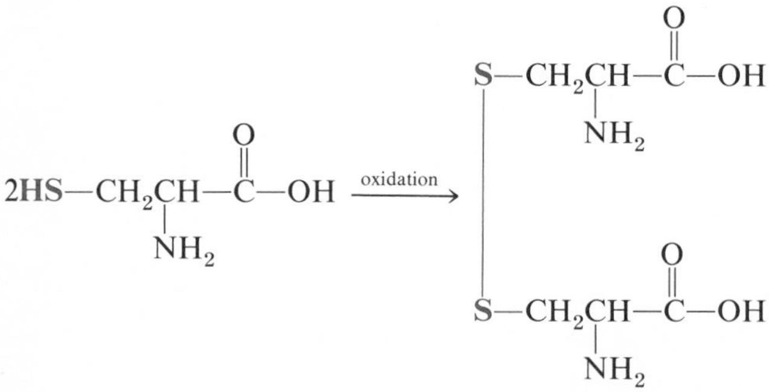

cysteine   cystine

A NOTE ON SMELLS   As a class, sulfur compounds are some of the foulest-smelling substances known. The defensive odor of the common American skunk comes from three odiferous sulfur compounds the animal produces:

---

*Oxygen-oxygen bonds exist in compounds such as hydrogen peroxide, H_2O_2, but such bonds are easily broken.

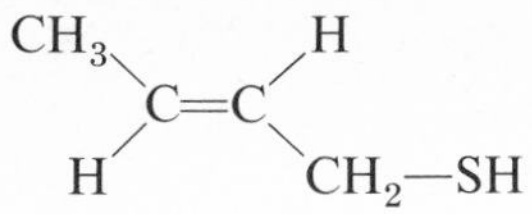

$$CH_3 \diagdown \hspace{-0.5em} C = C \diagup H$$
$$H \diagup \hspace{1em} \diagdown CH_2-SH$$

trans-2-butene-1-thiol

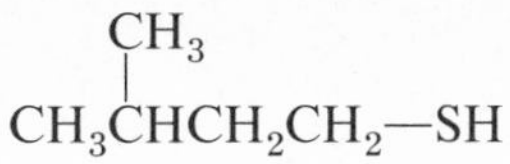

$$CH_3$$
$$CH_3CHCH_2CH_2-SH$$

3-methyl-1-butanethiol

$$CH_3 \diagdown \hspace{-0.5em} C = C \diagup H$$
$$H \diagup \hspace{1em} \diagdown CH_2-S-S-CH_3$$

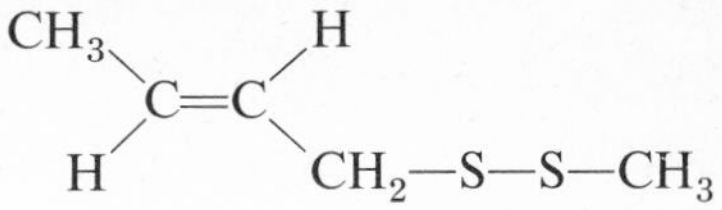

methyl-1-(*trans*-2-butenyl) disulfide

Since natural gas is odorless, a trace amount of ethanethiol

$$CH_3CH_2-SH$$

ethanethiol

is usually added to it to make gas leaks easily detectable.

A number of sulfur-containing compounds are essential to the survival of all living things. Foremost among these are the amino acids cysteine and methionine:

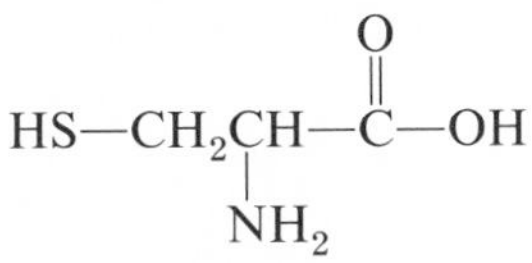

SULFUR IN METABOLISM AND NUTRITION

$$HS-CH_2CH-\overset{\overset{\displaystyle O}{\|}}{C}-OH$$
$$\underset{NH_2}{|}$$

cysteine

$$CH_3-S-CH_2CH_2CH-\overset{\overset{\displaystyle O}{\|}}{C}-OH$$
$$\underset{NH_2}{|}$$

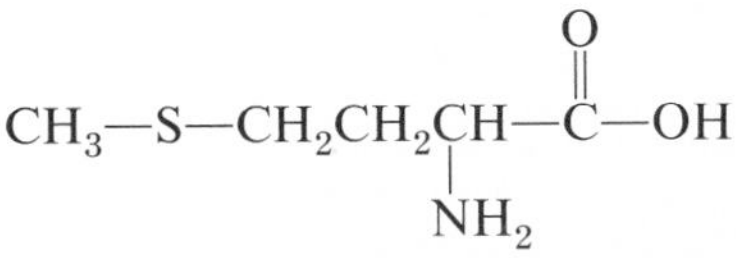

methionine

Cysteine serves a particularly important role in determining the structures of such materials as hair, fingernails, and animal horns (chapter 12). Methionine is used as the starting point for the formation of *S*-adenosylmethionine, which is involved in many biological nucleophilic displacement reactions (section 5-12).

The most abundant form of sulfur in nature is not either of these amino acids but sulfate ion, SO_4^{2-}. Microorganisms and plants are able to reduce sulfate to hydrogen sulfide. Another substrate [symbolized by (O)] is oxidized in this reaction:

$$SO_4^{2-} + H_2O \xrightarrow[\text{and plants}]{\text{microorganisms}} H_2S + (O)$$

This hydrogen sulfide is then used for the formation of cysteine

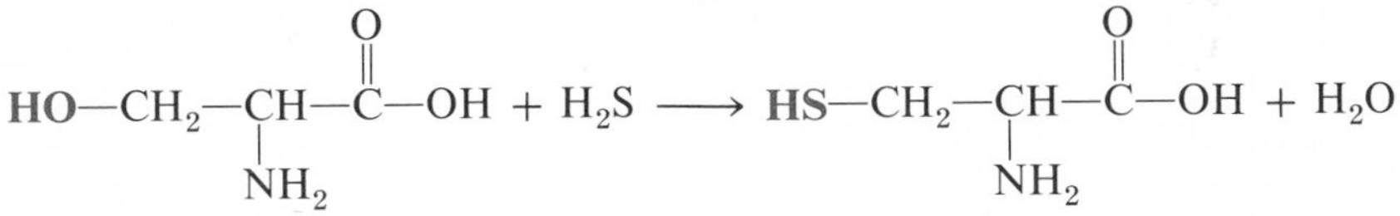

$$HO-CH_2-\overset{\overset{\displaystyle O}{\|}}{\underset{\underset{NH_2}{|}}{CH}-C}-OH + H_2S \longrightarrow HS-CH_2-\overset{\overset{\displaystyle O}{\|}}{\underset{\underset{NH_2}{|}}{CH}-C}-OH + H_2O$$

serine

cysteine

The mechanism of this reaction is not well understood. Methionine can be synthesized from cysteine by these organisms:

Cysteine $\longrightarrow$ methionine

Animals are not able to use sulfate as a source of sulfur for the synthesis of cysteine or methionine. Instead, they depend on their diet for the sulfur-containing amino acids. Optimum wool production by sheep requires a diet rich in the sulfur-containing amino acids because of the large amount of cystine present in wool.

5-12 NUCLEOPHILIC DISPLACEMENT REACTIONS IN BIOCHEMISTRY

Many important molecules in living systems have a methyl group attached to oxygen, nitrogen, or sulfur. Some typical examples are shown in figure 5-16. In all cases these molecules are synthesized by a nucleophilic displacement reaction. The donor of the methyl group is *S-adenosylmethionine,* which is by far the most important methyl donor in living systems. The formation of adrenaline from noradrenaline and *S*-adenosylmethionine is typical (figure 5-17). In this reaction the methylating agent is a sulfonium salt, and the leaving group is therefore a sulfide.

*Figure 5-16
Compounds whose
methyl groups arise
from
S-adenosylmethionine.*

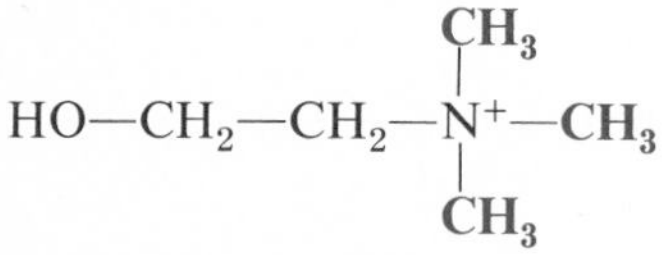

choline
precursor of the neuro-
transmitter acetylcholine

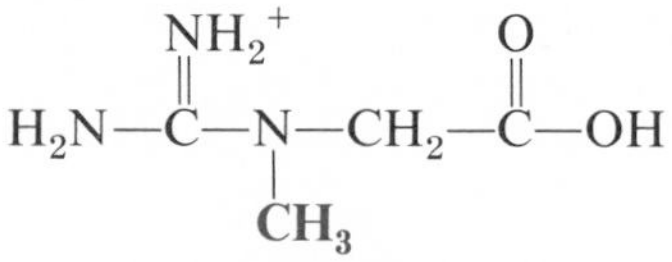

creatine
involved in energy production
in muscle

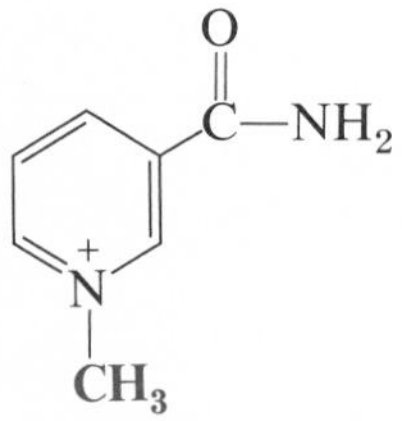

N-methylnicotinamide
found in urine; formed in the
liver from nicotinamide

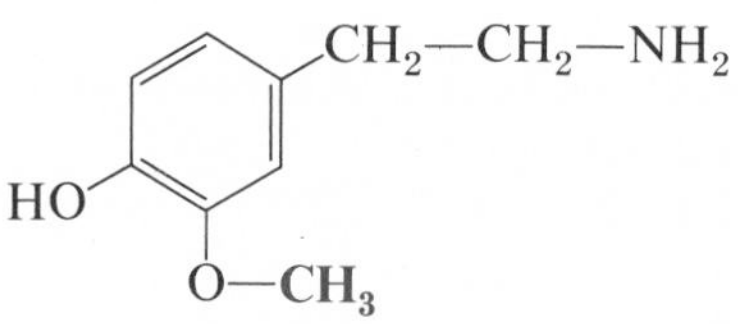

3-methoxytyramine

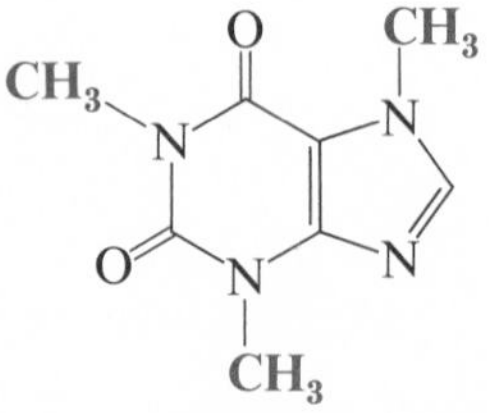

caffeine
central nervous system stimulant
found in coffee and tea

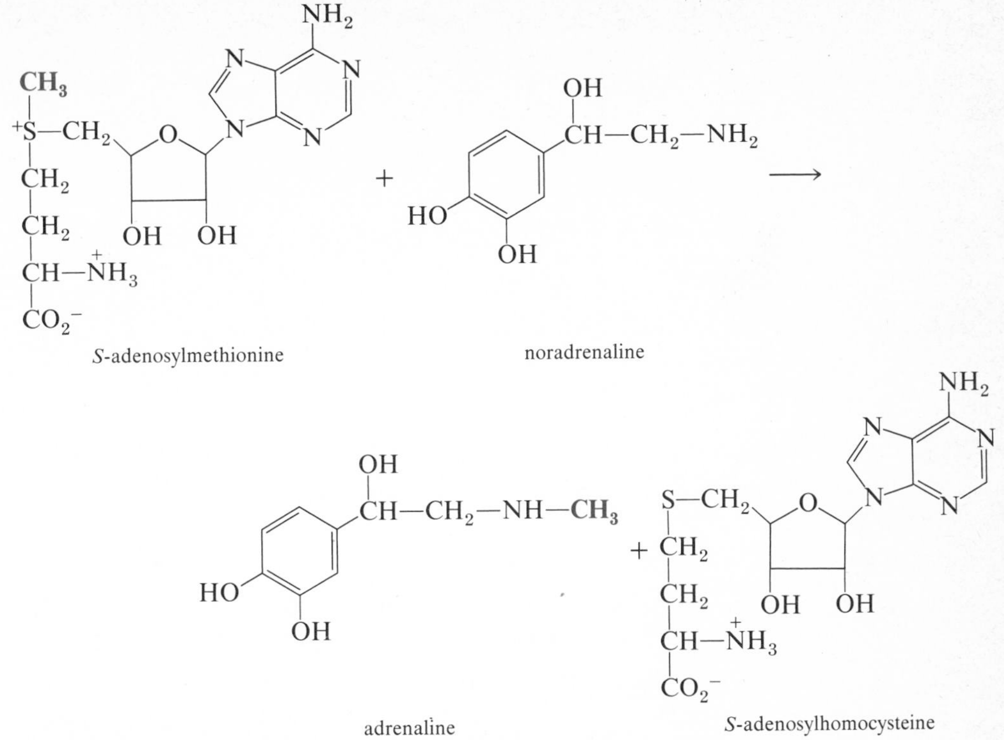

S-adenosylmethionine

noradrenaline

adrenaline

S-adenosylhomocysteine

S-Adenosylmethionine reacts with noradrenaline to form adrenaline and S-adenosylhomocysteine. The nitrogen atom of noradrenaline serves as a *nucleophile, and a sulfide serves as the leaving group in this nucleophilic displacement.*

Figure 5-17 A biological methylation reaction.

Why does nature choose such a complex methylating agent as S-adenosylmethionine instead of simple sulfonium salts like those discussed in section 5-11? Is there something special about the reactivity of the complex salt? Seemingly not. The reactivities of trimethylsulfonium ion and of S-adenosylmethionine are quite similar. The choice of structure has been dictated by the necessity for precise control of the methylation reaction. Each enzyme which uses S-adenosylmethionine has a binding site for this compound and a binding site for the substrate to which the methyl group is being transferred. The shapes of these binding sites are precisely complementary to the shapes of the molecules that bind. Because of the size and complexity of S-adenosylmethionine, this complementarity results in tight and rigid binding. This rigidity allows the enzyme to control the course and outcome of the displacement reaction precisely.

Even though there exist numerous enzymes whose sole function is to catalyze and control displacement reactions involving S-adenosylmethionine, nature can occasionally go awry because of reactions with reactive nucleophiles which are not moderated by enzymes. This seems to be true of *trigonelline,* which is synthesized, apparently quite by accident, from nicotinic acid (a common molecule) and *S*-adenosyl-methionine. Trigonelline has no known function in metabolism.

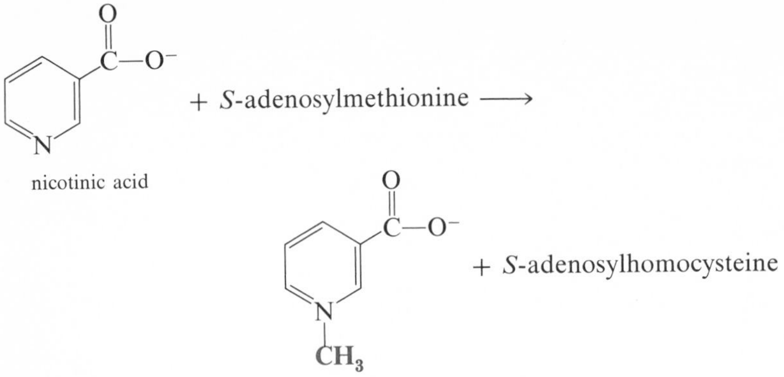

nicotinic acid

+ *S*-adenosylmethionine ⟶

+ *S*-adenosylhomocysteine

trigonelline

PROBLEMS **5-16** Give the structure of a compound which is (*a*) an ether; (*b*) a tertiary alcohol; (*c*) a primary halide; (*d*) a secondary mercaptan; (*e*) a disulfide; (*f*) a diol.

5-17 Give the mechanism and products of each of the following reactions: (*a*) 1-iodohexane + NaOH; (*b*) 2-iodo-2-methylbutane + H_2O; (*c*) 2-iodohexane + NaOH; (*d*) 1-hexanol + HI; (*e*) 2-hexanol + conc. H_2SO_4; (*f*) 2-hexanol + $H_2Cr_2O_7$.

5-18 Give the products and the stereochemistry of the following reactions: (*a*) 2(*R*)-chlorohexane + NaI; (*b*) 3(*R*)-bromo-3-methyl-hexane + H_2O.

5-19 Name each of the following compounds and characterize it as primary, secondary, or tertiary.

(*a*) $Cl-CH_2CH_2CH_2CHCH_3$
  |
  CH_3

(*b*) $CH_3CH_2CHCH_3$
  |
  I

(*c*) $HO-CHCH_2CH_2CHCH_2CH_3$
  | |
  CH_3 CH_2CH_3

(*d*) $HO-CH_2CH_2-OH$

(*e*) CH_3
  |
  CH_3C-Cl
  |
  CH_3

5-20 2-Bromo-3-methylpentane has two asymmetric carbon atoms and four different optical isomers (two pairs of enantiomers). Use molecular models to demonstrate that different diastereomers give different elimination products on reaction with sodium ethoxide (assume that both reactions follow Saytzeff's rule).

5-21 Cyanate ion, OCN^-, is a good nucleophile, but it usually forms a mixture of two different displacement products. Write two resonance forms for cyanate ion and show the products of reaction with bromoethane.

5-22 Arrange the following in order of increasing acidity: ethanethiol, water, hydronium ion, ethanol.

5-23 *S*-Adenosylethionine sometimes exists in living systems. Its structure is like that of *S*-adenosylmethionine but with an ethyl group rather than a methyl group attached to the sulfonium-ion sulfur. Give the product of reaction of *S*-adenosylethionine with noradrenaline (figure 5-17). Would *S*-adenosylethionine be more reactive or less reactive than *S*-adenosylmethionine? Why?

5-24 An important fate of primary and secondary alcohols in nature is oxidation to an aldehyde or ketone. Give the product which would be formed by oxidation of cholesterol (figure 5-7).

5-25 2,6-Dichloroheptane reacts with Na_2S to form a cyclic compound. The stereochemistry of the product is different when the starting material is the R,R isomer than when it is the R,S isomer. Give the structures of the products and explain this difference. Use molecular models if necessary.

5-26 Arrange the following in order of increasing reactivity toward bromoethane: H_2O, NaCl, $NaOCH_3$, NaI.

5-27 Arrange the following in order of increasing reactivity toward NaCN: ethanol, ethyl iodide, ethyl chloride, triethyl sulfonium chloride.

5-28 Arrange the following compounds in order of increasing amount of olefin formed on reaction with sodium ethoxide: *t*-butyl bromide, ethyl bromide, isopropyl bromide.

5-29 Give the structure of the principal product or products of each of the following reactions:

(*a*) 2-Methyl-2-butanol + conc. H_2SO_4 $\longrightarrow$

(*b*) —CH_2SH + $CH_3\overset{\underset{\displaystyle |}{Cl}}{C}HCH_3$ $\longrightarrow$

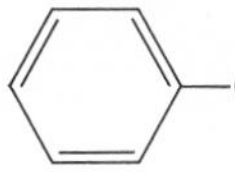

(*c*) —$CH_2\overset{\underset{}{\overset{\displaystyle OH}{|}}}{C}HCH_3$ + $Na_2Cr_2O_7$ $\xrightarrow{H^+}$

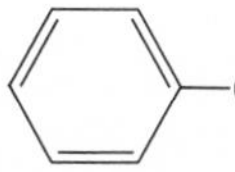

(*d*) 2-Bromo-2-methylheptane + $NaOCH_2CH_3$ $\longrightarrow$

(*e*) $ClCH_2CH_2CH_2Br$ + 1 equiv. KCN $\longrightarrow$

(*f*)  + KI + H_3PO_4 $\longrightarrow$

(g) $CH_3CH_2CH_2Br + NaSH \longrightarrow$

(h) $CH_3(CH_2)_{11}Br + KCN \longrightarrow$

(i) 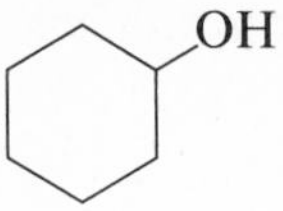
$+ Na \longrightarrow$

(j) 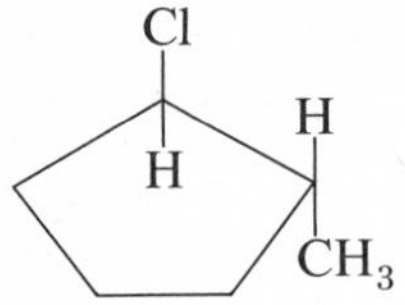
$+ NaOCH_2CH_3 \longrightarrow$

(k) 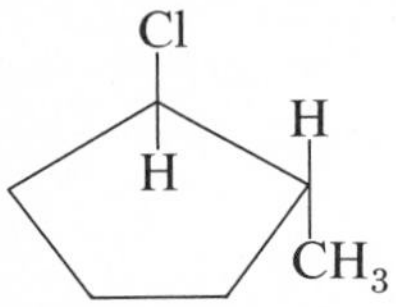
$+ KI \longrightarrow$

(l) 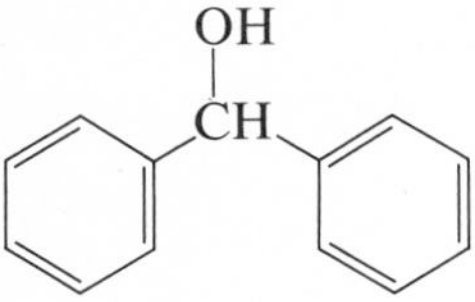
$+ KMnO_4 \xrightarrow{OH^-}$

(m) 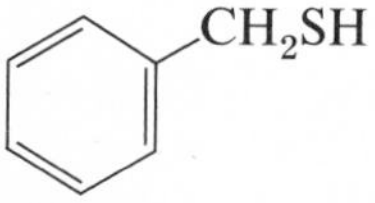
$+ NaOH \longrightarrow$

(n) 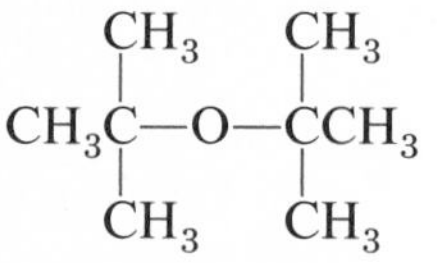
$+ H_2SO_4 \longrightarrow$

(o) $CH_3CH_2CH_2SH + Na \longrightarrow$

SUGGESTED READINGS

Evans, D. A., and C. L. Green: Insect Attractants of Natural Origin, *Chem. Soc. Rev.*, **2:** 75 (1973). An interesting account of structure-function relationships in insect sex attractants.

Fieser, L. F.: Steroids, p. 32 in *Organic Chemistry of Life: Readings from Scientific American,* Freeman, San Francisco, 1973. An introduction to the structures and functions of steroids.

Fox, C. F.: The Structure of Cell Membranes, p. 261 in *Organic Chemistry of Life: Readings from Scientific American,* Freeman, San Francisco, 1973. Good discussion of the role of polar and nonpolar groups in membrane structure.

Wilson, E. O.: Pheromones, p. 115 in *Organic Chemistry of Life: Readings from Scientific American,* Freeman, San Francisco, 1973. A description of chemicals used by insects for communication.

6

Amines

Simple inorganic compounds frequently provide us with the basis for understanding corresponding organic compounds. Our discussion of alcohols and ethers, for example, began with a discussion of water. Our discussion of mercaptans derives from the consideration of hydrogen sulfide. In the same way, our discussion of amines begins with a discussion of ammonia.

The chemistry of amines is similar in many ways to that of ammonia. Like ammonia, amines have an unshared pair of electrons on nitrogen; amines can act as bases, using that electron pair to form a bond to a proton, and they can act as nucleophiles, using that electron pair to form a bond to a carbon atom. The nitrogen atom of the amino group, rather than the carbon atom to which it is attached, will usually serve as the focus of our discussion of amines.

6-1 CLASSIFICATION AND NOMENCLATURE

This chapter is concerned with two slightly different types of compounds. The *amines* are organic analogs of ammonia, having three atoms bonded to nitrogen; the *ammonium salts* are organic analogs of ammonium ion, having four atoms bonded to a positively charged nitrogen

$$H-\overset{..}{\underset{|}{\overset{|}{N}}}-H \qquad H-\overset{\overset{H}{|}}{\underset{|}{N^+}}-H$$

ammonia   ammonium ion

CLASSIFICATION   Amines and ammonium salts are classified according to the number of carbon substituents attached to the amine nitrogen. *Primary amines* have one carbon substituent attached to nitrogen, *secondary amines* have two, *tertiary amines* have three, and *quaternary ammonium salts* have four (figure 6-1). Note that this classification system is based on the number of substituents attached to nitrogen, whereas the classification system for alcohols and alkyl halides is based on the number of substituents attached to carbon.

One or more of the carbon substituents in an amine may be a benzene ring (for example, aniline, in figure 6-1). Although the carbon atom attached to nitrogen in such compounds is at a higher oxidation level than that in the simple amines, the properties of such compounds are very similar to those of simple amines.

The amine nitrogen atom may also be part of an aliphatic ring (for example, piperidine and pyrrolidine in figure 6-1). Amines in

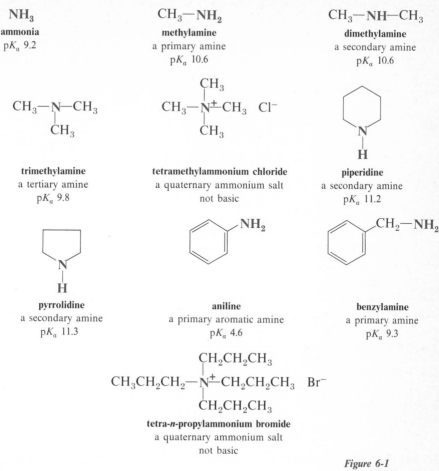

Figure 6-1
Some simple amines.
The pK_a values
given are those of
the conjugate acids
of the amines
(section 6-4).

which the nitrogen atom is part of an aromatic ring have properties quite different from those of the amines in figure 6-1 and will be considered in chapter 11.

NOMENCLATURE

Amines are frequently named by using functional class names. The alkyl groups attached to nitrogen are named using the names in table 3-2, and this is followed by the ending -*amine*. For example, CH_3—NH_2 is methylamine, and $CH_3CH_2CH_2$—NH_2 is *n*-propylamine. Some other examples of this system are shown in figure 6-1.

Quaternary ammonium salts are named by giving the names of the substituent groups, followed by the word *ammonium* and the name of the anion. Some examples are given in figure 6-1.

PROBLEM 6-1   Give the structure of (a) n-butylamine; (b) methylisopropylamine; (c) methyltriethylammonium bromide; (d) ethyl-n-propylamine.

6-2   THE STRUCTURE OF AMINES

The nitrogen atom in most amines is sp^3-hybridized. Three of the four hybrid orbitals form σ bonds to adjacent carbon or hydrogen atoms; the fourth hybrid orbital contains the unshared pair of electrons* (figure 6-2).

The sp^3-hybridized nitrogen atom, like the sp^3-hybridized carbon atom, is tetrahedral. The three substituents occupy three corners of the tetrahedron, and the unshared electron pair occupies the fourth. The angle between any two bonds to nitrogen is near 109° (the angles in methane are all 109.5°).

PROBLEM 6-2   Build models of ammonia and methane and compare their shapes. Do the same for tetramethylammonium ion and for neopentane, $(CH_3)_4C$.

PROBLEM 6-3   Draw and name all structural isomers of (a) C_3H_9N; (b) $C_4H_{11}N$.

AMMONIUM   The ammonium ion, NH_4^+, has an sp^3-hybridized nitrogen atom
SALTS   surrounded by four equivalent hydrogens. Like methane, the ammonium ion is tetrahedral (figure 6-3). Other ammonium ions are also tetrahedral. Tetramethylammonium ion, $(CH_3)_4N^+$, has an sp^3-hybridized nitrogen atom surrounded by four tetrahedrally arranged methyl groups. Tetramethylammonium ion has the same geometry and the same electronic structure as neopentane (figure 6-3).

PROBLEM 6-4   Draw the structures of all isomers of $C_4H_{12}N^+$.

INVERSION OF   As a result of the tetrahedral geometry of the nitrogen atom in amines,
AMINES   compounds such as methylethylamine, $CH_3-NH-CH_2CH_3$, are chiral. However, it is not possible to separate such compounds into enantiomers because the nitrogen atom inverts its configuration very

---

*The nitrogen atom in aniline and other aromatic amines is sp^2-hybridized, and the unshared pair of electrons occupies a p orbital rather than a hybrid orbital. The unshared pair of electrons on the nitrogen atom in aniline is involved in resonance interactions with the benzene ring (chapter 11); such interactions are possible only if the unshared electron pair occupies a p orbital.

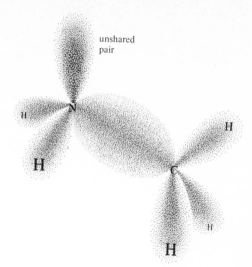

The nitrogen atom is sp³-hybridized. The carbon-nitrogen bond is a σ bond formed by an sp³ hybrid orbital of carbon and an sp³ hybrid orbital of nitrogen. The

unshared electron pair on nitrogen (shown here in brown) also occupies an sp³ hybrid orbital.

Figure 6-2
Structure of
methylamine.

rapidly. This inversion, which resembles an umbrella turning inside out, interconverts enantiomers of compounds such as methylethylamine (figure 6-4). Inversion is very easy and occurs millions of times a second.

Quaternary ammonium ions cannot undergo the type of inversion shown in figure 6-4; as a result, quaternary ammonium ions with four different carbon substituents do not undergo inversion and can be resolved. One of the first such compounds to be resolved was methylallylbenzylphenylammonium iodide

methylallylbenzylphenylammonium iodide

Note that although it is chiral, this compound does not have an asymmetric carbon atom. Instead, it has an asymmetric nitrogen atom.

Build a model of methylisopropylamine and show that inversion of the nitrogen atom interconverts the compound and its enantiomer.

PROBLEM 6-5

PROBLEM 6-6   Build a model of methyl-*s*-butylamine and show that inversion of the nitrogen atom does not lead to interconversion of enantiomers. Why is this true?

CYCLIC AMINES   Because of the tetrahedral geometry of the nitrogen atom in most amines, cyclic amines such as piperidine and pyrrolidine (figure 6-1) can exist in two different conformations, which can be interconverted by inversion of the nitrogen atom (figure 6-5). In some cases, this inversion interchanges cis and trans isomers.

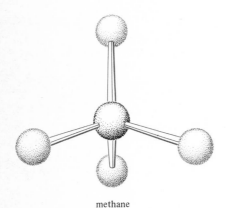

methane

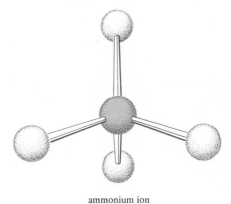

ammonium ion

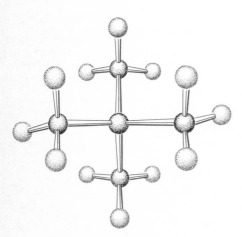

neopentane

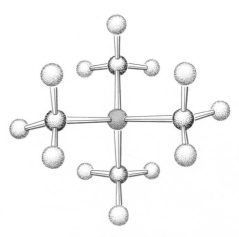

tetramethylammonium ion

Figure 6-3
Geometry of
ammonium salts.

The ammonium ion and methane both have a tetrahedral center atom and therefore have the same geometry.

Tetramethylammonium ion and neopentane both have the same geometry.

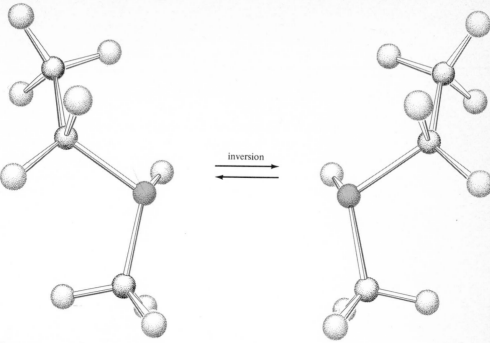

The inversion of the nitrogen atom of amines occurs millions of times a second. If the amine is chiral because of an asymmetric nitrogen atom, the inversion interconverts R and S enantiomers. Quaternary ammonium salts do not undergo this inversion.

Figure 6-4
Amine inversion.

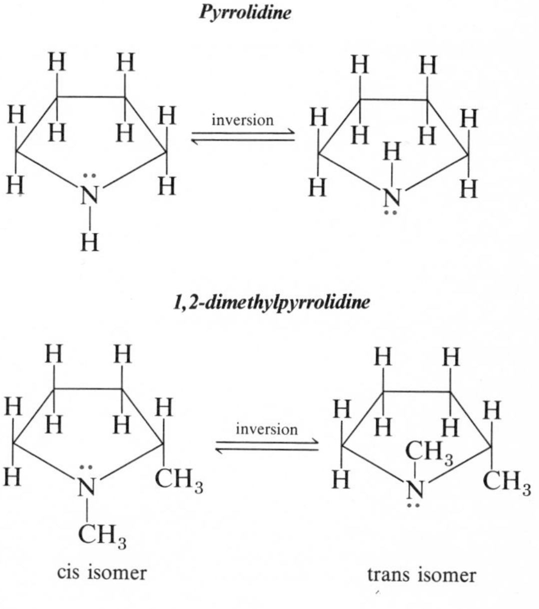

Figure 6-5
Nitrogen inversion in cyclic amines.
Inversion of the nitrogen atom in 1,2-dimethylpyrrolidine results in interconversion of cis and trans isomers.

Pyrrolidine

inversion

1,2-dimethylpyrrolidine

inversion

cis isomer   trans isomer

PROBLEM 6-7 Use molecular models to show that interconversion of the axial and equatorial conformations of the chair form of *N*-methylpiperidine can occur without interconversion of the two chair forms.

6-3 NITROGEN FIXATION

By far the largest fraction of the nitrogen near the earth's surface is tied up in the form of nitrogen gas, $N\equiv N$. The triple bond in nitrogen is very strong, with the result that the conversion of molecular nitrogen into ammonia and other forms suitable for incorporation into living systems requires a large amount of energy. Only a small fraction of living organisms have the ability to carry out this process, called *nitrogen fixation:**

$$N_2 + 6H \xrightarrow[\text{and plants}]{\text{microorganisms}} 2NH_3$$

The source of hydrogen for this process is water or other hydrogen-containing compounds.

Nitrogen fixation provides a source of the ammonia needed to sustain life. From this ammonia, living things synthesize other nitrogen-containing compounds used in metabolism. However, few higher plants and no animals are capable of carrying out this important process by themselves; most plants and animals depend on the nitrogen-fixing ability of various microorganisms and legumes (plants such as peas and beans).

The chemical reactions involved in nitrogen fixation by living organisms have been under study for a number of years, but many parts of this important process remain to be elucidated. The two enzymes involved in this process have been isolated and studied. One of the enzymes contains molybdenum, an unusual metal in biological systems. It has recently been shown that nitrogen-fixing microorganisms are capable of reducing acetylene to ethane

$$H-C{\equiv}C-H + 4H \xrightarrow[\text{organisms}]{\text{nitrogen-fixing}} \underset{\underset{H}{|}\;\underset{H}{|}}{\overset{\overset{H}{|}\;\overset{H}{|}}{H-C-C-H}}$$

This reaction has led many people to speculate on the possibility that hydrazine

$$\underset{\underset{H}{|}\;\underset{H}{|}}{H-N-N-H}$$

hydrazine

---

*In some instances, nitrogen fixation leads to the formation of nitrite ion, NO_2^-, or nitrate ion, NO_3^-, rather than ammonia.

is an intermediate in nitrogen fixation, but hydrazine has not been detected in any nitrogen-fixing system.

The microorganisms which fix nitrogen can also reduce other molecules, including HCN and ethylene. What is the product of the reaction of HCN? Of ethylene?

PROBLEM 6-8

Suggest an experiment which might be used to determine whether hydrazine is an intermediate in nitrogen fixation.

PROBLEM 6-9

Most plants are unable to synthesize their own ammonia from nitrogen gas and depend on ammonia obtained from their environment.* Modern agriculture has evolved sophisticated chemical methods for supplementing the natural supply of ammonia. Industrial nitrogen fixation is a strictly chemical process which converts nitrogen into ammonia by means of high temperature, high pressure, and a catalyst

INDUSTRIAL NITROGEN FIXATION

$$N\equiv N + 3H_2 \xrightarrow{\text{metal cat., high pressure, 500°}} 2NH_3$$

The ammonia produced by this process is used as fertilizer. This industrial nitrogen fixation has an enormous effect on both the economy and the ecology. The world total of nitrogen fixed per year by this process (presently about 30 million tons) is as large as the total amount of nitrogen fixed each year by nature prior to the advent of modern agriculture.

Nitrogen fixation results in the conversion of N_2 into NH_3 and eventually into various amines and other organic nitrogen compounds. If this were the only pathway involving nitrogen, the atmosphere would eventually become depleted and all the earth's nitrogen would be converted into organic compounds. However, this does not happen. Certain plants and microorganisms catalyze *denitrification,* the conversion of NH_3, NO_2^-, and NO_3^- into N_2. It is not known whether the rate of denitrification is keeping pace with the increasing rate of nitrogen fixation; if it is not, serious ecological consequences may result from what might be described as overfertilization. Some manifestations of this phenomenon are already apparent in certain lakes and rivers.

DENITRIFICATION

---

*The algae which congest many lakes also depend on external sources of ammonia. Most of it comes from fertilizer runoff from neighboring farms or yards.

NITROGEN METABOLISM IN HIGHER ANIMALS Ammonia itself is toxic to most higher animals and does not constitute an important part of their diet. Instead, they ingest most of their nitrogen in the form of amino acids (chapter 12). A complex series of reactions converts amino acids into other nitrogen-containing substances.

THE NITROGEN CYCLE The movement of nitrogen in nature is an important and complex process. Some of the important pathways for nitrogen transfer are shown in figure 6-6.

6-4 BASICITY OF AMINES

Ammonia is basic. In the presence of acid it is converted into ammonium ion

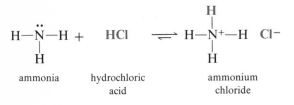

Figure 6-6
The nitrogen cycle of the atmosphere. The principal reservoir of nitrogen is the N_2 in the air. Organic forms of nitrogen come ultimately from this source.

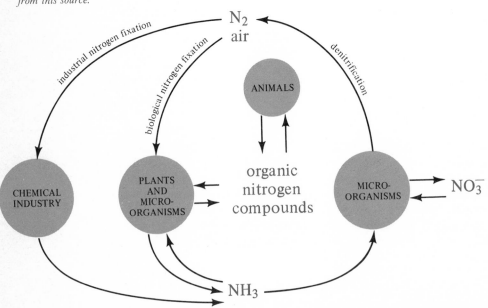

Amines also undergo this reaction

$$CH_3-\overset{\displaystyle ..}{\underset{\displaystyle H}{N}}-H + HNO_3 \rightleftharpoons CH_3-\overset{\displaystyle H}{\underset{\displaystyle H}{N^+}}-H \quad NO_3^-$$

| methylamine | nitric acid | methylammonium nitrate |

$$2 \left[\text{piperidine} \right] + H_2SO_4 \rightleftharpoons 2 \left[\text{piperidinium} \right] \quad SO_4^{2-}$$

| piperidine | sulfuric acid | piperidinium sulfate |

Quaternary ammonium salts, of course, are unable to react in this way.

QUANTITATIVE DETERMINATION OF BASICITY

The definition of the basicity of an amine is similar to that of the acidity of an alcohol (section 5-3). For the acid-base reaction

$$R-NH_3^+ \rightleftharpoons R-NH_2 + H^+$$

the acid dissociation constant of the conjugate acid* is defined as

$$K_a = \frac{[H^+][RNH_2]}{[RNH_3^+]}$$

in which the brackets indicate concentrations. As usual, the pK_a is given by

$$pK_a = -\log K_a$$

Some pK_a values are given in figure 6-1. The pK_a values for most amines fall in the range between 9 and 11. Thus, under the conditions existing in most cells (approximately pH 7), most amines exist not as free amines, RNH_2, but as ammonium ions, RNH_3^+.

---

*It is important to realize that in order to consider acid dissociation constants as we did in chapter 5, we must write this equation in terms of the dissociation of a proton from the conjugate acid of the amine, rather than as the acceptance of a proton by the amine itself.

AMINES
AS DRUGS
A number of amines are important drugs. A notable example is 1-aminoadamantane, also called *amantadine*

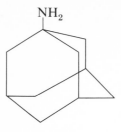

amantadine

This drug is of particular interest—chemically because of its highly symmetrical structure and biologically because it is an antiviral agent which acts against only one particular virus, the one causing Asian influenza.

Many amines are used as drugs, and many additional amines are present as part of the normal complement of molecules in the cell. Whether each amine is present as the amine or as its conjugate acid depends on its pK_a and on the pH of its surroundings. When the pH is below the pK_a, most of the amine exists as the conjugate acid; above the pK_a, most is free amine. Therapeutically it makes no difference whether the drug is administered in the amine form or the ammonium form; because of the buffering action of the digestive system, the final ratio of amounts of the two forms will be the same in either case.*

PROBLEM 6-10   Is amantadine chiral? Build a model to confirm your answer.

PROBLEM 6-11   Amantadine is usually administered in the form of amantadine hydrochloride, the product of reaction of amantadine with hydrochloric acid. Give the structure of amantadine hydrochloride.

ACIDITY OF
AMMONIUM
SALTS
Ammonium salts which have at least one hydrogen on nitrogen are slightly acidic, and in basic solution they can be transformed into amines. This reaction is simply the reverse of the reaction of amines with acids considered above. It occurs because hydroxide and alkoxides are more strongly basic than amines:

---

*Ordinarily if a drug is administered in the free-amine form, it will be converted quickly into the ammonium form. This conversion requires one equivalent of protons, and the absorption of these protons may produce irritation because of local pH changes in the gastrointestinal tract. For this reason, ammonium salts are usually administered.

$$CH_3-NH_3{}^+ \ Cl^- + NaOH \rightleftharpoons CH_3-NH_2 + NaCl + HOH$$

$$C_6H_5-\overset{\overset{\displaystyle H}{|}}{\underset{\underset{\displaystyle H}{|}}{N}}{}^+-CH_3 \ \ Cl^- + Na^+ \ {}^-O-CH_3 \rightleftharpoons$$

$$C_6H_5-\underset{\underset{\displaystyle H}{|}}{N}-CH_3 + NaCl + HO-CH_3$$

Which of the amines given in figure 6-1 would exist at pH 10 primarily **PROBLEM 6-12**
as the free amine rather than as the ammonium salt? Would any of
the amines in figure 6-1 exist as free amines at pH 7?

6-5 *ALKALOIDS*

In addition to the usual complement of amino acids and other amines,
many plants produce a number of nitrogen-containing substances
which are of little use to the plants themselves but have profound
effects on any animals that come in contact with them. These compounds, called *alkaloids,* are derived from amino acids and have a
great variety of structures. Although more than 4000 alkaloids have
been isolated and identified (a task which has been an important part
of organic chemistry for many years), the roles of these compounds
in the lives of the plants which produce them are largely unknown.
A number of alkaloids are shown in figure 6-7.

Many alkaloids have pronounced effects on humans; some of these
were among the very first drugs. Quinine, long used in the treatment
of malaria, is derived from the bark of the cinchona tree in South
America. It has been used in medicine at least since 1633 (when the
first written record of its use appeared), and it may have been used
by the Indians for many years before that. Morphine, the principal
active constituent of opium, is widely used for the relief of severe pain.
Early Babylonian and Egyptian writings contain many references to
its analgesic properties. Strychnine has been widely used as a rat poison
since the sixteenth century. In spite of its high toxicity, it has found
occasional use in medicine because of its strong stimulatory effect on
the central nervous system.

Since the beginning of recorded history, humans have sought to expand   *HALLUCI-*
their world by chemically expanding their minds. In primitive tribes,   *NOGENIC*
this was the province of the medicine man, who inherited from his   *DRUGS*

Figure 6-7
Alkaloids. *These*
compounds are
synthesized in plants
from amino acids.
Many alkaloids are
used as drugs.

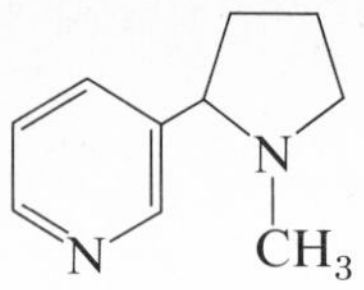

nicotine
from tobacco

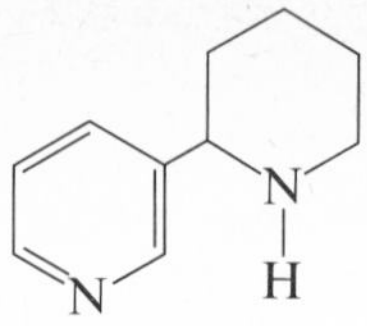

anabasine
from tobacco

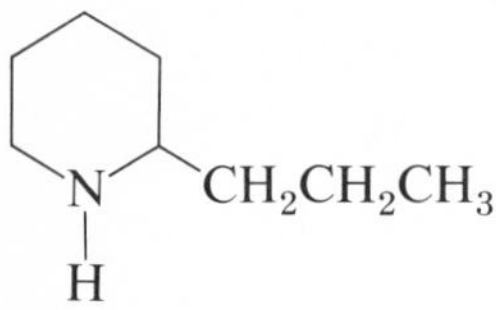

coniine
from hemlock; this
is the poison that
killed Socrates

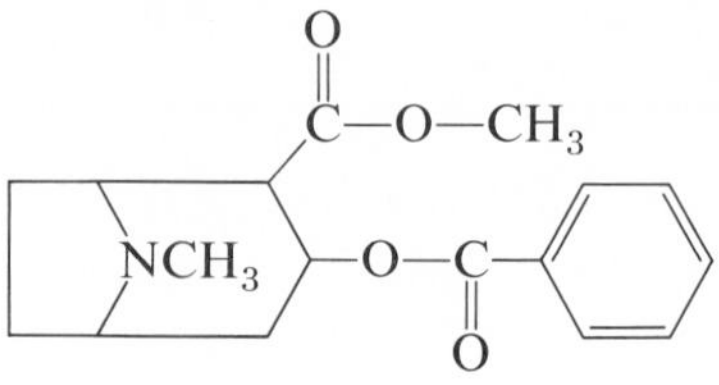

cocaine
from cocoa leaves;
a central nervous system
stimulant that is also
used as a local anesthetic

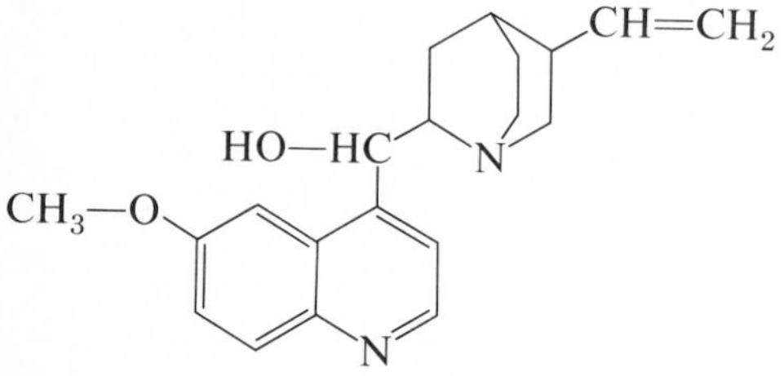

quinine
long used in
the treatment of
malaria

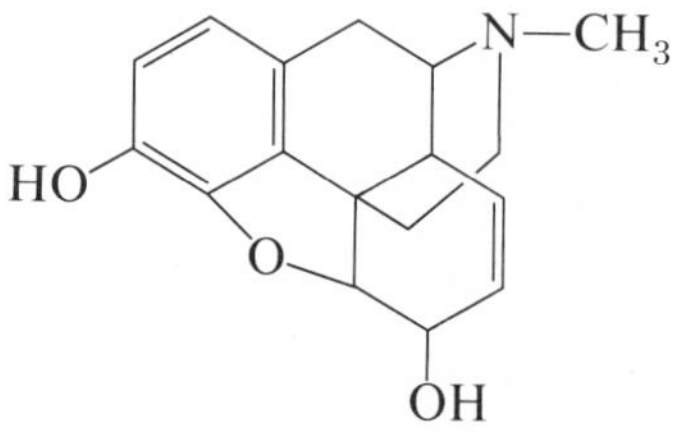

morphine
the principle analgesic
component of opium; widely
used and abused (heroin is
made from opium)

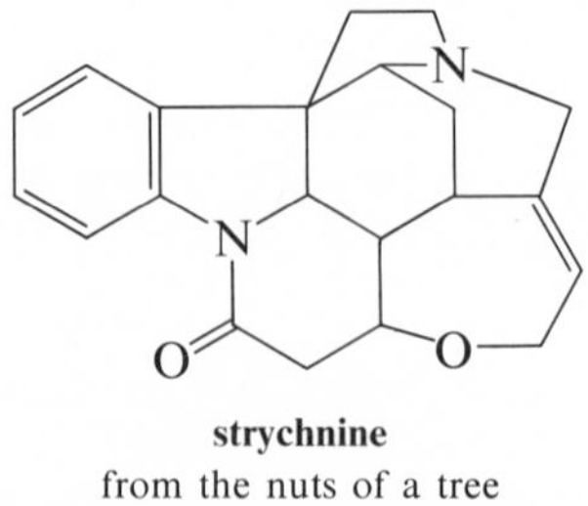

strychnine
from the nuts of a tree
native to India; strong
nervous system stimulant;
very toxic

predecessor the methods for obtaining such chemicals from plants. More recently, Havelock Ellis, Aldous Huxley, and others have investigated the effects of these potent substances on the mind.

Most of these so-called *hallucinogens* are alkaloids and are similar in structure to a class of compounds involved in the transmission of nerve impulses, the *catecholamines,* which are amines having a dihydroxybenzene, or *catechol,* ring.

catechol

Naturally occurring catecholamines include serotonin, an important neurotransmitter, norepinephrine, a neurotransmitter in the sympathetic nervous system, and dopamine, the neurotransmitter from which norepinephrine is made (figure 6-8). The hallucinogenic drugs have structures which are strikingly similar to those of the catecholamines: all have an aromatic ring connected to a two-carbon chain, which is in turn connected to an amino group. Almost all currently popular hallucinogenic drugs, as well as a number of other important drugs, show this same structural feature.

Δ^1-Tetrahydrocannabinol,*

Δ^1-tetrahydrocannabinol

the active ingredient in marihuana, is not an amine and does not fit into this structural class. Biological studies indicate that its mechanism of action is different from that of the hallucinogens.

6-6 REACTIONS OF AMINES

The first and most important characteristic of amines is the presence of an unshared pair of electrons. It figures importantly in nearly every reaction of amines.

---

*The prefix Δ^1 indicates that the double bond is at carbon atom 1.

Catecholamines

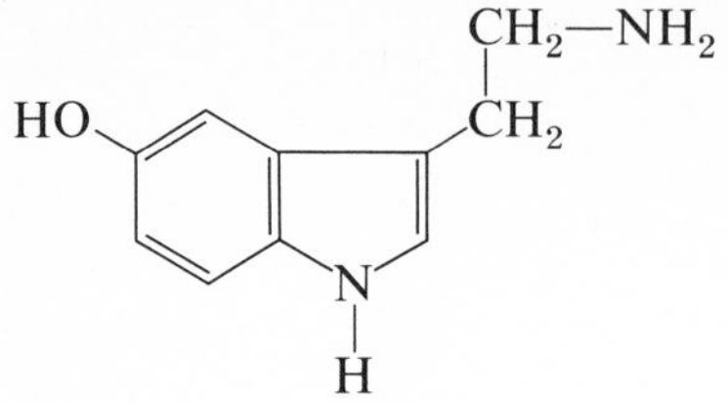

serotonin
derived from tryptophan

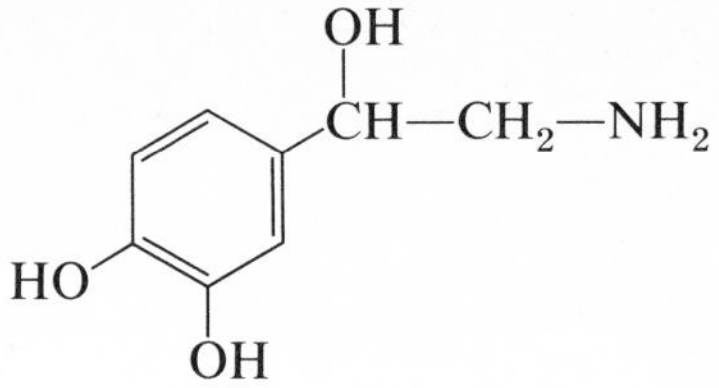

norepinephrine
derived from dopamine

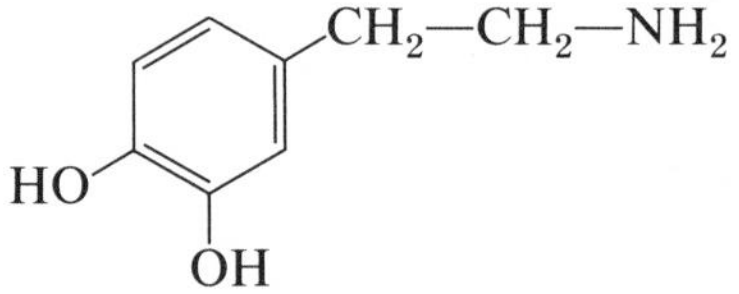

dopamine
derived from phenylalanine,
via tyrosine and dopa

Figure 6-8
Catecholamines and hallucinogens.

The two groups of compounds have similar structures, and it seems likely that hallucinogens work (at least in part) by interfering with the functioning of the catecholamines.

AMINES AS NUCLEOPHILES

Amines are excellent nucleophiles and undergo a variety of displacement reactions similar to those considered in chapter 5. For example,

$$CH_3CH_2-NH_2 + CH_3-Br \longrightarrow CH_3CH_2-NH_2^+-CH_3 \quad Br^-$$

The details of this reaction are shown in mechanism 6-1.

However, this reaction frequently takes a more complicated course than that shown in mechanism 6-1. In the first place, the ammonium salt which is shown as the product of the above reaction is not nucleophilic, but it can act as an acid and donate a proton to a base. Frequently this base is another molecule of the starting amine

$$CH_3CH_2-NH_2^+-CH_3 + CH_3CH_2-NH_2 \rightleftharpoons$$
$$CH_3CH_2-NH-CH_3 + CH_3CH_2-NH_3^+$$

The secondary amine produced in this last reaction is nucleophilic and can react with another molecule of alkyl halide, forming a new ammonium salt

$$CH_3CH_2-NH-CH_3 + CH_3-Br \longrightarrow CH_3CH_2-\underset{\underset{CH_3}{|}}{N}H^+-CH_3 \quad Br^-$$

After loss of a proton, this product, too, can react with an alkyl halide. In this case, the product is a quaternary ammonium salt, and no further

Hallucinogens

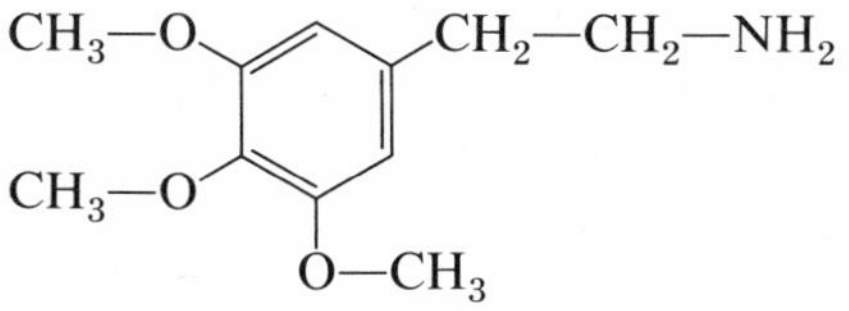

mescaline
the active principle of peyote; known to the Indians of Central America since pre-Columbian times

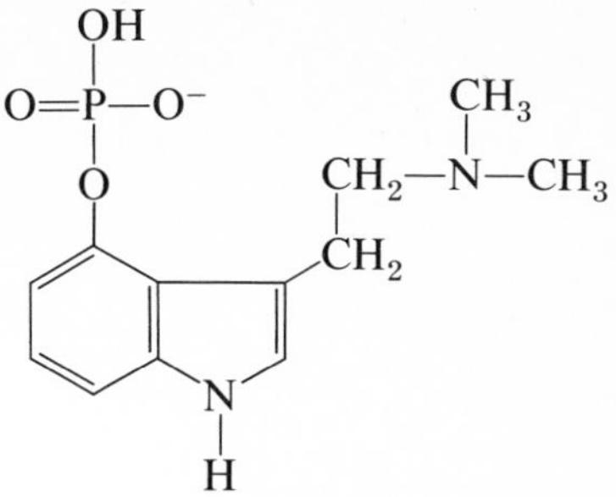

psilocybin
from Mexican mushrooms; once used in rituals by Indians in southern Mexico

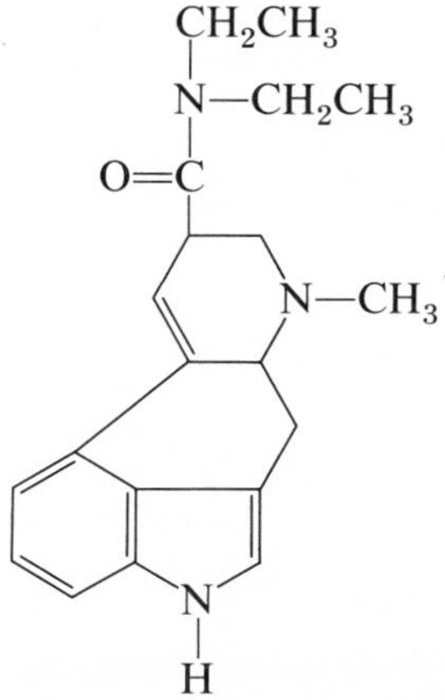

LSD (*d*-lysergic acid diethylamide)
does not occur in nature, although the parent compound, *d*-lysergic acid, occurs in the fungus *ergot,* which sometimes infests cereal grains; extremely potent—doses as small as 0.025 milligram produce an effect in susceptible individuals

reaction is possible

$$CH_3CH_2{-}\underset{\underset{CH_3}{|}}{N}{-}CH_3 + CH_3{-}Br \longrightarrow CH_3CH_2{-}\overset{\overset{CH_3}{|}}{\underset{\underset{CH_3}{|}}{N^+}}{-}CH_3 \quad Br^-$$

These complications limit the usefulness of reactions of alkyl halides

Mechanism 6-1
Nucleophilic
Displacement by an
Amine

$$CH_3\overset{..}{-}NH_2 + CH_3\overset{}{-}I$$

Amines have an unshared pair of electrons which is readily available to form a bond to an electron-poor carbon atom.

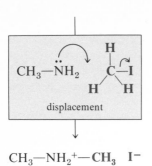

displacement

The key step in this reaction involves two simultaneous changes: (1) forming a new bond between carbon and the nucleophilic nitrogen and (2) breaking a bond between that carbon atom and the leaving group.

$$CH_3-NH_2{}^+-CH_3 \quad I^-$$

The product of the reaction is an ammonium salt. The ammonium ion contains one more carbon substituent on nitrogen than the starting amine does.

with amines. In most cases, reaction of an alkyl halide with an amine leads to a mixture of products

$$\text{(aniline)} + CH_3CH_2-I \longrightarrow$$

$$I^- {}^+NH_2-CH_2CH_3 + I^- {}^+\underset{CH_2CH_3}{NH}-CH_2CH_3 + CH_3CH_2-\underset{CH_2CH_3}{N^+}-CH_2CH_3 \quad I^-$$

The proportions of the various products are seldom easy to predict. The use of a large excess of alkyl halide results in complete conversion of the amine into a quaternary ammonium salt. *Choline,* which is the natural precursor of the neurotransmitter acetylcholine, can be synthesized by this method

$$HO-CH_2CH_2-NH_2 + 3CH_3-I \longrightarrow HO-CH_2CH_2-\underset{\underset{CH_3}{|}}{\overset{\overset{CH_3}{|}}{N^+}}-CH_3 \quad I^- + 2HI$$

ethanolamine   (excess)   choline

The oxygen of the hydroxyl group is not sufficiently reactive to displace iodide from methyl iodide under the reaction conditions. *Decamethonium,* a drug which interferes with the functioning of choline, can be made by a similar procedure

$$NH_2-(CH_2)_{10}-NH_2 + 6CH_3-I \longrightarrow CH_3-\overset{\overset{\displaystyle CH_3}{|}}{\underset{\underset{\displaystyle CH_3}{|}}{N^+}}-(CH_2)_{10}-\overset{\overset{\displaystyle CH_3}{|}}{\underset{\underset{\displaystyle CH_3}{|}}{N^+}}-CH_3 \quad 2I^- + 4HI$$

(excess)   decamethonium

Ammonia can also act as a nucleophile in displacement reactions. If a large excess of ammonia is used, a primary amine will be formed

(excess)

Give the mechanism of the last reaction above. Explain why the use of a large excess of ammonia results in the formation of a primary amine.

PROBLEM 6-13

Ammonium salts which have one or more hydrogens attached to the tetravalent nitrogen transfer a proton readily to a variety of bases (section 6-4)

ELIMINATION REACTIONS

Quaternary ammonium ions have no hydrogen atoms attached to the tetravalent nitrogen and are unable to undergo this reaction. Instead, upon heating in the presence of base, they undergo an elimination reaction analogous to that of secondary and tertiary alkyl halides (section 5-7):

$$H-CH_2CH_2-\overset{\overset{\displaystyle CH_2CH_3}{|}}{\underset{\underset{\displaystyle CH_2CH_3}{|}}{N^+}}-CH_2CH_3 \quad Cl^- + OH^- \xrightarrow{\text{heat}}$$

$$CH_2{=}CH_2 + \overset{\overset{\displaystyle CH_2CH_3}{|}}{\underset{\underset{\displaystyle CH_2CH_3}{|}}{N}}-CH_2CH_3 + H_2O + Cl^-$$

The mechanism of this reaction is the same as mechanism 5-3, by which olefins are formed from alkyl halides in the presence of base. Another example is

$$
\begin{array}{c}
CH_3 \\
|^{} \\
CH_3-N^+-CH_3 \quad Cl^- \\
|^{} \\
CH-CH_2-H \ + \ Na^+ \ \ ^-O-CH_3 \xrightarrow{\text{heat}}
\end{array}
$$

$$
CH=CH_2 \qquad\qquad
\begin{array}{c}
CH_3 \\
| \\
+ \ CH_3-N \ + \ HO-CH_3 \ + \ NaCl \\
| \\
CH_3
\end{array}
$$

Although the mechanism of this reaction is similar to that of the reaction of alkyl halides with base, the two differ in one important respect: Saytzeff's rule is not followed in elimination reactions of quaternary ammonium salts. Instead, elimination reactions of quaternary ammonium salts form a greater amount of the product resulting from loss of a hydrogen atom from the carbon bearing the *most* hydrogen atoms.*

$$
\begin{array}{c}
CH_3 \\
| \\
CH_3-N^+-CH_3 \ + \ OH^- \xrightarrow{\text{heat}} \\
| \\
CH_3CH_2CH_2CHCH_3
\end{array}
$$

$$
CH_3CH_2CH_2CH=CH_2 \ + \ CH_3CH_2CH=CH-CH_3 \ + \
\begin{array}{c}
CH_3 \\
| \\
CH_3-N-CH_3 \\
\end{array}
\ + \ H_2O
$$

major product   minor product

Elimination reactions like these have frequently been used to determine the structures of unknown amines. The amine is first reacted with an excess of methyl iodide. The ammonium salt so formed is subjected to heat in the presence of base, forming an olefin. The determination of the structure of *piperidine,* a compound isolated from black pepper, involved two such methylation and elimination sequences (figure 6-9).

PROBLEM 6-14   For each of the following compounds, give the structures of the two products which would be formed on heating in the presence of sodium

---

*It should be borne in mind that neither this elimination rule (called the *Hoffman rule*) nor the Saytzeff elimination rule is absolute; in both cases appreciable amounts of more than one product may be formed. The rules indicate which possible product is formed in the greater amount.

hydroxide, and indicate which of the two would be formed in greater amount.

(a)
$$CH_3-N^+-CH_3$$
with CH_3 above and CH_3 below, attached to a cyclohexane ring

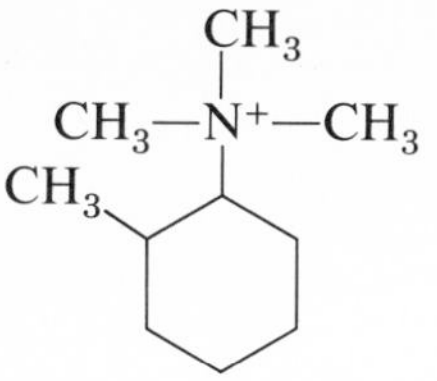

(b)
$$CH_3-N^+-CH_2CH_3$$
with CH_3 above, attached to a cyclohexane ring

---

piperidine

$$2\,CH_3I \longrightarrow$$

OH⁻
heat

$$CH_3I \longleftarrow$$

OH⁻
heat

(1,3-pentadiene). The rearrangement of 1,4-pentadiene into 1,3-pentadiene takes place spontaneously during the elimination reaction.

The determination of this structure involved the conversion of the unknown compound (piperidine) by a sequence of methylations and eliminations into a compound of known structure

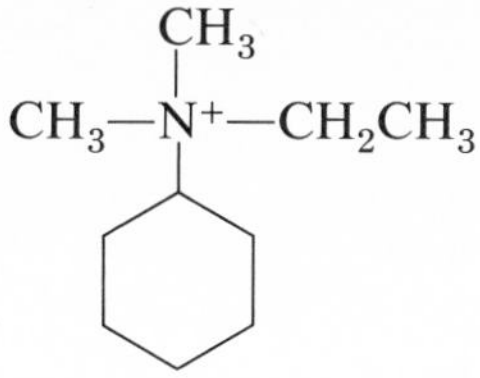

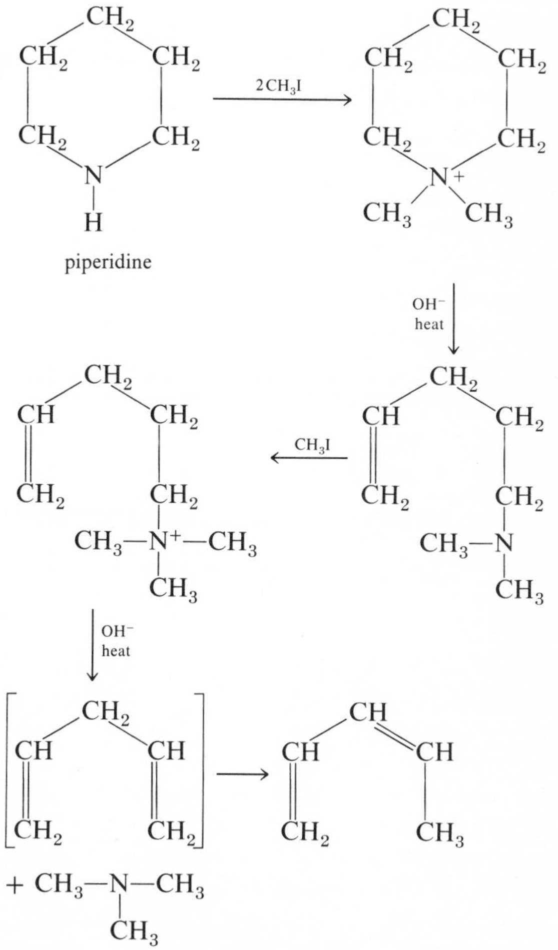

Figure 6-9
Structure of piperidine.

A few elimination reactions of amines analogous to those of quaternary ammonium salts are known. *Cinnamic acid,* the starting point for the synthesis of a variety of color pigments in plants, is made by this route from phenylalanine

phenylalanine   cinnamic acid

The enzyme which catalyzes this reaction, *phenylalanine ammonia-lyase,* occurs in a wide variety of plants. This reaction is unlike those discussed earlier in that it involves elimination of ammonia from a primary amine, rather than elimination of a tertiary amine from a quaternary ammonium salt. The means by which phenylalanine ammonia-lyase achieves this is unknown.

6-7 AMINES IN NATURE

Amines are widespread in nature. Many of them are produced by the *decarboxylation* of amino acids, a reaction in which the —COOH group of the amino acid is replaced by a single hydrogen. The decarboxylation of dihydroxyphenylalanine (often called *dopa*) to dihydroxyphenylethylamine (called *dopamine*) is typical

dopa   dopamine

Dopamine is a catecholamine (figure 6-8) and an important transmitter in the central nervous system. It also serves as the precursor of norepinephrine and epinephrine.

Parkinson's disease, a widespread disease of adults in their fifties and sixties, is caused at least in part by a lack of dopamine in the central nervous system. Administration of dopamine to patients is usually ineffective because dopamine cannot pass the blood-brain barrier. Instead, large doses of dopa are given. This compound passes through the digestive system, into the bloodstream, and then through the blood-brain barrier, after which it is converted by the enzyme *dopa decarboxylase* into dopamine. The discovery of the effectiveness of dopa against parkinsonism has restored many thousands of formerly debilitated patients to reasonably normal lives.

Not all amines which occur in nature are desirable substances. Many   *TOXIC AMINES*
simple amines are toxic to living things (at high concentrations ammonia itself is toxic to mammals). Food spoilage is largely the result of
decarboxylation reactions like that which produces dopamine from
dopa. These reactions are catalyzed by enzymes present in airborne
bacteria. The names of the reaction products are indicative of their
unpleasant smells

$$NH_3{}^+-CH_2CH_2CH_2CH_2CH-\overset{\overset{\displaystyle O}{\|}}{C}-O^- \xrightarrow{\text{decarboxylation}}$$
$$\underset{NH_3{}^+}{|}$$

lysine

$$NH_3{}^+-CH_2CH_2CH_2CH_2CH_2-NH_3{}^+ + CO_2$$
cadaverine

$$NH_3{}^+-CH_2CH_2CH_2CH-\overset{\overset{\displaystyle O}{\|}}{C}-O^- \xrightarrow{\text{decarboxylation}}$$
$$\underset{NH_3{}^+}{|}$$

ornithine

$$NH_3{}^+-CH_2CH_2CH_2CH_2-NH_3{}^+ + CO_2$$
putrescine

Another important amine of this group is spermidine,

$$NH_3{}^+-CH_2CH_2CH_2CH_2-NH_2{}^+-CH_2CH_2CH_2-NH_3{}+$$
spermidine

which is synthesized by microorganisms from putrescine and *S*-adeno-
sylmethionine via a decarboxylation of the latter compound, followed
by nucleophilic displacement. This reaction is unusual in that the
nucleophilic displacement on the sulfonium salt occurs not on the
methyl group but on the primary carbon atom of the aminopropyl
group (figure 5-17). Little is known about the biological roles of
spermidine and the related compound spermine

$$NH_3{}^+-CH_2CH_2CH_2-NH_2{}^+-CH_2CH_2CH_2CH_2-NH_2{}^+-CH_2CH_2CH_2-NH_3{}^+$$
spermine

Both compounds appear to be involved in the maintenance of cell-wall
structure in microorganisms, but despite their widespread occurrence
in higher animals, their functions are not known.

Write out the reactions involved in the synthesis of spermidine in   *PROBLEM 6-15*
nature.

PROBLEM 6-16   Suggest a scheme for the formation of spermine in nature analogous to that of spermidine.

PROBLEMS   **6-17**   Give the mechanism and the products of (*a*) triethylamine + 2-bromopropane; (*b*) *n*-butyltrimethylammonium chloride + NaOH.

6-18   Classify and name the following compounds:

(*a*) $CH_3CH_2CH-NH_2$
           |
           CH_3

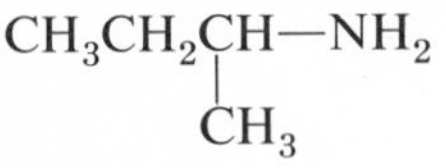

(*b*) $CH_3CH_2-NH-CH_3$

(*c*) $CH_3CH-N-CHCH_3$
         | | \
        CH_3 CH_3 CH_3

(*d*) $CH_3-N-CCH_3$
              CH_3
         | \
     CH_3CH_2 CH_3

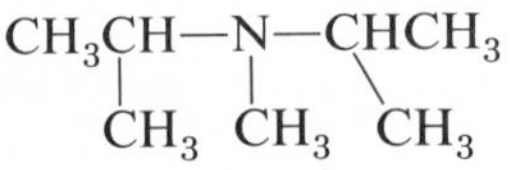

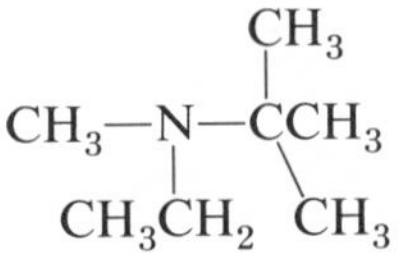

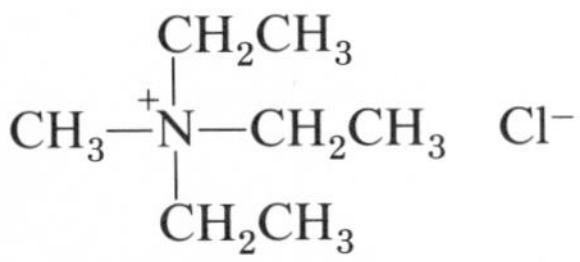

(*e*) $CH_3-\overset{+}{N}-CH_2CH_3$ Cl^-
            CH_2CH_3 (above)
            CH_2CH_3 (below)

(*f*) $CH_3CH_2CH_2CH_2-\overset{+}{N}-CH_3$ Br^-
            CH_2CH_3 (above)
            CH_3 (below)

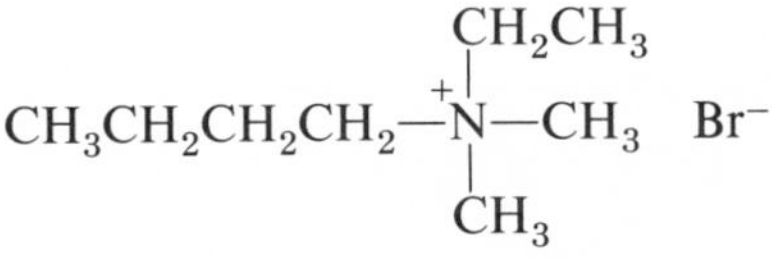

6-19   The same methylation-elimination sequence was used in the proof of the structure of coniine (figure 6-7) as in the proof of the structure of piperidine (figure 6-9). Give the structure of the final hydrocarbon produced in the reactions of coniine.

6-20   Complete each of the following acid-base equilibria and indicate whether the predominant species at equilibrium are those on the left, those on the right, or both: (*a*) piperidine + H_3O^+; (*b*) piperidine + H_2O; (*c*) $(CH_3CH_2CH_2)_3N$ + NH_4^+; (*d*) $(CH_3CH_2)_2NH$ + HCl; (*e*) NH_3 + H_2O.

6-21   Although amines are good nucleophiles, it is not possible to make amines by the reaction of ammonia with an alcohol in acidic solution. Why not?

6-22   Ammonium salts such as

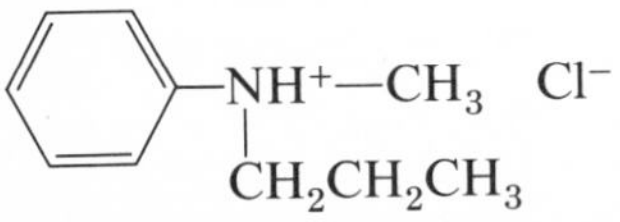

$$\text{(phenyl ring)}-NH^+-CH_3 \quad Cl^-$$
$$\qquad\qquad\qquad CH_2CH_2CH_3$$

cannot be resolved even in strongly acidic solution, conditions under which they are converted essentially quantitatively into the conjugate acid form shown here. Why can't they be resolved?

6-23   Give the mechanism of the reaction

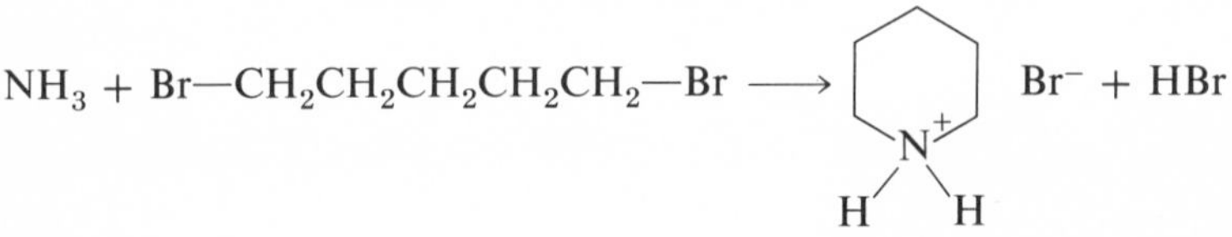

$$NH_3 + Br-CH_2CH_2CH_2CH_2CH_2-Br \longrightarrow \text{(piperidinium ring)} \quad Br^- + HBr$$
$$\overset{+}{N}$$
$$H \quad H$$

6-24 Amino groups are more nucleophilic than hydroxyl groups. Near room temperature, only amino groups participate in nucleophilic displacements. Give the product or products of each of the following reactions:

(*a*) Dopamine (figure 6-8) + CH_3CH_2I (excess) $\longrightarrow$

(*b*) Dopamine (excess) + CH_3I (limited amount) $\longrightarrow$

(*c*) Norepinephrine (figure 6-8) + CH_3I (excess) $\longrightarrow$

(*d*) Coniine (figure 6-7) + $BrCH_2CH_2CH_2CH_2CH_2Br$ $\longrightarrow$

(*e*) Cocaine (figure 6-7) + CH_3I $\longrightarrow$

(*f*) Morphine (figure 6-7) + H^+ $\longrightarrow$

6-25 Give the structure of the principal product or products of each of the following reactions:

(*a*)

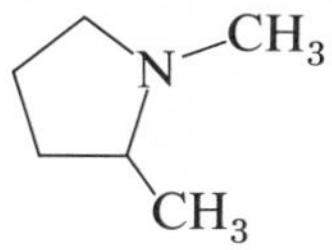

$+ CH_3CH_2{-}I \longrightarrow$ two stereoisomers

(*b*)

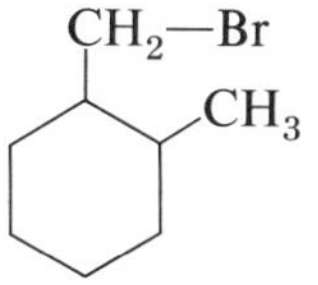

trans
isomer

(*c*) $CH_3CH_2CH_2Cl + CH_3{-}NH_2 \longrightarrow$

(excess)

(*d*) $(CH_3CH_2CH_2CH_2)_3NH^+$ $Cl^- + NaOH \xrightarrow{\text{heat}}$

(*e*) $(CH_3CH_2CH_2CH_2)_4N^+$ $Cl^- + NaOH \xrightarrow{\text{heat}}$

Barron, F., M. E. Jarvik, and S. Bunnell, Jr.: The Hallucinogenic Drugs, p. 85 in *Organic Chemistry of Life: Readings from Scientific American,* Freeman, San Francisco, 1973. The structures, functions, and physiology of a number of hallucinogens.

Delwiche, C. C.: The Nitrogen Cycle, *Sci. Am.,* September 1970, p. 136. Discusses nitrogen fixation, denitrification, and related matters.

Grinspoon, L.: Marihuana, *Sci. Am.,* December 1969, p. 17.

Robinson, T.: Alkaloids, p. 41 in *Organic Chemistry of Life: Readings from Scientific American,* Freeman, San Francisco, 1973. The variety of alkaloids, their history, and their uses.

SUGGESTED READINGS

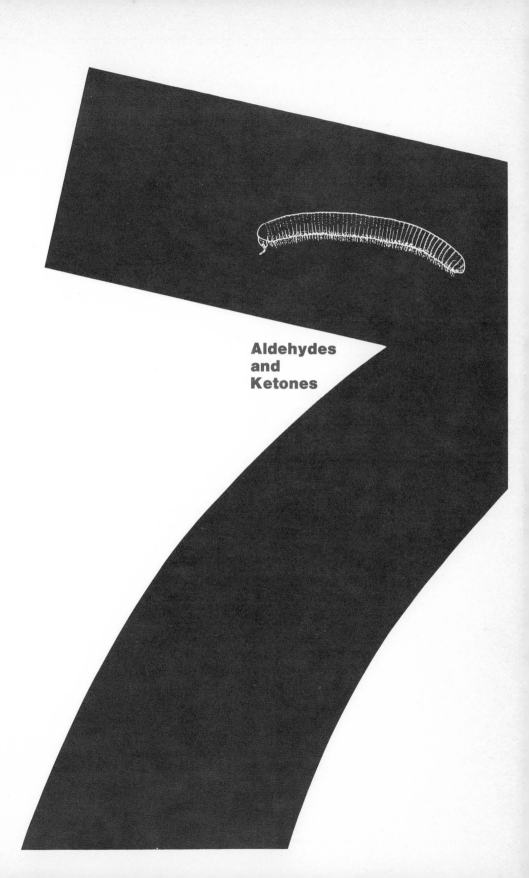

**Aldehydes
and
Ketones**

This is the first of three chapters concerned with the chemistry of the *carbonyl group,* $>C=O$. Aldehydes and ketones are compounds having only carbon or hydrogen attached to the carbonyl carbon atom. Ketones have two carbon atoms attached to the carbonyl carbon, whereas aldehydes have at least one hydrogen attached to the carbonyl carbon (figure 7-1).

Although aldehydes and ketones can undergo a wide variety of reactions, all these reactions fit into one of three classes:

1 Reactions in which the first step is attack by a nucleophile on the carbonyl carbon atom

2 Reactions in which the first step is protonation of the carbonyl oxygen and the second step is nucleophilic attack on the carbonyl carbon

3 Reactions in which the first step is removal of a proton from the carbon adjacent to the carbonyl carbon

7-1 THE STRUCTURE OF THE CARBONYL GROUP

The structure of formaldehyde, the simplest aldehyde, is shown in figure 7-2. Formaldehyde is planar, with 120° bond angles around the sp^2-hybridized carbon atom.

Like all double bonds, the carbon-oxygen double bond is composed of a σ bond and a π bond. Because the electronegativity of oxygen is greater than that of carbon, the carbon-oxygen double bond is a polar bond; the electron density is greater near oxygen than it is near carbon. Because of this polarity, nucleophiles, which are electron-rich, attack the carbon atom and electrophiles, which are electron-poor, attack oxygen.

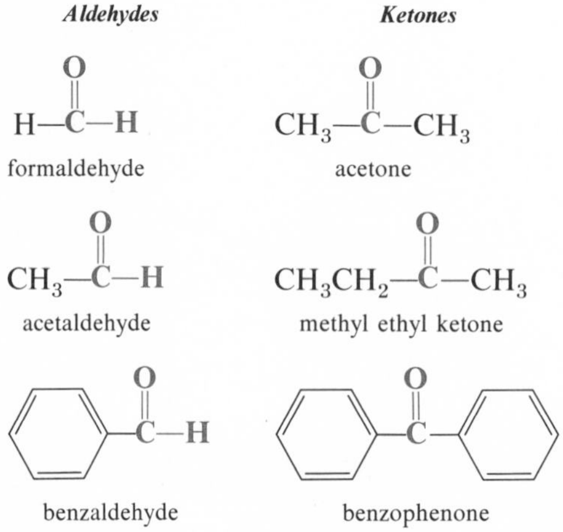

Figure 7-1
Aldehydes and ketones. Aldehydes have at least one hydrogen attached to the carbonyl carbon. Ketones have two carbons attached to the carbonyl carbon.

Aldehydes

$$H-\overset{\overset{\textstyle O}{\|}}{C}-H$$

formaldehyde

$$CH_3-\overset{\overset{\textstyle O}{\|}}{C}-H$$

acetaldehyde

benzaldehyde

Ketones

$$CH_3-\overset{\overset{\textstyle O}{\|}}{C}-CH_3$$

acetone

$$CH_3CH_2-\overset{\overset{\textstyle O}{\|}}{C}-CH_3$$

methyl ethyl ketone

benzophenone

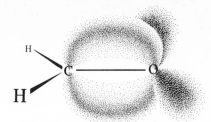

Both carbon and oxygen are
sp^2-hybridized. Lines indicate σ bonds.
The carbon-oxygen π bond is shown in
red. The oxygen atom has two unshared

pairs of electrons. Because of the
electronegativity of oxygen, the
carbon-oxygen bonds are polarized, with
the electron density higher near oxygen.

*Figure 7-2
Formaldehyde.*

The polarity of the carbonyl group can also be shown by the use of
resonance (figure 7-3). The carbonyl group is more like the first resonance structure in figure 7-3 than it is like the second, but the second
clearly shows the polarity of the carbonyl group.

*RESONANCE OF
THE CARBONYL
GROUP*

Draw two resonance structures for acetone. Indicate which atoms in
acetone must lie in the plane of the carbonyl group.

PROBLEM 7-1

7-2   NOMENCLATURE OF ALDEHYDES AND KETONES

From a chemical point of view, aldehydes and ketones are similar
and may be considered together, but from the point of view of nomenclature, the two are quite different and must be considered separately.

Aldehydes and carboxylic acids (chapter 9) are frequently named by
a set of common names which have different roots from the names
of other compounds which we have considered. The common names
of some simple aldehydes are given in table 7-1, together with their
IUPAC names. The roots of the common names are given in color.

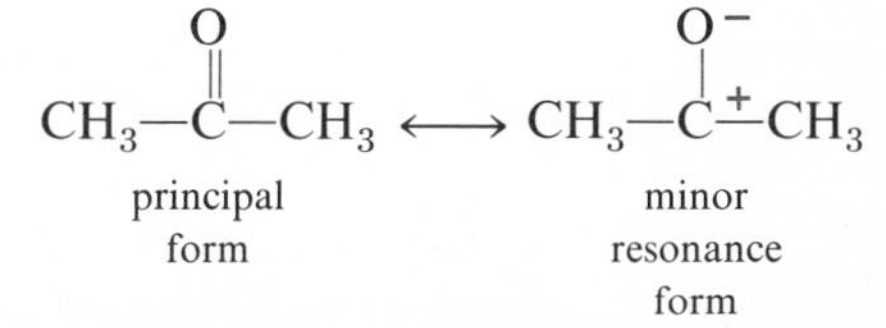

*COMMON NAMES
OF ALDEHYDES*

$$CH_3{-}\overset{\overset{\textstyle O}{\|}}{C}{-}CH_3 \longleftrightarrow CH_3{-}\overset{\overset{\textstyle O^-}{|}}{\underset{+}{C}}{-}CH_3$$

principal
form

minor
resonance
form

The polarization of the carbonyl group
can be pictured in terms of a resonance
structure with charges on oxygen and

carbon. This resonance form contributes
only slightly to the structure of the
carbonyl group.

*Figure 7-3
Resonance of the
carbonyl group.*

Table 7-1
Structures and Names of Some Aldehydes

Structure	Common Name	IUPAC Name
$H-\overset{\overset{\displaystyle O}{\|}}{C}-H$	**form**aldehyde	methanal*
$CH_3-\overset{\overset{\displaystyle O}{\|}}{C}-H$	**acet**aldehyde	ethanal*
$CH_3CH_2-\overset{\overset{\displaystyle O}{\|}}{C}-H$	**propion**aldehyde	propanal*
$CH_3CH_2CH_2-\overset{\overset{\displaystyle O}{\|}}{C}-H$	**butyr**aldehyde	butanal*
$CH_3\overset{\displaystyle}{\underset{\overset{\displaystyle \|}{CH_3}}{CH}}-\overset{\overset{\displaystyle O}{\|}}{C}-H$	**isobutyr**aldehyde	2-methylpropanal
$CH_3CH_2CH_2CH_2-\overset{\overset{\displaystyle O}{\|}}{C}-H$	**valer**aldehyde	pentanal
$CH_3\underset{\overset{\displaystyle \|}{CH_3}}{CH}CH_2-\overset{\overset{\displaystyle O}{\|}}{C}-H$	**isovaler**aldehyde	3-methylbutanal
$CH_3CH_2\underset{\overset{\displaystyle \|}{CH_3}}{CH}CH_2\underset{\overset{\displaystyle \|}{CH_3}}{CH}-\overset{\overset{\displaystyle O}{\|}}{C}-H$	—	2,4-dimethylhexanal
$CH_3CH_2CH_2CH{=}CHCH_2CH_2-\overset{\overset{\displaystyle O}{\|}}{C}-H$	—	4-octenal
$CH_3CH_2CH_2\underset{\overset{\displaystyle \|}{CH_3}}{\overset{\overset{\displaystyle CH_3CH_2}{\|}}{CH}}CHCH_2-\overset{\overset{\displaystyle O}{\|}}{C}-H$	—	3-ethyl-4-methylheptanal

*Although these names are technically correct, they are seldom used.

Carboxylic acids (chapter 9) may be named by attaching the ending *-ic acid* to the root.

Positions are assigned to substituents by Greek letters, starting with α (alpha) for a substituent on the carbon attached to the carbonyl carbon, β (beta) for a substituent on the next carbon, then γ (gamma), δ (delta), ε (epsilon), etc. Examples of this nomenclature are

$$\underset{\text{α-chloropropionaldehyde}}{CH_3-\underset{\underset{Cl}{|}}{CH}-\overset{\overset{O}{\|}}{CH}} \qquad \underset{\text{γ-hydroxybutyraldehyde}}{HO-CH_2CH_2CH_2-\overset{\overset{O}{\|}}{CH}}$$

Common names of aldehydes are useful only for simple compounds; complex compounds require a more systematic approach.

The IUPAC names of aldehydes are analogous to the IUPAC names of other types of compounds. The rules are as follows:

IUPAC NAMES OF ALDEHYDES

1 Find the longest continuous carbon chain which has the aldehyde group as one end. The root of the name is the name of the hydrocarbon with that number of carbons (table 3-1). The aldehyde ending *-al* replaces the hydrocarbon ending *-e*.

2 Number the carbons of the main chain. The aldehyde carbon is given the number 1.

3 Name and locate substituents on the main chain as usual, using the substituent-group names given in table 3-2.

IUPAC names of some aldehydes are given in table 7-1.

Simple ketones can be named by naming the two groups attached to the carbonyl carbon (table 3-2), followed by the word *ketone*. Some examples are given in table 7-2. This system is satisfactory for naming simple compounds, but it becomes cumbersome for complex compounds. As with aldehydes, positions of substituents can be indicated by the Greek letters α, β, γ, etc., but for ketones containing two different substituent groups, such names are usually ambiguous and should be avoided.

COMMON NAMES OF KETONES

The rules for constructing IUPAC names of ketones are similar to those used for other compounds:

IUPAC NAMES OF KETONES

1 Find the longest chain of carbon atoms containing the carbonyl group. The root of the name is the name of the hydrocarbon having this number of carbon atoms, with the ketone ending *-one* substituted for the alkane ending *-e.*

Table 7-2
Structures and Names of Some Ketones

Structure	Common Name	IUPAC Name
$CH_3-\overset{\overset{\displaystyle O}{\|\|}}{C}-CH_3$	dimethyl ketone (acetone)	2-propanone
$CH_3-\overset{\overset{\displaystyle O}{\|\|}}{C}-CH_2CH_3$	methyl ethyl ketone	2-butanone
$CH_3-\overset{\overset{\displaystyle O}{\|\|}}{C}-\underset{\underset{\displaystyle CH_3}{\|}}{C}HCH_3$	methyl isopropyl ketone	3-methyl-2-butanone
$CH_3CH_2CH_2-\overset{\overset{\displaystyle O}{\|\|}}{C}-CH_2CH_2CH_2CH_3$	n-propyl n-butyl ketone	4-octanone
$CH_3\underset{\underset{\displaystyle CH_3}{\|}}{C}H-\overset{\overset{\displaystyle O}{\|\|}}{C}-\underset{\underset{\displaystyle CH_3}{\|}}{C}HCH_3$	diisopropyl ketone	2,4-dimethyl-3-pentanone
(cyclobutanone ring with O and two CH₃ groups)	—	2,2-dimethyl-cyclobutanone
(cyclohexanone ring with O)	—	cyclohexanone

2 Number the main chain starting at the end nearer the carbonyl group. Place the number of the carbonyl carbon atom immediately before the root.

3 Number and identify substituents on the main chain in the usual manner.

Some examples of ketone names are given in table 7-2.

Ketones in which the carbonyl carbon is part of a ring are named by dropping the final -e and adding the suffix -one to the name of the ring. The carbonyl carbon atom is assigned the number 1 when numbering the ring.

7-3 ALDEHYDES AND KETONES IN NATURE

Aldehydes and ketones undergo a variety of reactions—perhaps a greater variety than any other group of compounds we have seen. Nature takes advantage of this variety; aldehydes and ketones are central to the metabolism of every organism.

The reactions of enolates (section 7-10) are among the principal reactions by which carbon-carbon bonds are formed in nature. Sugars, which almost invariably contain a carbonyl group (chapter 13), are usually synthesized and degraded by means of such reactions. In addition, the formation of hemiacetals and acetals (section 7-5) is a key to understanding the structures of sugars.

Aldehydes and ketones react readily with primary amines to form imines (section 7-4), compounds with carbon-nitrogen double bonds. Since this reaction is readily reversible, imines are often used in nature to hold things together temporarily in cases where a permanent bond is not necessary.

In addition, a variety of aldehydes and ketones occur in nature as flavoring and coloring agents, as vitamins, and in many other forms. Some naturally occurring aldehydes and ketones are shown in figure 7-4.

7-4 REACTIONS OF ALDEHYDES AND KETONES WITH NUCLEOPHILES

Because of the polarization of the carbonyl group, reactive nucleophiles are able to attack the carbonyl carbon atom. A variety of such reactions will be considered in this section. In section 7-5 we will consider reactions of less reactive nucleophiles, reactions which are initiated by protonation of the carbonyl oxygen atom.

Cyanide ion is a good nucleophile and readily displaces halide ions from many alkyl halides (section 5-6). Similarly, cyanide is an excellent nucleophile toward the carbonyl carbon atom of aldehydes and ketones. The product of reaction with cyanide is called a *cyanohydrin*

CYANOHYDRIN FORMATION

$$CH_3-\overset{\displaystyle O}{\overset{\|}{C}}-CH_3 + HC\equiv N \xrightleftharpoons{Na^+ \ ^-C\equiv N} CH_3-\overset{\displaystyle OH}{\underset{\underset{\displaystyle C\equiv N}{|}}{\overset{|}{C}}}-CH_3$$

acetone   acetone
                                                                    cyanohydrin

Figure 7-4
Some naturally
occurring aldehydes
and ketones.

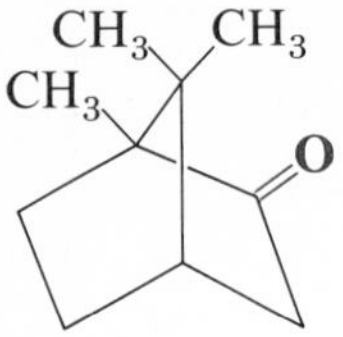

camphor
used in
medicine

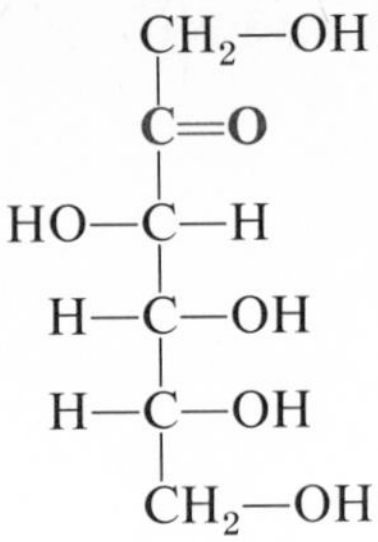

fructose
a sugar (chapter 13)

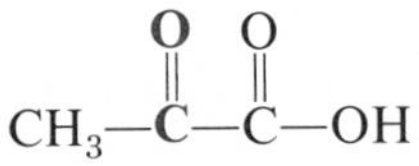

pyruvic acid
important intermediate in metabolism

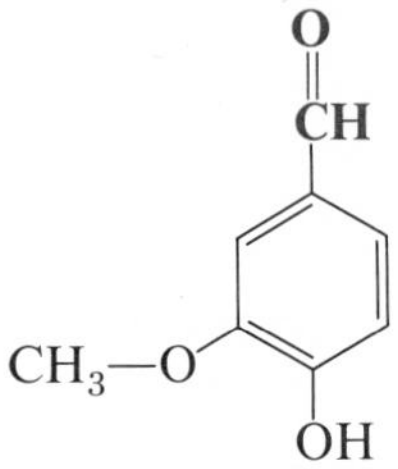

vanillin
odor constituent
of vanilla

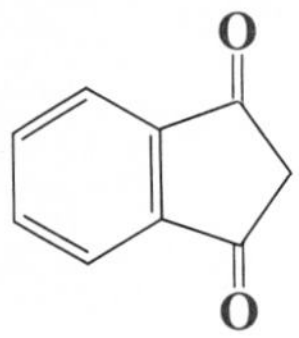

indane 1,3-dione
anticoagulant

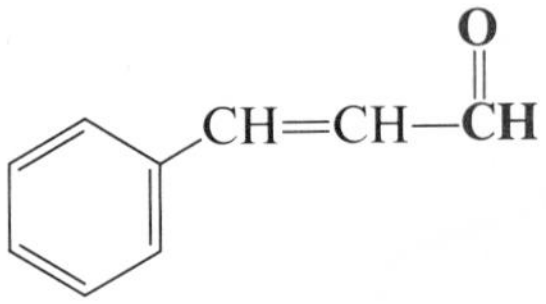

cinnamaldehyde
from cinnamon

The mechanism of this reaction is given in mechanism 7-1. Cyano-hydrin formation is reversible, and, as a result, the reaction must be carried out with a large excess of HCN in order to convert the alde-hyde or ketone completely into the cyanohydrin. Only a trace of NaCN is required because it functions as a catalyst and is not con-sumed during the reaction.

Cyanohydrins are rare, though not unknown, in living systems. Benzaldehyde cyanohydrin occurs as the principal component of the defensive secretion of certain millipedes. As this compound is sprayed toward an approaching enemy, it is mixed with a catalyst which rapidly decomposes the cyanohydrin into HCN and benzaldehyde

$$CH_3-\overset{\overset{\displaystyle O}{\|}}{CH} + \ ^-C{\equiv}N \ +$$

$$HC{\equiv}N$$

Reaction of an aldehyde or ketone with HCN occurs in two steps. Both HCN and ^-CN must be present for the reaction to occur.

Mechanism 7-1
Cyanohydrin
Formation

$$CH_3-\overset{\overset{\displaystyle O}{\|}}{\underset{\underset{\displaystyle ^-C{\equiv}N}{|}}{CH}}$$

nucleophilic
attack

In the first step, cyanide ion acts as a nucleophile and attacks the carbonyl carbon atom. A new carbon-carbon bond is formed, and simultaneously the carbon-oxygen double bond is transformed into a carbon-oxygen single bond. In this process the carbonyl oxygen atom acquires a pair of electrons.

$$\left[CH_3-\overset{\overset{\displaystyle O^-}{|}}{\underset{\underset{\displaystyle C{\equiv}N}{|}}{CH}} \right]$$

alkoxide
intermediate

The intermediate formed as a result of this attack no longer has a carbon-oxygen double bond, but instead is an alkoxide ion. This intermediate is basic and rapidly removes a proton from the HCN present.

proton
transfer
from
HCN

This proton transfer from HCN neutralizes the alkoxide-ion intermediate and generates a cyanide ion to replace the one consumed in the first step.

$$CH_3-\overset{\overset{\displaystyle OH}{|}}{\underset{\underset{\displaystyle C{\equiv}N}{|}}{CH}} + \ ^-C{\equiv}N$$

cyanohydrin

*The product of the reaction is called a **cyanohydrin**.*

benzaldehyde
cyanohydrin
(not very toxic)

$\xrightarrow{\text{catalyst}}$

benzaldehyde

$+ \ HC{\equiv}N$

hydrogen
cyanide
(very toxic)

The HCN thus formed is highly toxic and is usually successful in repelling predators.

PROBLEM 7-2   Give the mechanism of the reaction of pentanal with HCN in the presence of NaCN.

REACTION WITH PRIMARY AMINES   Primary amines are reactive nucleophiles and readily attack carbonyl carbon atoms of aldehydes and ketones. The product of the reaction is an *imine,* a compound with a carbon-nitrogen double bond

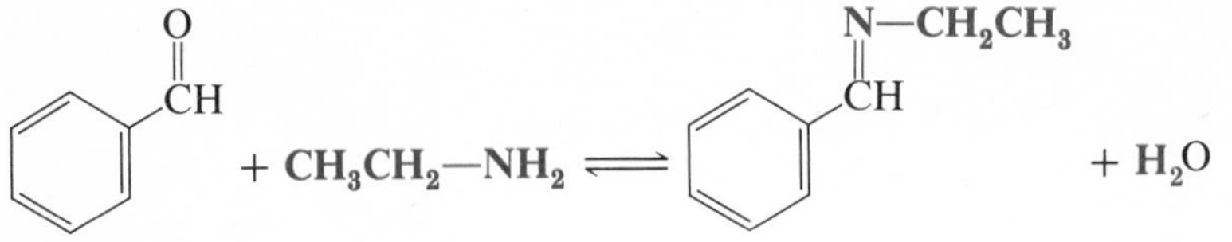

aldehyde   primary amine   imine

The mechanism of this reaction is given in mechanism 7-2.

Like cyanohydrin formation, imine formation is an equilibrium reaction. The equilibrium can be shifted to the right by using a large excess of amine or by removing the water formed.

PROBLEM 7-3   Give the mechanism of the reaction

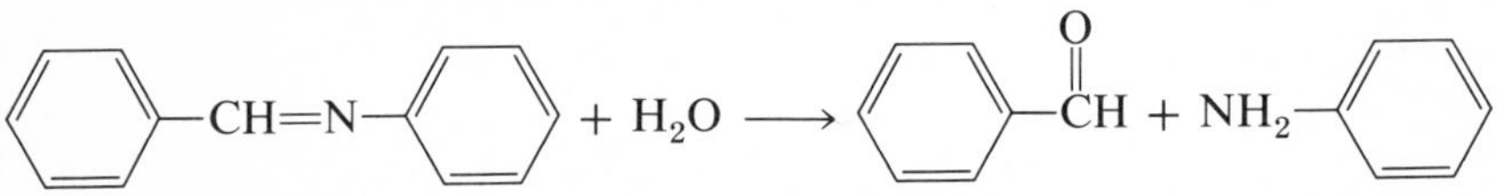

IMINES IN NATURE   Many imines occur in biochemistry. The bond between opsin and retinal (figure 3-7) is an imine bond. Imines occur as intermediates in many reactions; the synthesis of proline in nature, for example, occurs by way of an imine intermediate (figure 7-5).

HYDROXYLAMINE AND OTHER REACTIVE AMINES   There is a special class of amines for which imine formation is particularly easy. These compounds are primary amines in which the nitrogen is attached to oxygen or nitrogen rather than to carbon. The most important members of this class are

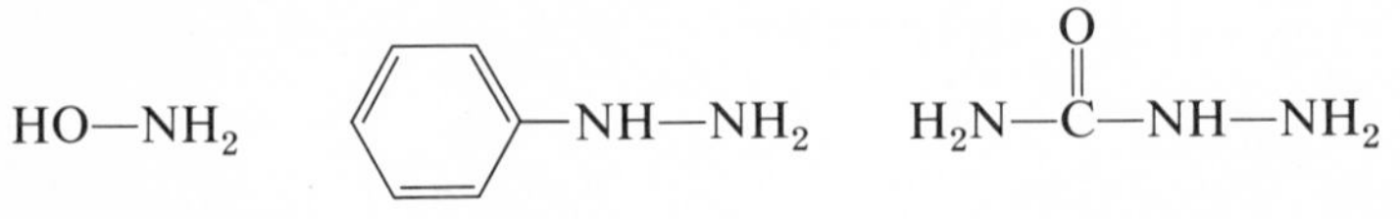

hydroxylamine   phenylhydrazine   semicarbazide

The exceptional usefulness of these compounds results because the equilibrium for imine formation is much more on the side of the imine

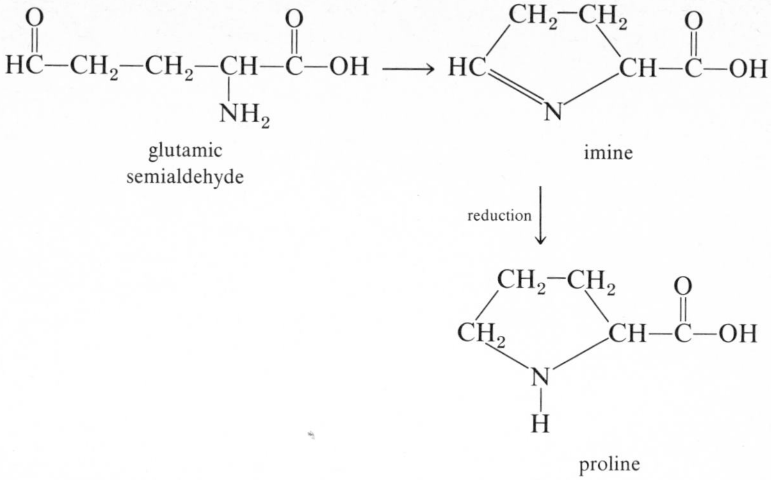

glutamic
semialdehyde

imine

reduction

proline

The synthesis of the amino acid proline in nature involves the reaction of glutamic semialdehyde with itself to form   a cyclic imine, followed by reduction of this imine to form proline.   *Figure 7-5*
A natural imine.

than it is with the simple primary amines discussed previously. Reactions of these amines with aldehydes and ketones have been widely used for the identification of aldehydes and ketones because the imine products formed are easily isolated and characterized. These imines are formed by mechanism 7-2.

Imines derived from these special amines all have special names. Those derived from hydroxylamine are called *oximes;* those derived from phenylhydrazine are called *phenylhydrazones;* those derived from semicarbazide are called *semicarbazones.* Reactions and product names are illustrated by the following examples:

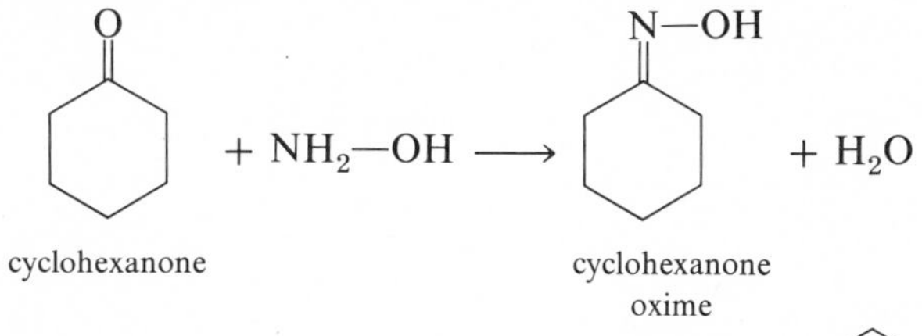

cyclohexanone

cyclohexanone
oxime

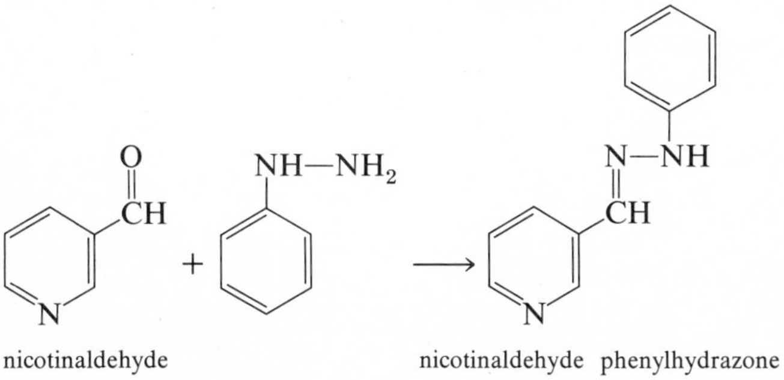

nicotinaldehyde

nicotinaldehyde phenylhydrazone

Mechanism 7-2
Imine Formation

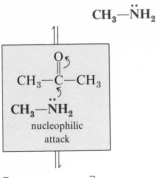

Primary amines have an unshared electron pair which enables them to act as nucleophiles.

O
‖
CH₃—C—CH₃

CH₃—N̈H₂

$$CH_3-\overset{\overset{\displaystyle O}{\|}}{C}-CH_3$$

$$CH_3-\ddot{N}H_2$$
nucleophilic
attack

As in cyanohydrin formation, the first step in the reaction is nucleophilic attack on the carbonyl carbon, with simultaneous breaking of the carbon-oxygen π bond.

$$\left[CH_3-\overset{\overset{\displaystyle O^-}{|}}{C}-CH_3 \right.$$
$$\left. CH_3-\overset{+}{N}H_2 \right]$$
unstable
intermediate

The intermediate formed as a result of this attack has an acidic ammonium group and a basic alkoxide anion.

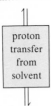
proton
transfer

As a result, proton transfer from nitrogen to oxygen occurs very rapidly.

$$\left[CH_3-\overset{\overset{\displaystyle OH}{|}}{C}-CH_3 \right.$$
$$\left. CH_3-\ddot{N}H \right]$$
carbinolamine
intermediate

The product thus formed is a moderately stable intermediate called a **carbinolamine.**

proton
transfer
from
solvent

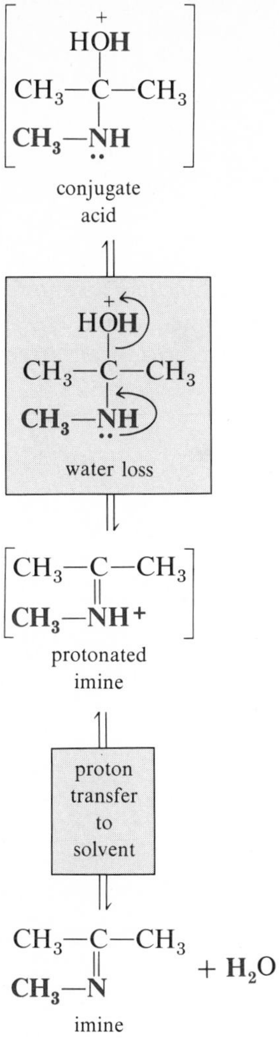

$$
\left[
\begin{array}{c}
\overset{+}{\text{H}}\text{OH} \\
\text{CH}_3-\overset{|}{\underset{|}{\text{C}}}-\text{CH}_3 \\
\text{CH}_3-\underset{\cdot\cdot}{\text{NH}}
\end{array}
\right]
$$

conjugate
acid

Both the oxygen atom and the nitrogen atom
of the carbinolamine intermediate are slightly
basic. Protonation of the oxygen of the carbinolamine forms a conjugate acid, which can
react further.

$$
\begin{array}{c}
\overset{+}{\text{H}}\text{OH} \\
\text{CH}_3-\overset{|}{\underset{|}{\text{C}}}-\text{CH}_3 \\
\text{CH}_3-\underset{\cdot\cdot}{\text{NH}}
\end{array}
$$

water loss

The protonated carbinolamine contains a very
good leaving group—water—and it decomposes readily by the loss of this water. At the
same time, a carbon-nitrogen π bond is formed
by using the unshared electrons on nitrogen.

$$
\left[
\begin{array}{c}
\text{CH}_3-\overset{|}{\underset{||}{\text{C}}}-\text{CH}_3 \\
\text{CH}_3-\text{NH}^+
\end{array}
\right]
$$

protonated
imine

The product of this step is a protonated imine,
which rapidly loses a proton.

proton
transfer
to
solvent

$$
\begin{array}{c}
\text{CH}_3-\overset{|}{\underset{||}{\text{C}}}-\text{CH}_3 \\
\text{CH}_3-\text{N}
\end{array}
\quad + \text{ H}_2\text{O}
$$

imine

The final product of the reaction is an imine.

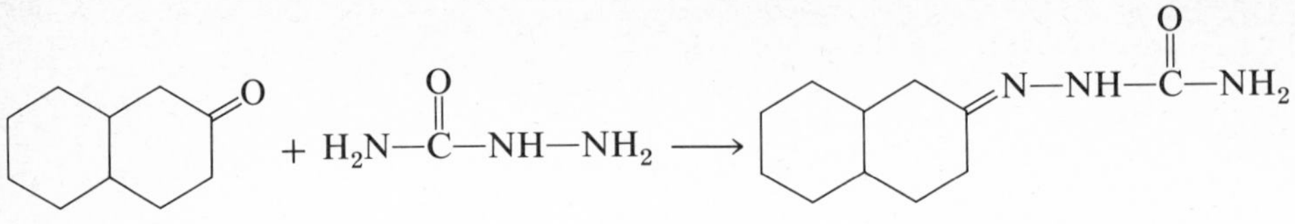

β-decalone   β-decalone semicarbazone

PROBLEM 7-4 Two different stereoisomers are formed when 2-methylcyclohexanone reacts with hydroxylamine. Only one isomer is formed when cyclohexanone reacts with hydroxylamine. Give the structures of all three products.

HEMIACETALS Alkoxide ions, the conjugate bases of alcohols, are excellent nucleo-
AND philes. A mixture of alcohol and alkoxide will react with an aldehyde
HEMIKETALS to form a small amount of a *hemiacetal*

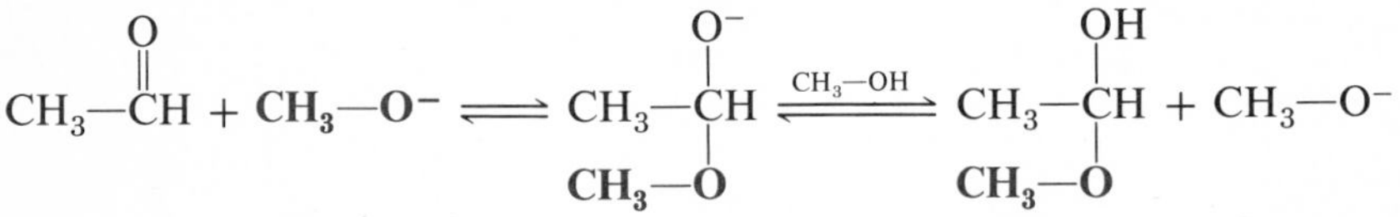

aldehyde   alkoxide   hemiacetal

The corresponding compound formed by reaction of a ketone with an alcohol is called a *hemiketal*. This reaction is not of great importance for simple aldehydes and ketones because the equilibrium usually lies far on the side of the free aldehyde, and little hemiacetal is formed at equilibrium. With sugars, however, hemiacetal and hemiketal formations are strongly favored (chapter 13).

Hemiacetal formation can also occur in acidic solution. In that case, further reaction of the hemiacetal can take place, leading to formation of an acetal (section 7-5).

7-5 ACID-CATALYZED NUCLEOPHILIC ATTACK ON THE CARBONYL GROUP

Reactive nucleophiles like those discussed in the previous section are able to attack the carbonyl groups of aldehydes and ketones without the help of a catalyst. Alcohols, on the other hand, are poor nucleophiles and do not attack carbonyl groups readily. Reaction of such poor nucleophiles with carbonyl groups is facilitated by acid catalysis.

The carbonyl oxygen atom of an aldehyde or ketone is very slightly basic. The protonated form is stabilized by resonance

$$CH_3-\overset{\overset{\displaystyle O}{\parallel}}{C}-CH_3 + H^+ \rightleftharpoons \left[CH_3-\overset{\overset{\displaystyle +O-H}{\parallel}}{C}-CH_3 \longleftrightarrow CH_3-\overset{\overset{\displaystyle O-H}{|}}{\overset{+}{C}}-CH_3 \right]$$

acetone   conjugate acid

However, even in strongly acidic solution only a very small amount of the conjugate acid is present at equilibrium. The conjugate acid of acetone has a pK_a of -7.2. Thus, even when acetone is dissolved in 1 M acid, only about 0.00001 percent of the compound exists as the conjugate acid. This conjugate acid is sufficiently reactive, however, to react readily with alcohols.

Aldehydes react with alcohols in the presence of an acid catalyst to form *acetals*

$$CH_3CH_2-\overset{\overset{\displaystyle O}{\parallel}}{CH} + 2CH_3CH_2CH_2-OH \overset{H^+}{\rightleftharpoons} CH_3CH_2\overset{\overset{\displaystyle O-CH_2CH_2CH_3}{|}}{\underset{\underset{\displaystyle O-CH_2CH_2CH_3}{|}}{CH}} \quad + H_2O$$

aldehyde   alcohol (excess)   acetal

The mechanism of this reaction is shown in mechanism 7-3. A hemi-acetal intermediate is formed. Because the equilibrium for this reaction lies on the left, the reaction must be carried out with a large excess of alcohol, together with an efficient means for removing water.

Ketones react with alcohols in the presence of an acid catalyst to form *ketals,* which are analogous to acetals. The equilibrium for this reaction is even more unfavorable than that for acetal formation.

cyclohexanone   (excess)   cyclohexanone
                                           dimethyl
                                           acetal

Under most conditions, acetals and ketals are stable compounds. However, in the presence of aqueous acid, hydrolysis of the acetal or ketal occurs, regenerating the carbonyl compound. The mechanism

Mechanism 7-3
Acetal Formation

$$CH_3-\overset{\overset{\textstyle O}{\|}}{C}H + H^+ +$$

$$2CH_3-\ddot{O}H$$

The carbonyl oxygen atom is slightly basic, and the aldehyde exists in equilibrium with a small concentration of its conjugate acid.

proton
transfer

$$\left[CH_3-\overset{\overset{\textstyle +OH}{\|}}{C}H \leftrightarrow CH_3-\overset{\overset{\textstyle OH}{|}}{\overset{+}{C}}H \right]$$

resonance-stabilized
conjugated acid

The concentration of the protonated aldehyde is small, but it is very reactive.

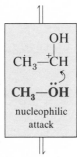

nucleophilic
attack

Because of the positive charge on carbon, methanol can act as a nucleophile and form a carbon-oxygen bond.

protonated
hemiacetal

The product formed as a result of this nucleophilic attack has a very acidic proton, which is lost rapidly to the solvent.

proton
transfer

$+ H^+$

hemiacetal
intermediate

*The compound resulting from loss of the proton is a **hemiacetal**. Although hemiacetals are marginally stable and may occasionally be isolated, under most conditions the reaction continues.*

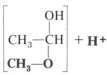

proton
transfer
from
solvent

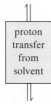

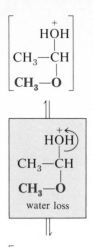

$$\left[\begin{array}{c} \overset{+}{\text{H}}\text{OH} \\ | \\ \text{CH}_3-\text{CH} \\ | \\ \text{CH}_3-\text{O} \end{array} \right]$$

In this acidic solution the hemiacetal exists to a small extent in a protonated form.

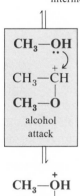

$$\overset{+}{\text{H}}\text{OH}$$
$$|$$
$$\text{CH}_3-\text{CH}$$
$$|$$
$$\text{CH}_3-\text{O}$$
water loss

This form can decompose by loss of water.

$$\left[\begin{array}{cc} \text{CH}_3-\text{CH} & \text{CH}_3-\overset{+}{\text{CH}} \\ \| & | \\ \text{CH}_3-\overset{+}{\text{O}} \leftrightarrow & \text{CH}_3-\text{O} \end{array} \right]$$
resonance-stabilized
intermediate

The product of the loss of water has a structure similar to the conjugate acid of the original aldehyde, but it cannot be neutralized by loss of a proton. Instead, it will react with methanol.

$$\text{CH}_3-\overset{..}{\text{O}}\text{H}$$
$$\overset{+}{}$$
$$\text{CH}_3-\overset{+}{\text{CH}}$$
$$|$$
$$\text{CH}_3-\text{O}$$
alcohol
attack

Attack by an unshared electron pair of oxygen on the positively charged carbon atom results in formation of a new carbon-oxygen bond.

$$\text{CH}_3-\overset{+}{\text{O}}\text{H}$$
$$|$$
$$\text{CH}_3-\text{CH}$$
$$|$$
$$\text{CH}_3-\text{O}$$

The protonated intermediate formed as a result of this attack rapidly loses a proton to the solvent.

proton
transfer
to
solvent

$$\text{CH}_3-\text{O}$$
$$|$$
$$\text{CH}_3-\text{CH} + \text{H}^+ + \text{H}_2\text{O}$$
$$|$$
$$\text{CH}_3-\text{O}$$
acetal

The final product of the reaction is a stable compound called an ***acetal***.

of this reaction is mechanism 7-3, taken backward

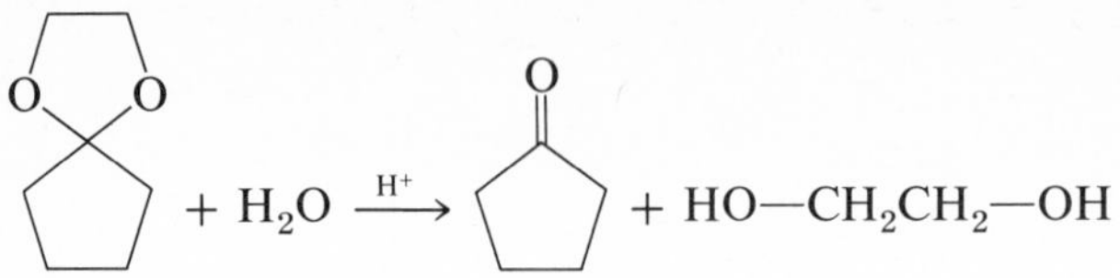

$$\text{(dioxaspiro structure)} + H_2O \xrightarrow{H^+} \text{(cyclopentanone)} + HO-CH_2CH_2-OH$$

PROBLEM 7-5   Give the mechanism of the reaction of cyclohexanone with methanol in the presence of acid, which was shown above.

7-6   THE GRIGNARD REACTION

Until now we have said very little about reactions which form new carbon-carbon bonds, even though such reactions are very important for the synthesis of complex molecules. Two common methods for making carbon-carbon bonds will be discussed in this chapter, the Grignard reaction (this section) and the reactions of enolates (section 7-10). Both reactions involve electron-rich carbon atoms acting as nucleophiles and attacking carbonyl carbon atoms.

FORMATION OF GRIGNARD REAGENTS

Grignard reagents can be made from a variety of chlorides, bromides, and iodides, including primary, secondary, tertiary, and aromatic halides. Reaction of the halide with magnesium in ether or tetrahydrofuran solvent results in cleavage of the carbon-halogen bond and formation of a carbon-magnesium bond and a magnesium-halogen bond

$$CH_3CH_2-I + Mg \xrightarrow[\text{tetrahydrofuran}]{\text{ether or}} CH_3CH_2-Mg-I$$

ethyl   ethylmagnesium
iodide   iodide, a Grignard
          reagent

Grignard reagents are stable in solution, but they cannot be isolated. We have represented the structure of the Grignard reagent above as ethylmagnesium iodide, but this representation is not strictly correct. The exact nature of the Grignard reagent is still a matter of speculation.

REACTIONS OF GRIGNARD REAGENTS

The reactivity of the Grignard reagent is that of a negatively charged carbon species, and its structure is sometimes represented in resonance terms as

$$CH_3CH_2-\mathbf{MgBr} \longleftrightarrow CH_3CH_2^- \; {}^+\mathbf{MgBr}$$

major contributor   minor resonance form

Grignard reagents decompose on contact with solvents containing an —OH group. The products are a hydrocarbon and a magnesium salt

$$CH_3CH_2—MgBr + H_2O \longrightarrow CH_3CH_3 + MgBrOH$$

The great usefulness of Grignard reagents in organic chemistry is a result of their reactivity toward carbonyl groups. Addition of a Grignard reagent to an aldehyde or ketone results in nucleophilic attack by the negatively charged carbon of the Grignard reagent on the carbonyl carbon of the aldehyde or ketone and formation of a magnesium alkoxide

After reaction of the Grignard reagent with the aldehyde is complete, water is added to hydrolyze the magnesium alkoxide. The product of this hydrolysis is an alcohol

Reaction of a Grignard reagent with an aldehyde forms a secondary alcohol after hydrolysis. Reaction with a ketone forms a tertiary alcohol. The following sequence of reactions is typical:

$$\text{(structure with OMgBr)} + H_2O \longrightarrow \text{(structure with OH)} + HOMgBr$$

tertiary
alcohol

Reaction of a Grignard reagent with carbon dioxide followed by hydrolysis forms a carboxylic acid:

$$\text{Ph—}CH_2\text{—Cl} + Mg \longrightarrow \text{Ph—}CH_2\text{—MgCl}$$

$$\text{Ph—}CH_2\text{—MgCl} + CO_2 \longrightarrow \text{Ph—}CH_2\overset{\displaystyle O}{\overset{\|}{—C}}\text{—OMgCl}$$

$$\text{Ph—}CH_2\overset{\displaystyle O}{\overset{\|}{—C}}\text{—OMgCl} + H_2O \longrightarrow \text{Ph—}CH_2\overset{\displaystyle O}{\overset{\|}{—C}}\text{—OH} + HOMgCl$$

a carboxylic acid

PROBLEM 7-6   Give the structure of the product of the following sequence of reactions: (*a*) a Grignard reagent is formed by reaction of 3-bromooctane with magnesium; (*b*) this reagent is reacted with propionaldehyde; (*c*) water is added.

7-7   REDUCTION OF ALDEHYDES AND KETONES

The reductions of aldehydes and ketones by complex metal hydrides, most notably lithium aluminum hydride, $LiAlH_4$, and sodium borohydride, $NaBH_4$, are further examples of the reaction of the carbonyl group with nucleophiles; in this case the nucleophile is the *hydride ion*, H^-. However, the whole discussion to follow has something of an "as if" character about it: these complex metal hydrides do not really liberate free H^- in solution, even though the structures of the products which are formed are those which would be expected if H^- were the reactive species.

Lithium aluminum hydride is a salt composed of Li^+ cations and

AlH_4^- anions:

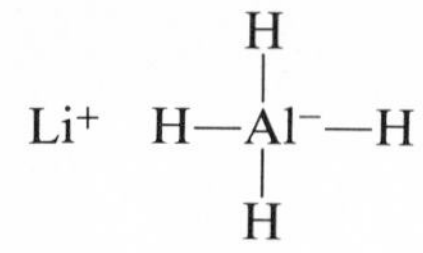

$$Li^+ \quad H\!-\!\overset{\displaystyle H}{\underset{\displaystyle H}{\overset{|}{\underset{|}{Al^-}}}}\!-\!H$$

lithium aluminum hydride

When lithium aluminum hydride reacts with carbonyl compounds, the AlH_4^- anion acts as a donor of hydride ion, H^-, and the electron-deficient AlH_3 becomes bound to the electron-rich carbonyl oxygen. Addition of water to the complex thus formed yields an alcohol

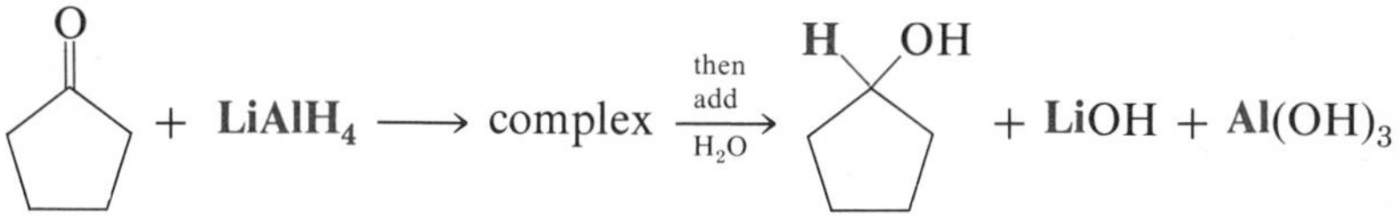

The mechanism of this reaction is shown in mechanism 7-4. This reaction is in many ways analogous to the reaction of Grignard reagents with aldehydes and ketones; in both cases, a compound with carbon or hydrogen bound to a metal reacts with a carbonyl group, placing the carbon or hydrogen on the carbonyl carbon and the metal on the carbonyl oxygen. Subsequent hydrolysis liberates an alcohol.

Sodium borohydride, like lithium aluminum hydride, reduces aldehydes and ketones

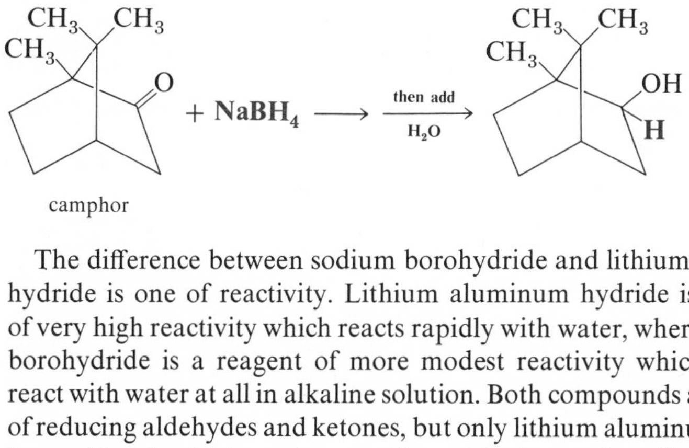

camphor

The difference between sodium borohydride and lithium aluminum hydride is one of reactivity. Lithium aluminum hydride is a reagent of very high reactivity which reacts rapidly with water, whereas sodium borohydride is a reagent of more modest reactivity which does not react with water at all in alkaline solution. Both compounds are capable of reducing aldehydes and ketones, but only lithium aluminum hydride is capable of reducing esters, amides, and other carboxylic acid derivatives (chapter 10).

Give the mechanism of the reaction of camphor with $NaBH_4$ shown above. Use molecular models to show why only one stereoisomer is formed in the reaction. **PROBLEM 7-7**

USE OF SODIUM BOROHYDRIDE IN BIOCHEMISTRY Reducing agents such as lithium aluminum hydride and sodium borohydride do not occur as natural components of living systems. Reductions in living systems are accomplished by much more subtle means (section 7-9). Sodium borohydride, however, has found favor with biochemists as a mild reducing agent that can be used in aqueous solution. (Lithium aluminum hydride cannot be used, as it decomposes violently on contact with water.) An example is shown in figure 7-6.

7-8 OXIDATION OF ALDEHYDES

Ketones cannot be oxidized except under very vigorous conditions. Aldehydes, on the other hand, can be oxidized easily to the corresponding carboxylic acids. This difference between aldehydes and ketones exists because oxidation of the carbonyl carbon atom of a ketone would require breaking a carbon-carbon bond, whereas oxidation of an aldehyde requires breaking only a carbon-hydrogen bond.

The most common reagents for oxidizing aldehydes to carboxylic acids are chromic acid, $H_2Cr_2O_7$, in acidic solution and potassium permanganate, $KMnO_4$, in acidic or basic solution. For example,

cyclopentane carboxaldehyde + $KMnO_4$ $\xrightarrow{H^+}$ cyclopentane carboxylic acid + $KMnO_3$

The mechanisms of these oxidations have many features in common with other reactions of aldehydes we have considered. In particular, they all involve nucleophilic attack by an oxygen on the carbonyl carbon atom. The mechanism of aldehyde oxidation by acidic potassium permanganate is shown in mechanism 7-5.

Oxidation also occurs in basic solution. The products of the reaction in basic solution are the same as those in acidic solution

(benzaldehyde) + $KMnO_4$ + OH^- ⟶ (benzoate) + $KMnO_3$ + H_2O

Mechanism 7-4
*Lithium Aluminum
Hydride Reduction
of Aldehydes and
Ketones*

$$CH_3-\overset{\overset{\displaystyle O}{\|}}{C}-CH_3 \; +$$

$$Li^+ \quad AlH_4{}^-$$

The starting materials are the carbonyl compound and $AlH_4{}^-$ ion.

reduction

Two bonds are made and two are broken in the key step of this reaction: an aluminum-hydrogen bond is broken, and a carbon-hydrogen bond is formed. At the same time, the carbonyl group π bond breaks, and an oxygen-aluminum bond is formed. This is the step in which the actual reduction takes place.

$$\left[CH_3-\overset{\overset{\displaystyle O-AlH_3{}^-}{|}}{\underset{\underset{\displaystyle CH_3}{|}}{C}}-H \right]$$

reduced intermediate

The product of this step is an alkoxide–AlH_3 complex. Although under certain circumstances this intermediate may remain in solution, the usual fate of the complex is to react further. The aluminum atom in the complex is capable of donating three H^- to additional molecules of aldehyde or ketone.

three more
reduction
steps

In general, three additional molecules of aldehyde or ketone will react sequentially with the above complex. Such addition continues until no further hydrogens are left on aluminum.

This final tetrasubstituted aluminum complex is stable and remains in solution until water or alcohol is added.

hydrolysis

Water is not added until after the reduction steps are completed.

$$4CH_3-\overset{\overset{\displaystyle OH}{|}}{CH}-CH_3$$

$$+ \; Al(OH)_3$$
$$+ \; OH^-$$

Addition of water results in formation of a primary or secondary alcohol, depending on whether the starting compound was an aldehyde or a ketone.

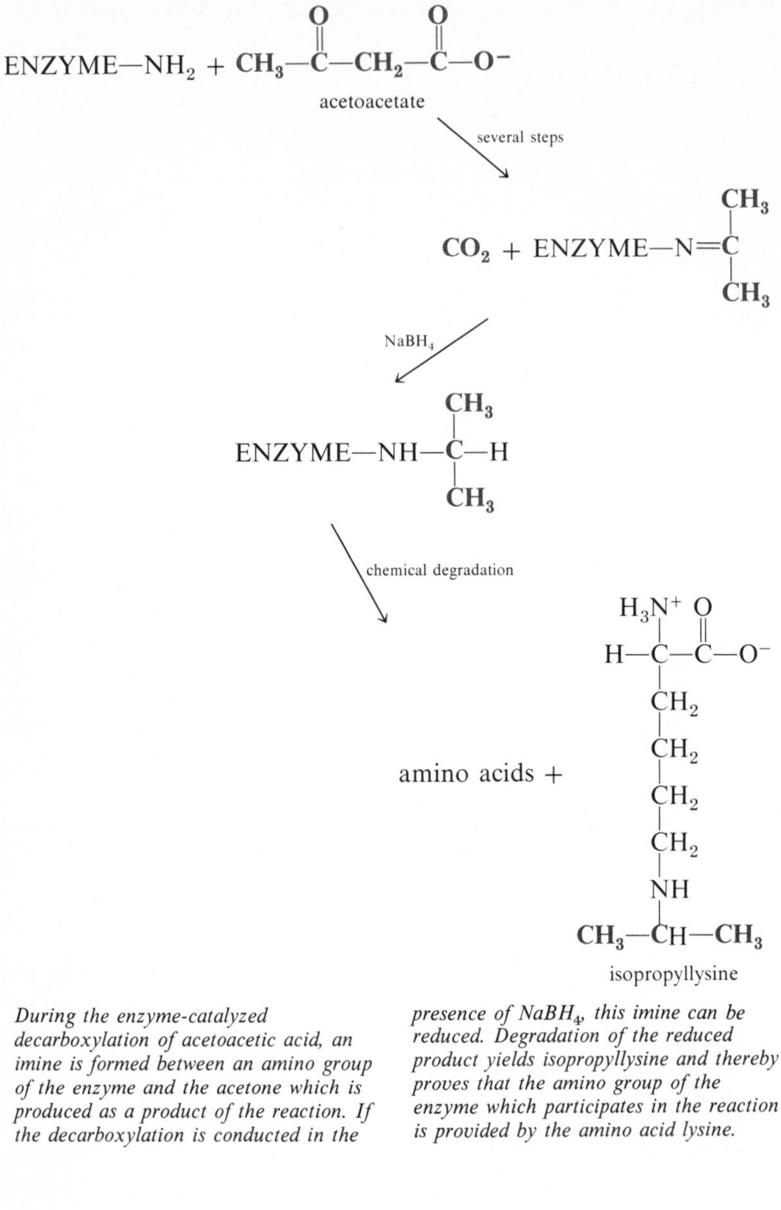

$$\text{ENZYME—NH}_2 + \text{CH}_3\text{—}\overset{\overset{\displaystyle O}{\|}}{\text{C}}\text{—CH}_2\text{—}\overset{\overset{\displaystyle O}{\|}}{\text{C}}\text{—O}^-$$

acetoacetate

several steps ↓

$$\text{CO}_2 + \text{ENZYME—N}\overset{\displaystyle CH_3}{=}\overset{\displaystyle |}{\underset{\displaystyle CH_3}{\text{C}}}$$

NaBH₄ ↙

$$\text{ENZYME—NH—}\overset{\overset{\displaystyle CH_3}{|}}{\underset{\underset{\displaystyle CH_3}{|}}{\text{C}}}\text{—H}$$

chemical degradation ↓

$$\text{amino acids} + \quad \overset{\overset{\displaystyle H_3N^+}{|}}{\text{H—C}}\text{—}\overset{\overset{\displaystyle O}{\|}}{\text{C}}\text{—O}^-$$

$$\text{CH}_2$$
$$\text{CH}_2$$
$$\text{CH}_2$$
$$\text{CH}_2$$
$$\text{NH}$$
$$\text{CH}_3\text{—CH—CH}_3$$

isopropyllysine

Figure 7-6
Sodium borohydride trapping of a natural imine. *During the enzyme-catalyzed decarboxylation of acetoacetic acid, an imine is formed between an amino group of the enzyme and the acetone which is produced as a product of the reaction. If the decarboxylation is conducted in the presence of NaBH₄, this imine can be reduced. Degradation of the reduced product yields isopropyllysine and thereby proves that the amino group of the enzyme which participates in the reaction is provided by the amino acid lysine.*

but there is evidence that the mechanism is somewhat more complex in this case.

PROBLEM 7-9   Give the mechanism of the oxidation of octanal by $KMnO_4$ in acidic solution.

7-9 OXIDATIONS AND REDUCTIONS IN BIOCHEMISTRY

In organic chemistry, oxidation is equivalent to the removal of hydrogen from carbon, and reduction is addition of hydrogen to carbon. Thus, reducing agents, such as lithium aluminum hydride, add a hydrogen to the carbonyl carbon of aldehydes and ketones; oxidizing agents, such as potassium permanganate, remove a hydrogen from the carbonyl carbon of aldehydes.

Oxidations and reductions in biochemistry also involve removal or addition of hydrogen, but the oxidizing and reducing agents are different from those in organic chemistry. The oxidizing agent in many biochemical oxidations is *nicotinamide adenine dinucleotide* (NAD$^+$), and the reducing agent is the reduced form of this same substance (NADH). These structures and some typical biological oxidation-reduction reactions are shown in figure 7-7.

7-10 ENOLIZATION

Hydrogen atoms attached to carbon are not ordinarily acidic to any extent. That is, even strong bases are unable to remove protons bound to carbon; the reaction

$$\underset{}{\diagdown}C-H + base \rightleftharpoons\!\!\!/\!\!\!\rightleftharpoons \underset{}{\diagdown}C^- + base-H^+$$

cannot be observed. There is one important exception to this generalization: the hydrogen atoms on the carbon *adjacent to* the carbonyl carbon atom of aldehydes and ketones (as well as esters and other carboxylic acid derivatives) are appreciably acidic.* Under the influence of base, loss of one of these protons may occur, forming an *enolate anion*

$$CH_3-\overset{\overset{\displaystyle O}{\|}}{C}-CH_3 + {}^-OH \rightleftharpoons \left[CH_3-\overset{\overset{\displaystyle O}{\|}}{C}-CH_2{}^- \longleftrightarrow CH_3-\overset{\overset{\displaystyle O^-}{|}}{C}=CH_2 \right] + H_2O$$

acetone   enolate anion

The enolate anion is stabilized by resonance. The enhanced acidity of aldehydes and ketones is due to this resonance stabilization. Although the above equilibrium favors the ketone, rather than the enolate, the extraordinary reactivity of the enolate more than compensates for its low concentration.

---

*A hydrogen atom attached to a carbon-carbon triple bond is also slightly acidic, but we will have no occasion to consider such compounds.

Mechanism 7-5
Permanganate
Oxidation of
Aldehydes

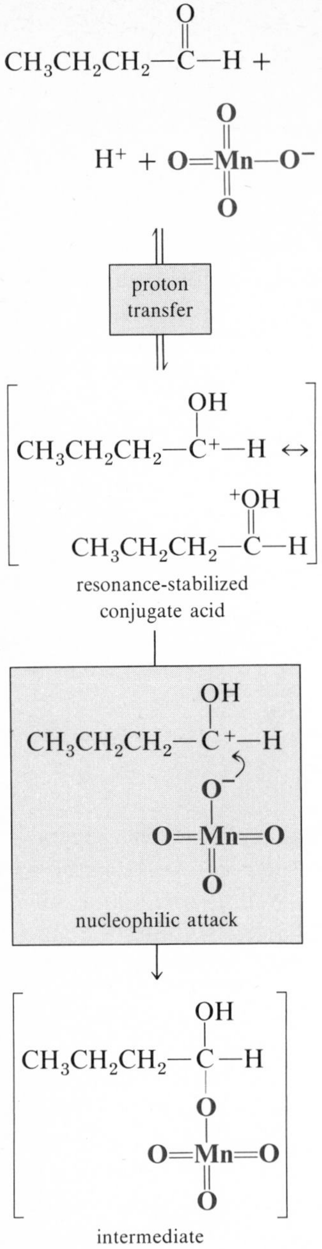

$$CH_3CH_2CH_2-\overset{\overset{\displaystyle O}{\|}}{C}-H +$$

The oxidation of butyraldehyde by potassium permanganate can be conducted in slightly acidic solution.

$$H^+ + O=\overset{\overset{\displaystyle O}{\|}}{\underset{\underset{\displaystyle O}{\|}}{Mn}}-O^-$$

proton transfer

$$\left[CH_3CH_2CH_2-\overset{\overset{\displaystyle OH}{|}}{C^+}-H \quad \leftrightarrow \quad CH_3CH_2CH_2-\overset{\overset{\displaystyle {}^+OH}{\|}}{C}-H \right]$$

resonance-stabilized
conjugate acid

Under these conditions, the aldehyde is in equilibrium with a very small concentration of protonated aldehyde.

$$CH_3CH_2CH_2-\overset{\overset{\displaystyle OH}{|}}{C^+}-H$$
$$O^-$$
$$O=\overset{\overset{\displaystyle }{}}{\underset{\underset{\displaystyle O}{\|}}{Mn}}=O$$

nucleophilic attack

In the first step an oxygen of the MnO_4^- ion acts as a nucleophile and attacks the carbonyl carbon atom of the protonated aldehyde.

$$\left[\begin{array}{c} CH_3CH_2CH_2-\overset{\overset{\displaystyle OH}{|}}{\underset{\underset{\displaystyle O}{|}}{C}}-H \\ O=\overset{\overset{\displaystyle }{}}{\underset{\underset{\displaystyle O}{\|}}{Mn}}=O \end{array} \right]$$

intermediate

The product of this attack is a hemiacetal-like species. This intermediate is unlike any other intermediate we have seen in this chapter, in that it is capable of losing a proton from the carbonyl carbon atom.

The oxidation step is actually an elimination reaction. Water acts as a base, removing a proton at the same time that an oxygen-manganese bond is broken. A carbon-oxygen π bond is formed as a result of this elimination.

The products of the reaction are a carboxylic acid, produced by oxidation of the aldehyde, and MnO_3^-, produced by reduction of MnO_4^-. The MnO_3^- which is produced usually undergoes disproportionation, yielding MnO_2 and MnO_4^-.

Most aldehydes and ketones are capable of forming enolates. In fact, many ketones, such as

2-butanone   cyclohexyl ethyl ketone

may form either of two enolates with different structures, but reaction occurs only *once in each molecule*. Some molecules lose one of the enolizable hydrogens, and some lose another, but no molecules lose two enolizable hydrogens. Such doubly charged enolate anions would be extremely unstable because of repulsion of the two negative charges.

A few aldehydes and ketones are not capable of forming enolates at all because they lack α hydrogens. Examples include*

formaldehyde   benzaldehyde   benzophenone

*Note that we have removed an α hydrogen of an aldehyde or ketone. The hydrogen which is attached to the carbonyl carbon of an aldehyde cannot be removed by this reaction.

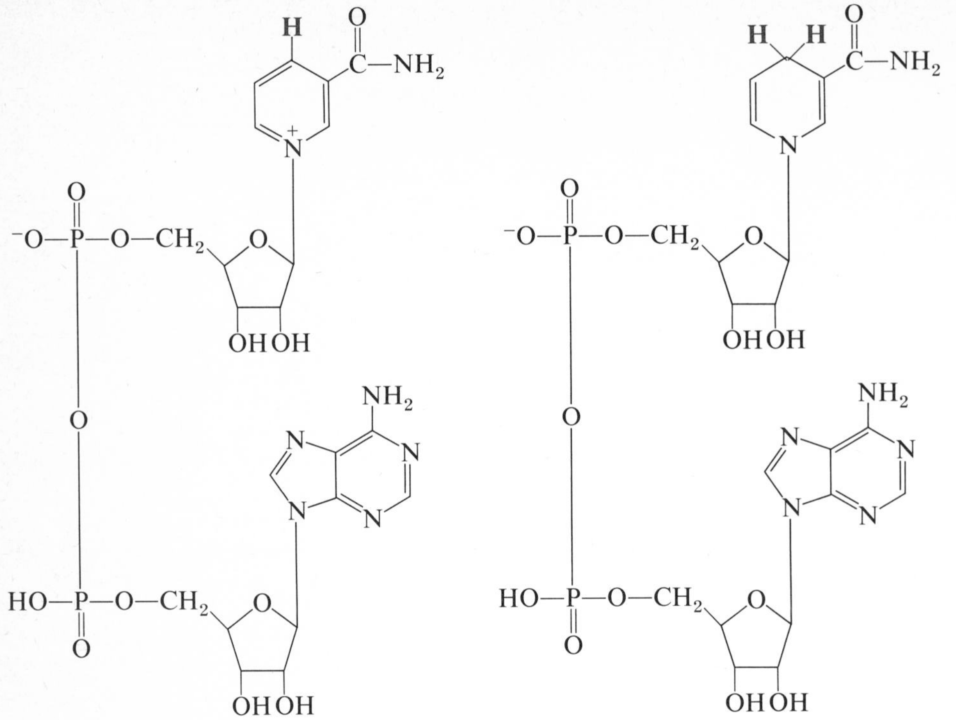

$$CH_3-CH_2-OH + NAD^+ \underset{\text{dehydrogenase}}{\overset{\text{alcohol}}{\rightleftharpoons}} CH_3-\overset{\overset{\displaystyle O}{\|}}{C}H + NADH + H^+$$

$$CH_3-\overset{\overset{\displaystyle O}{\|}}{C}H + H_2O + NAD^+ \underset{\text{dehydrogenase}}{\overset{\text{aldehyde}}{\rightleftharpoons}} CH_3-\overset{\overset{\displaystyle O}{\|}}{C}-OH + NADH + H^+$$

Figure 7-7
Biological oxidations and reductions. *The oxidizing agent is usually nicotinamide adenine dinucleotide (NAD+), and the reducing agent is reduced nicotinamide adenine dinucleotide (NADH). When reduction of a compound by NADH occurs, a hydride ion, that is, a hydrogen nucleus and a pair of electrons, is transferred from NADH to the substrate. When NAD+ oxidizes a compound, a hydride ion is transferred from that compound to NAD+, forming NADH.*

PROBLEM 7-10 Draw the structures of *two* enolates which are formed when 2-penta-none reacts with NaOH.

ALKYLATION OF ENOLATES Enolate anions are very good nucleophiles. Both resonance forms of the enolate anion have an unshared pair of electrons which can act as a nucleophile in reactions with a number of compounds. The two resonance structures also suggest that two different products might be formed when an enolate reacts with an alkyl halide. The following reaction occurs by way of the enolate:

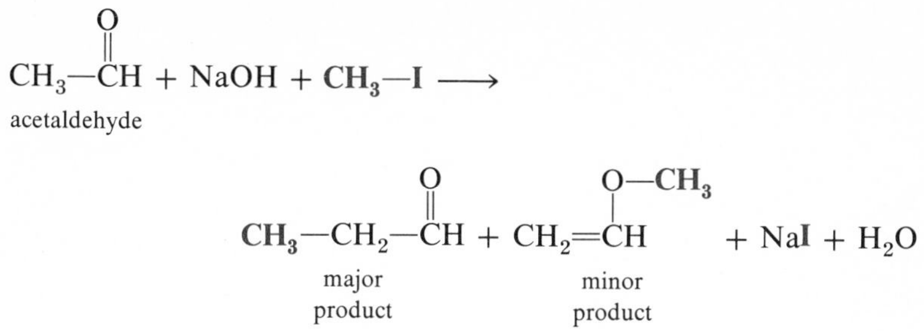

$$\underset{\text{acetaldehyde}}{CH_3-\overset{\displaystyle O}{\overset{\|}{C}}H} + NaOH + CH_3-I \longrightarrow$$

$$CH_3-CH_2-\overset{\displaystyle O}{\overset{\|}{C}}H + CH_2{=}\overset{\displaystyle O-CH_3}{\overset{\displaystyle |}{C}}H \quad + NaI + H_2O$$

$$\underset{\substack{\text{major} \\ \text{product}}}{} \qquad \underset{\substack{\text{minor} \\ \text{product}}}{}$$

The first of the two products shown here is the type usually formed in the greater amount. We will not consider the formation of the second product in detail.

Formation and alkylation of an enolate are shown in mechanism 7-6. This reaction is ordinarily carried out with an excess of base. When the reaction of acetaldehyde above is carried out, the product can react with sodium hydroxide to form another enolate, which can react with methyl iodide to form a new aldehyde

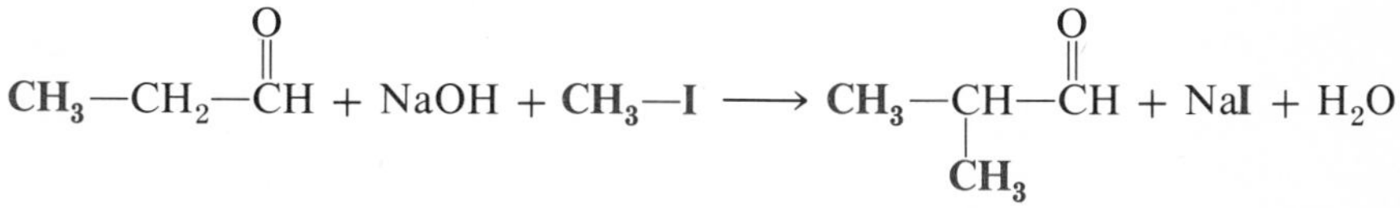

$$CH_3-CH_2-\overset{\displaystyle O}{\overset{\|}{C}}H + NaOH + CH_3-I \longrightarrow CH_3-\underset{\displaystyle \underset{\displaystyle CH_3}{|}}{C}H-\overset{\displaystyle O}{\overset{\|}{C}}H + NaI + H_2O$$

This product can also undergo enolization and further reaction

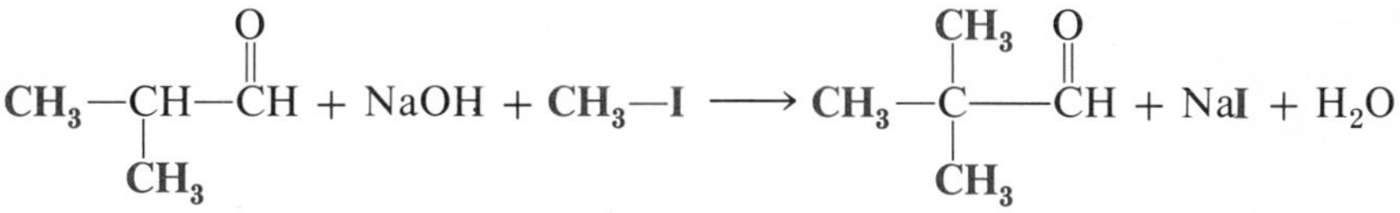

$$CH_3-\underset{\displaystyle \underset{\displaystyle CH_3}{|}}{C}H-\overset{\displaystyle O}{\overset{\|}{C}}H + NaOH + CH_3-I \longrightarrow CH_3-\underset{\displaystyle \underset{\displaystyle CH_3}{|}}{\overset{\displaystyle \overset{\displaystyle CH_3}{|}}{C}}-\overset{\displaystyle O}{\overset{\|}{C}}H + NaI + H_2O$$

This last product is not capable of further enolization and is therefore the final product of the reaction.

Thus, the reaction of aldehydes and ketones with alkyl halides in the presence of base ordinarily results in the replacement of all α hydrogens by alkyl groups. In the case of acetone, all six α hydrogens can be replaced

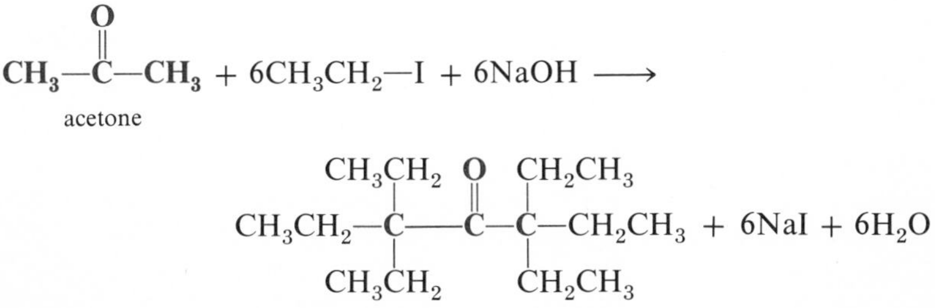

$$\underset{\text{acetone}}{CH_3-\overset{\displaystyle O}{\overset{\|}{C}}-CH_3} + 6CH_3CH_2-I + 6NaOH \longrightarrow$$

$$CH_3CH_2-\underset{\displaystyle \underset{\displaystyle CH_3CH_2}{|}}{\overset{\displaystyle \overset{\displaystyle CH_3CH_2}{|}}{C}}-\overset{\displaystyle O}{\overset{\|}{C}}-\underset{\displaystyle \underset{\displaystyle CH_2CH_3}{|}}{\overset{\displaystyle \overset{\displaystyle CH_2CH_3}{|}}{C}}-CH_2CH_3 + 6NaI + 6H_2O$$

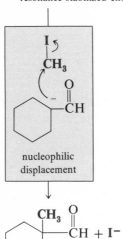

Mechanism 7-6
Alkylation of an
Enolate in Basic
Solution

When an aldehyde or ketone is dissolved in basic solution, loss of an α proton may occur.

proton removal by ⁻OH

resonance-stabilized enolate

Only a small amount of enolate is present at any one time, but this enolate is very reactive. Eventually all of the starting carbonyl compound is consumed via reaction of the enolate.

nucleophilic displacement

The enolate anion is a good nucleophile and readily displaces iodide from methyl iodide, forming a new carbon-carbon bond. The mechanism of this step is exactly like that of the displacement reactions discussed in chapter 5.

The product of the reaction is a new aldehyde or ketone.

PROBLEM 7-11 Give the mechanism of the reaction of acetone with the first molecule of ethyl iodide in the reaction at the bottom of page 199.

PROBLEM 7-12 The enolate formed by diketones such as

$$CH_3-\overset{\overset{O}{\|}}{C}-CH_2-\overset{\overset{O}{\|}}{C}-CH_3$$

is particularly well stabilized by resonance. Give the resonance forms of this enolate.

In alkaline solution an enolate can react with another molecule of aldehyde or ketone to form an *aldol,* a β-hydroxyaldehyde or β-hydroxyketone

$$CH_3-\overset{\overset{O}{\|}}{CH} + CH_3-\overset{\overset{O}{\|}}{CH} \xrightarrow{\text{NaOH}} CH_3-\overset{\overset{OH}{|}}{CH}-CH_2-\overset{\overset{O}{\|}}{CH}$$

an aldol

The base in this reaction functions as a catalyst. The reaction is reversible and can also occur in acidic solution. The mechanism is shown in mechanism 7-7.

Ketones are less reactive than aldehydes, and aldol condensation of ketones ordinarily does not work as well as aldol condensation of aldehydes. The aldol condensation of acetone reaches equilibrium when only a small amount of condensation has occurred

$$2CH_3-\overset{\overset{O}{\|}}{C}-CH_3 \xrightleftharpoons{\text{NaOH or } H^+} CH_3-\overset{\overset{OH}{|}}{\underset{\underset{CH_3}{|}}{C}}-CH_2-\overset{\overset{O}{\|}}{C}-CH_3$$

acetone, 99%   1%

Aldol condensations may often be carried to completion by dehydration of the first condensation product. If the condensation is conducted in basic solution at high temperature, elimination of water occurs, and an α,β-unsaturated ketone is formed

$$CH_3-\overset{\overset{O}{\|}}{CH} \xrightarrow[\text{heat}]{\text{NaOH}} CH_3-CH=CH-\overset{\overset{O}{\|}}{CH} + H_2O$$

Although the dehydration step is reversible, the equilibrium often favors the unsaturated product. By this means it is possible to carry out condensations in high yield.

Give the mechanism of the aldol condensation of acetone in the presence of KOH.

Aldol condensations with a mixture of two compounds are sometimes useful. If one of the compounds is not capable of enolization, that compound can be used in excess and only the other compound will undergo enolization

(excess)

Mechanism 7-7
Base-catalyzed
Aldol Condensation

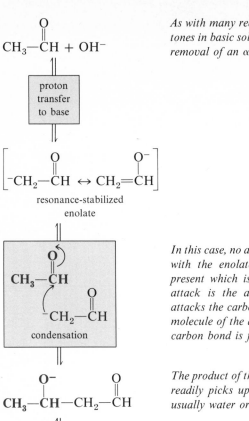

As with many reactions of aldehydes and ketones in basic solution, this one is initiated by removal of an α proton.

In this case, no alkyl halide is present to react with the enolate, and the only compound present which is susceptible to nucleophilic attack is the aldehyde itself. The enolate attacks the carbonyl carbon atom of another molecule of the aldehyde, and a new carbon-carbon bond is formed.

The product of this step is an alkoxide, which readily picks up a proton from the solvent, usually water or an alcohol.

The final product of the reaction is a β-hydroxyaldehyde.

Aldehydes are more reactive than ketones toward nucleophilic attack in the aldol condensation. As a result, the product shown above is formed in good yield. Dehydration of the first-formed aldol product occurs easily because the final product is *conjugated;* that is, the new double bond is adjacent to other double bonds. Another example of this same reaction is

$$NO_2—\bigcirc\!\!\!-\overset{\overset{\displaystyle O}{\|}}{C}H + CH_3—\overset{\overset{\displaystyle O}{\|}}{C}—\bigcirc \quad \xrightarrow{\text{NaOH}}$$

$$NO_2—\bigcirc\!\!\!-CH\!=\!CH—\overset{\overset{\displaystyle O}{\|}}{C}—\bigcirc \quad + H_2O$$

Many of the processes by which carbon-carbon bonds are formed in living systems are simply aldol condensations in disguise. Likewise, a number of the processes by which carbon-carbon bonds are broken are aldol condensations in reverse. These condensations and retrogressions are for the most part enzyme-catalyzed. The enzyme provides the necessary acid or base catalyst and holds the substrates in the proper orientation for reaction. The mechanisms of these enzyme-catalyzed reactions are generally quite similar to the mechanisms of the organic reactions which we have already discussed, but some may differ in detail.

ALDOL CONDENSATIONS IN BIOCHEMISTRY

Deoxyribose 5-phosphate is synthesized in nature by the reversible aldol condensation of acetaldehyde with glyceraldehyde phosphate. The last compound is at the same time an aldehyde, an alcohol, and a phosphate, but only the aldehyde function need concern us now:

$$^{2-}O_3PO—CH_2—\overset{\overset{\displaystyle OH}{|}}{C}H—\overset{\overset{\displaystyle O}{\|}}{C}H + CH_3—\overset{\overset{\displaystyle O}{\|}}{C}H \rightleftharpoons {}^{2-}O_3PO—CH_2—\overset{\overset{\displaystyle OH}{|}}{C}H—\overset{\overset{\displaystyle OH}{|}}{C}H—CH_2—\overset{\overset{\displaystyle O}{\|}}{C}H$$

glyceraldehyde   acetaldehyde   deoxyribose 5-phosphate
phosphate

The importance of reversed aldol condensations is shown by the enzyme-catalyzed cleavage of fructose 1,6-diphosphate to glyceraldehyde phosphate and dihydroxyacetone phosphate. This reaction is a central reaction in the biochemical degradation of six-carbon sugars

$$^{2-}O_3PO—CH_2—\overset{\overset{\displaystyle OH}{|}}{C}H—\overset{\overset{\displaystyle OH}{|}}{C}H—\underset{\underset{\displaystyle OH}{|}}{\overset{\overset{\displaystyle O}{\|}}{C}}H—\overset{\overset{\displaystyle O}{\|}}{C}—CH_2—OPO_3^{2-} \longrightarrow$$

fructose diphosphate

$$^{2-}O_3PO—CH_2—\overset{\overset{\displaystyle OH}{|}}{C}H—\overset{\overset{\displaystyle O}{\|}}{C}H + HO—CH_2—\overset{\overset{\displaystyle O}{\|}}{C}—CH_2—OPO_3^{2-}$$

glyceraldehyde   dihydroxyacetone
phosphate   phosphate

This reaction also illustrates the important point that aldol condensations in biochemistry frequently involve enolate formation on a carbon which also carries a hydroxyl group. In the above case, the reverse reaction, the condensation of glyceraldehyde phosphate with dihydroxyacetone phosphate, occurs by loss of a proton from the carbon of dihydroxyacetone phosphate which bears the —OH group, followed by condensation of the enolate with the carbonyl carbon of glyceraldehyde phosphate.

ACETOGENINS   Many of the brilliant colors that occur in nature are due to pigments called *acetogenins*. These compounds are synthesized in nature from polyketones by a long series of aldol condensations, which, although complex, are similar to the aldol condensations we have discussed. In these cases the aldol condensation is believed to occur while the polyketone is attached to an enzyme. This attachment generally occurs through a thiol ester linkage (chapter 10) between the polyketone chain and the enzyme. A typical example of acetogenin formation is the formation of orsellinic acid, a compound which occurs commonly in

Figure 7-8
Some acetogenins.

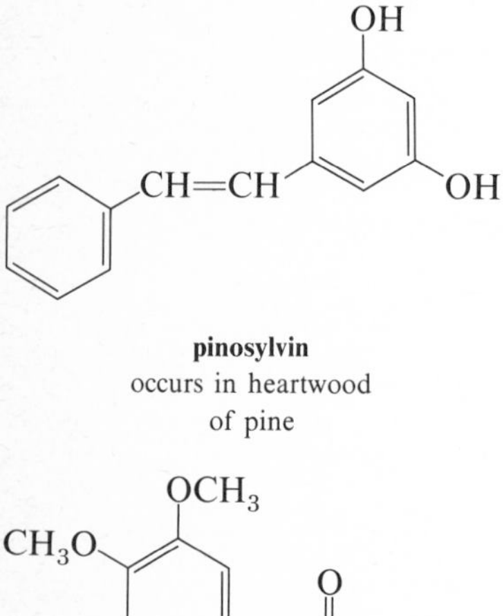

pinosylvin
occurs in heartwood
of pine

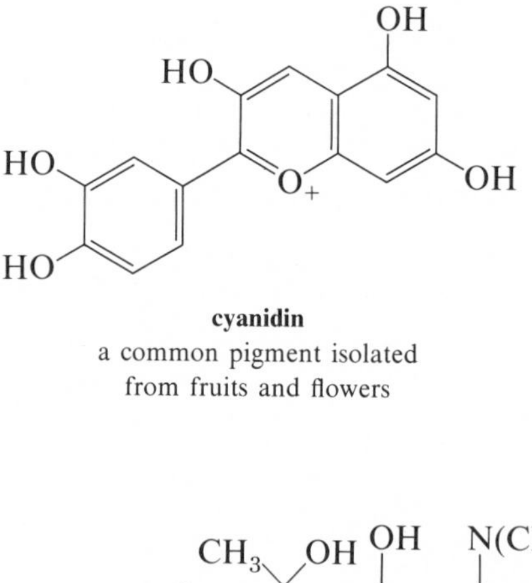

cyanidin
a common pigment isolated
from fruits and flowers

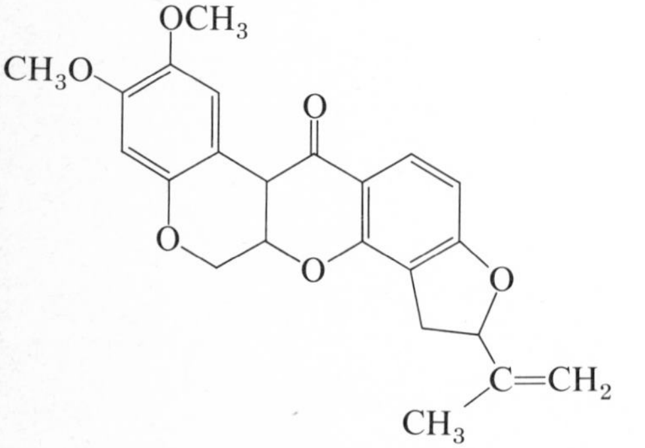

rotenone
a naturally occurring
insecticide

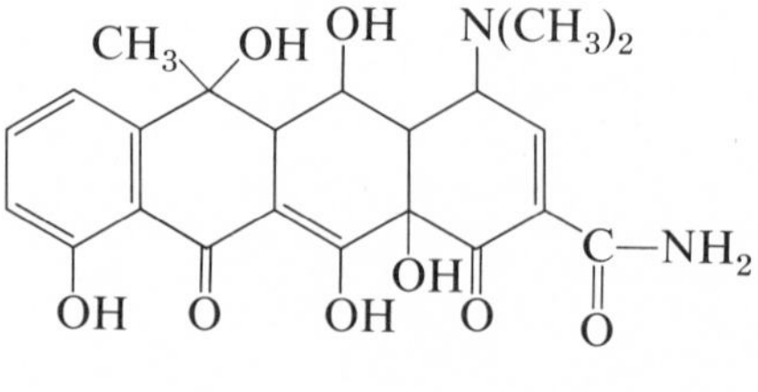

Terramycin
an important antibiotic

lower organisms and fungi, from an enzyme-bound eight-carbon precursor:

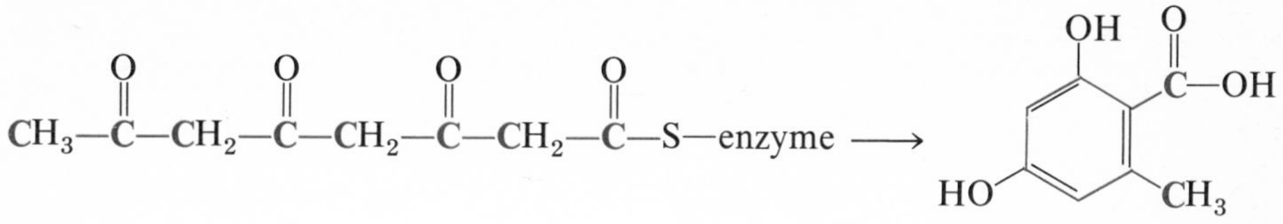

$$CH_3-\overset{O}{\underset{||}{C}}-CH_2-\overset{O}{\underset{||}{C}}-CH_2-\overset{O}{\underset{||}{C}}-CH_2-\overset{O}{\underset{||}{C}}-S-\text{enzyme} \longrightarrow$$

orsellinic acid

The number of known acetogenins exceeds a thousand. Some other simple examples are shown in figure 7-8.

PROBLEMS

7-14 Give the structure of a compound which is (*a*) an aldehyde; (*b*) a ketone; (*c*) an imine; (*d*) a hemiacetal; (*e*) an acetal; (*f*) a ketal; (*g*) a cyanohydrin; (*h*) a Grignard reagent; (*i*) an aldehyde that is not capable of undergoing enolate formation; (*j*) a ketone that is capable of forming two different enolates.

7-15 Give the mechanism and product of each of the following reactions: (*a*) butanal + HCN + NaCN; (*b*) butanal + isopropyl-amine; (*c*) butanal + 1-propanol + H⁺; (*d*) butanal + LiAlH₄, with H₂O added at the end; (*e*) butanal + KMnO₄ + H⁺; (*f*) butanal + methyl iodide + NaOH; (*g*) butanal + NaOH.

7-16 Give the structure of (*a*) 2-methyl-3-hexanone; (*b*) 5-ethylhep-tanal; (*c*) formaldehyde; (*d*) α-phenylpropionaldehyde; (*e*) cyclohep-tanone.

7-17 Show why secondary amines do not form stable imines.

7-18 The following compound is formed by the reaction of a diketone with itself in the presence of NaOH. What is the structure of the diketone?

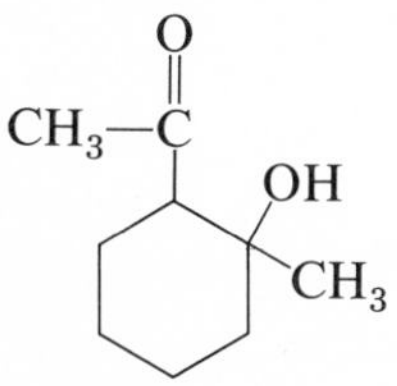

7-19 Although chloral, $CCl_3-CH{=}O$, is a powerful hypnotic, the hypnotic properties of chloral are due not to the compound itself but to the alcohol formed by reduction of chloral in the body. What is the structure of this alcohol?

7-20 Name the following compounds:

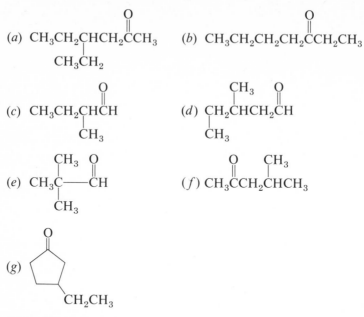

(a) $CH_3CH_2CHCH_2CCH_3$ with CH_3CH_2 substituent and O on the carbonyl

(b) $CH_3CH_2CH_2CH_2CCH_2CH_3$ with O on the carbonyl

(c) CH_3CH_2CHCH with CH_3 substituent and O on the carbonyl

(d) CH_2CHCH_2CH with CH_3 substituent and O on the carbonyl

(e) $CH_3C\!\!-\!\!CH$ with CH_3 and CH_3 substituents and O

(f) $CH_3CCH_2CHCH_3$ with O and CH_3 substituent

(g) cyclopentanone with CH_2CH_3 substituent

7-21 Mercaptans have chemical properties which are often similar to those of alcohols. Give the product of the reaction of benzaldehyde with ethanethiol in the presence of acid.

7-22 Many ketones are capable of forming two aldol condensation products because of the possibility of forming two different enolates. Choose a ketone for which this might occur, and give the structures of the two aldol condensation products. Show also the products which would be formed by loss of water from these aldol condensation products.

7-23 The aldol condensation of acetone with benzaldehyde can form a compound $C_{17}H_{14}O$. Give the structure of this compound.

7-24 Give the structure of the principal product or products of each of the following reactions:

(a) 2-Cyclohexenone + NH_2NHCNH_2 (with O on the carbonyl) $\longrightarrow$

(b) 2-Hexanone + $CH_3CH_2CH_2OH$ $\xrightarrow{H^+}$

(c) Hexanal + CH_3NH_2 $\longrightarrow$

(d) 3-Methylcyclobutanone + HCN $\xrightarrow{NaCN}$

(e) 2-Pentanone + [benzene ring]–MgBr $\longrightarrow$ $\xrightarrow[H_2O]{then\ add}$

(*f*) 3-Methyl-2-octanone + LiAlH$_4$ $\longrightarrow$ $\xrightarrow[\text{H}_2\text{O}]{\text{then add}}$

(*g*) Propionaldehyde + CH$_3$CH$_2$I + NaOH $\longrightarrow$
(excess)   (excess)

(*h*) + CH$_3$I + NaOH $\longrightarrow$
(excess)   (excess)

(*i*) HC(CH$_2$)$_6$CH + NaOCH$_2$CH$_3$ $\longrightarrow$
(with two C=O groups shown)

(*j*) + H$_2$O $\xrightarrow{\text{H}^+}$

(*k*) Valeraldehyde + KMnO$_4$ $\xrightarrow{\text{OH}^-}$

(*l*) + CO$_2$ $\longrightarrow$ $\xrightarrow[\text{H}_2\text{O}]{\text{then add}}$

(*m*) Decanal + H$_2$Cr$_2$O$_7$ $\longrightarrow$

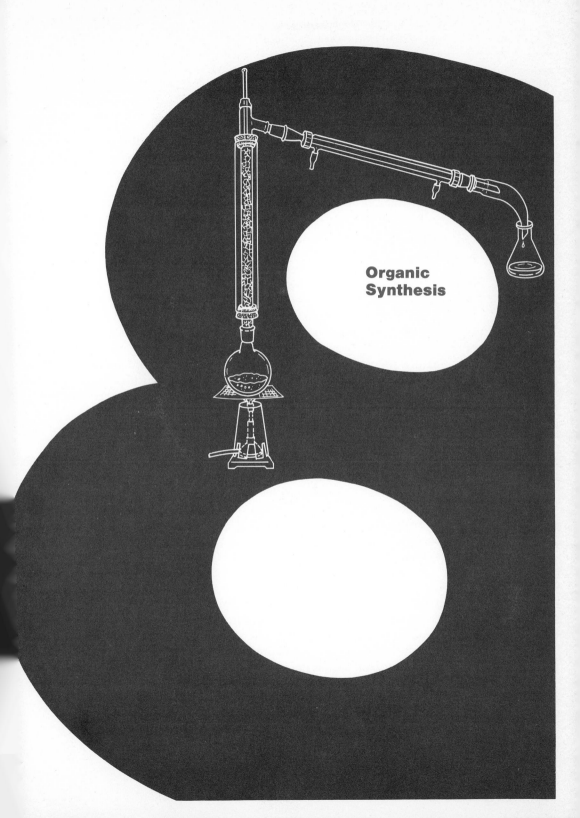

**Organic
Synthesis**

One of the principal applications of organic chemistry is in preparing organic compounds. These compounds are used by scientists for research and by the chemical industry for preparing commercial chemical products. In this chapter we will discuss the design and execution of organic syntheses.*

8-1   THE GROUND RULES

The starting point for designing a synthesis is the desired product. Once this product is identified, the chemist must work backward until a starting compound is reached which can be converted into that product by a reasonable sequence of chemical reactions.

THE PRODUCT   Synthetic goals are chosen for two reasons: because a compound is needed for studies in research or in the chemical industry, or simply, as Mallory said about Everest, "because it is there."

Organic compounds are needed for a wide variety of studies in science and industry. Many chemists synthesize compounds in order to study the mechanisms of organic reactions. Others synthesize compounds in order to study their medicinal or biological properties or to compare them with materials of unknown structure isolated from nature. Industrial chemists synthesize compounds in order to examine their potential uses in plastics, synthetic fabrics, insecticides, drugs, and many other products. The majority of organic syntheses are performed because the synthetic product is needed for further studies.

However, it is the syntheses which are undertaken for their own sake that often produce the most spectacular results. The Woodward-Eschenmoser synthesis of vitamin B_{12}, discussed in chapter 1, is one of the most noteworthy. Although this synthesis would never be used for the routine production of this vitamin, it produced many significant advances in the theory and practice of organic chemistry, including a revolutionary new theory of how the properties of orbitals affect the courses of chemical reactions and a number of new reactions which may be useful in the synthesis of other complex molecules.

THE STARTING   Ultimately, every synthesis must begin with an available organic
COMPOUND   compound—one that can either be purchased or easily synthesized from another available compound. The starting compound can be chosen only by beginning with the desired product and thinking back through a sequence of chemical reactions until an available compound is

---

*This chapter is optional. It may be covered following chapter 7 or later in the course, or it may be omitted entirely.

reached. This process of working backward is the most important and most difficult part of an organic synthesis because every compound can be made from a variety of other compounds. Each step in the backward process thus opens up several possible routes which must be evaluated.

A practicing chemist usually begins the search for a starting compound with the catalogs of commercial chemical companies, which produce over 50,000 compounds. However, these compounds vary greatly in cost and availability, and not all of them are suitable starting materials.

The choice of starting material is also dictated by the purpose of the synthesis. A laboratory chemist who needs only a few grams of a product can afford to buy a more expensive starting material if it will enable the product to be synthesized in a shorter time. An industrial chemist who must ultimately plan for the production of many pounds or even tons of a compound must be more aware of cost than a laboratory chemist.

PLANNING THE SYNTHESIS

The chemist's ability to design a successful synthesis depends on how many reactions he or she knows and on how well he or she understands the scope and limitations of those reactions. The successful chemist has spent many hours in the library reading about the reactions to be conducted, the methods of purifying the product, and the ways of proving its structure.

A good synthesis has two characteristics: (1) it uses the smallest possible number of reaction steps, and (2) each reaction gives a high yield of the desired product. After each reaction step in a synthesis, the product must be purified and its structure proved. Even though a variety of methods exist for both processes, these operations often take longer than carrying out the reactions. Thus, a premium is usually placed on short syntheses.

However, a short synthesis is a good synthesis only if all reactions proceed in satisfactory yield. A reaction which yields only 10 to 20 percent of the desired product often rules out a potential synthetic route.

OUR APPROACH TO SYNTHESIS

An introductory textbook cannot discuss all the reactions used in modern organic synthesis; in fact, it is not even possible to give a complete picture of the advantages and disadvantages of the reactions which are included. Nevertheless, a discussion of organic synthesis is useful and interesting because it provides a different perspective on organic chemistry.

Although the choice of a starting compound for a synthesis is important, it is a choice which the beginner is not prepared to make.

For that reason, most problems in organic synthesis will state the starting compound as well as the product.

8-2 THE TYPES OF ORGANIC REACTIONS

For the purpose of constructing syntheses, it is useful to divide organic reactions into two classes: (1) reactions in which carbon-carbon bonds are made or broken (called *carbon-skeleton changes*) and (2) reactions in which the carbon backbone is not changed (called *functional-group interchanges*). These two types of reactions are discussed in the next two sections.

Because an organic synthesis focuses on the product of the synthesis rather than on the starting compound, it is important to remember not only all the reactions which we have discussed but also all the methods which can be used to synthesize a given type of compound. Table 8-1 lists the principal reactions given in this book which are used for the synthesis of various types of compounds.

FUNCTIONAL-GROUP INTERCHANGES
Various methods are available for interconverting functional groups. For example, alcohols can easily be made from olefins, halides, aldehydes, ketones, and esters. Olefins can be made from alcohols, halides, and amines. Although not every possible functional group can be converted directly into every other possible functional group, the variety of available reactions is sufficient to ensure that most functional groups can be interconverted in two steps, if not in one.

CARBON-SKELETON CHANGES
In the course of organic syntheses it is often necessary to build up complex molecules from simpler molecules by the formation of new carbon-carbon bonds. Since methods for the construction of new carbon-carbon bonds are more limited in number and scope than the functional-group interchanges, syntheses must be planned with particular attention to formation of carbon-carbon bonds.

Two of the most important reactions by which new carbon-carbon bonds are formed are the Grignard reaction (sections 7-6 and 10-6) and reactions of enolates (sections 7-10 and 10-9). Both types of reactions involve carbonyl groups; in the first reaction a new bond is formed to the carbonyl carbon atom, and in the second reaction a new bond is formed to the α carbon atom.

Other methods for forming carbon-carbon bonds include the reaction of cyanide ion with alkyl halides (section 5-6) and with aldehydes and ketones (section 7-4), the reaction of carbon dioxide with Grignard reagents (section 7-6), and the Friedel-Crafts reaction of aromatic compounds (section 11-4). These reactions are limited in scope but occasionally useful.

Table 8-1
Methods for the Synthesis of Various Types of Compounds

Compound Type	Method	Discussed on Page
alkanes	hydrogenation of alkenes	72
	hydrolysis of Grignard reagents	189
alkenes	dehydration of alcohols	128
	elimination of alkyl halides	122
	elimination of quaternary ammonium salts	163
alcohols	addition of water to olefins	70
	displacement of halides with hydroxide	117
	reduction of aldehydes, ketones, and esters	190, 257
	Grignard reactions*	188, 255
	aldol condensation*	201
	ester hydrolysis	243
alkyl halides	addition to olefins	67
	displacement with primary alcohol	127
amines	reaction of amines with alkyl halides	160
	reduction of imines	194
	reduction of amides	259
aldehydes and ketones	oxidation of alcohols	132
	aldol condensation*	201
	Friedel-Crafts reactions*	293
carboxylic acids	hydrolysis of esters, amides, and nitriles	243
	oxidation of primary alcohols	132
	oxidation of aldehydes	192
	reaction of Grignard reagents with CO_2*	190

*Carbon-skeleton change.

8-3 PLANNING A SYNTHESIS

In this section we will show how to plan organic syntheses, starting with very simple examples not involving carbon-skeleton changes and progressing to more difficult syntheses.

SYNTHESES NOT REQUIRING CARBON-SKELETON CHANGE

Example 1

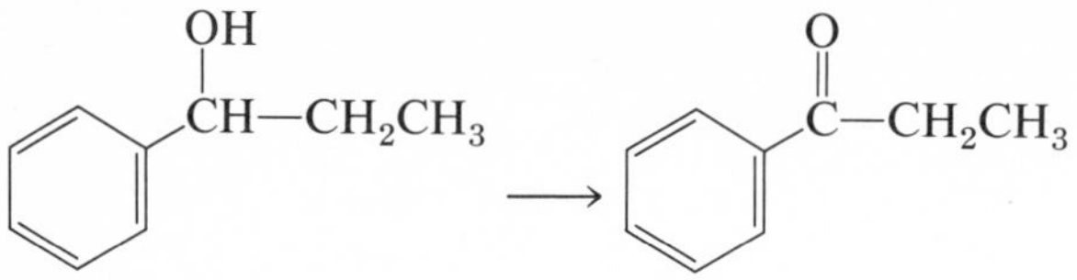

The first step in approaching a synthesis is to determine whether a carbon-skeleton change occurs. In this case, no such change is required, and it is necessary only to find a means for converting an alcohol into a ketone with the same carbon skeleton. This is easily accomplished by oxidation with permanganate or dichromate (section 5-9).

Solution Step 1 Oxidation with $KMnO_4$ in the presence of acid or base produces the desired ketone.

Example 2

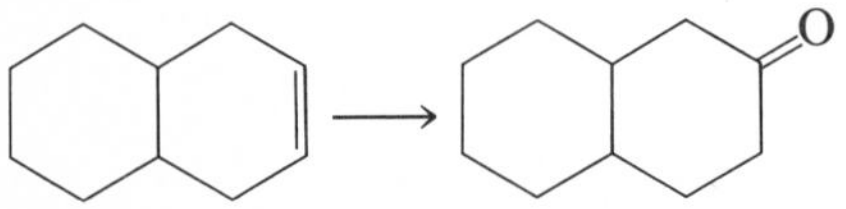

Again, no change in the carbon skeleton is required. However, an olefin cannot be converted directly into a ketone. The olefin must be first converted into an alcohol, which can then be oxidized to a ketone.*

Solution Step 1 Sulfuric acid–catalyzed addition of water to the olefin produces an alcohol.

Step 2 Oxidation with $KMnO_4$ in the presence of acid or base produces the desired ketone.

Example 3

$$CH_3CH_2CH_2CH_2CH{=}CH_2 \longrightarrow CH_3CH_2CH_2CH_2\underset{\underset{CH_3CH_2—O}{|}}{C}HCH_3$$

Even though the product of this synthesis has more carbon atoms than the starting material, no carbon-skeleton changes are involved because no new carbon-carbon bonds are formed.

Ethers are often made (section 5-10) by reactions of alkoxides with alkyl halides. Thus, the desired ether can be made by conversion of the starting olefin into an alkyl halide and reaction of that halide with sodium ethoxide.

*In this and subsequent examples, write out the structure of the product formed after each step of the synthesis.

Step 1 Addition of HBr to the olefin produces an alkyl halide. *Solution 1*

Step 2 Reaction of the alkyl halide with sodium ethoxide produces the desired ether.

However, the second step in this synthesis works poorly because elimination competes with displacement (section 5-7). A better synthetic method is conversion of the starting olefin to an alcohol, conversion of the alcohol to an alkoxide, and reaction of the alkoxide with ethyl iodide.

Step 1 Sulfuric acid–catalyzed addition of water to the olefin produces an alcohol. *Solution 2*

Step 2 Reaction of the alcohol with sodium produces an alkoxide.

Step 3 Reaction of the alkoxide with ethyl iodide produces the desired ether.

Because of the low yield expected in the second step of solution 1, this second synthesis is preferred even though it involves an additional step.

Example 4

Replacement of a hydroxyl group by an amino group cannot be achieved by direct displacement. The alcohol must be converted into an aldehyde, a ketone, or a halide before reaction with the amine. In this case, the best method is conversion into a ketone, formation of an imine, and reduction.

Step 1 Oxidation of the alcohol with $KMnO_4$ in the presence of acid or base produces a ketone. *Solution*

Step 2 Reaction of the ketone with *t*-butylamine produces an imine.

Step 3 Reduction of the imine with $NaBH_4$ or $LiAlH_4$, followed by addition of water, produces the desired amine.

Give a method for making each of the following compounds from cyclohexanone: (*a*) bromocyclohexane; (*b*) cyclohexene; (*c*) dimethyl-cyclohexylamine; (*d*) *trans*-1,2-dichlorocyclohexane; (*e*) cyclohexane; (*f*) cyclohexyl methyl sulfide; (*g*) cyclohexylamine. ***PROBLEM 8-1***

Example 5

The first step in working a problem like this is to identify the new carbon-carbon bonds in the product and determine their relationship to the functional groups present in the starting material. In this case, a new carbon-carbon bond has been formed between the carbonyl carbon of the starting compound and the center carbon of a three-carbon unit.*

Alkanes are often made by catalytic hydrogenation of alkenes. In this case, an alkene might have been produced by dehydration of the alcohol formed by a reaction of the starting ketone with a Grignard reagent. Thus, a possible synthesis is

Solution Step 1 Reaction of the ketone with isopropylmagnesium bromide in ether, followed by addition of water, produces an alcohol.

Step 2 Dehydration of the alcohol with concentrated sulfuric acid produces an alkene.

Step 3 Reduction of the alkene with hydrogen in the presence of a platinum or palladium catalyst produces the desired alkane.

Even though the sulfuric acid–catalyzed elimination may give a mixture of alkenes, they all give the same alkane on hydrogenation.

Example 6

In this case, the product has a new carbon-carbon bond to the α carbon of the starting material, and an enolate reaction is the best candidate for the synthesis. The solution given is superior to reaction of benzyl chloride with the enolate, which may give two successive displacement reactions.

Solution Step 1 Heating the starting ketone with benzaldehyde and sodium hydroxide produces an unsaturated ketone.

---

*Actually it is not possible in this case to determine that the new bond is formed to the carbonyl carbon; because of the structures involved, it might have been formed to an α carbon instead.

Step 2 Reduction of the unsaturated ketone with hydrogen in the presence of a platinum catalyst produces the desired ketone.

Example 7

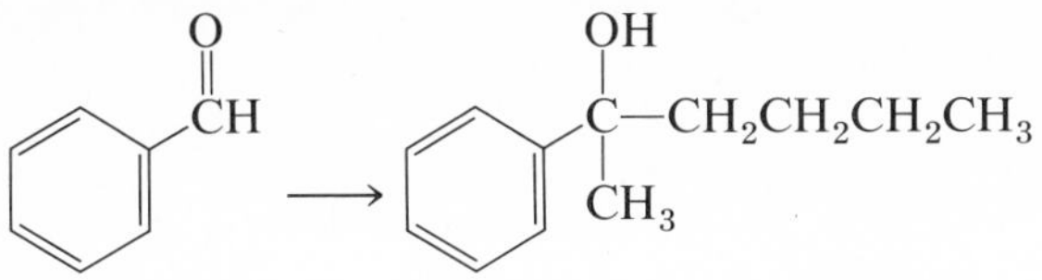

In the course of this synthesis, two new bonds are formed to the carbonyl carbon atom. This can be most easily accomplished by two separate Grignard reactions. The alcohol produced in the first Grignard reaction is oxidized to a ketone before the second Grignard reaction is carried out. Two different syntheses are possible; neither has any significant advantage over the other.

Step 1 Reaction of the starting aldehyde with methylmagnesium bromide in ether, followed by addition of water, produces an alcohol.

Solution 1

Step 2 Oxidation of the alcohol with $KMnO_4$ in the presence of acid or base produces a ketone.

Step 3 Reaction of the ketone with *n*-butylmagnesium bromide in ether, followed by addition of water, produces the desired alcohol.

Same as solution 1, except that the two Grignard reactions are performed in the opposite order: *n*-butylmagnesium bromide first, then methylmagnesium bromide.

Solution 2

Although we cannot list all the available compounds which might be useful for synthesis, a number of important ones are given in table 8-2.

WHEN THE STARTING MATERIAL IS NOT GIVEN

all organic compounds containing four or fewer carbon atoms
cyclopentanone
cyclopentene
cyclohexanone
cyclohexene
benzene
toluene
benzaldehyde
benzoic acid
bromobenzene
styrene

Table 8-2
Useful Compounds for Organic Synthesis

PROBLEMS **8-2** Show how each of the following compounds can be synthesized from 2-propanol: (*a*) propene; (*b*) 2-chloropropane; (*c*) acetone; (*d*) isopropylamine; (*e*) isopropyl ether; (*f*) isopropyl methyl ether; (*g*) propane; (*h*) isobutyric acid.

8-3 Show how each of the following syntheses can be carried out. Give the reagents needed for each step, and show the structure of the product formed after each step.

(*a*) $CH_3CH_2CH_2CH_2CH=CH_2 \longrightarrow CH_3CH_2CH_2CH_2\overset{\overset{\displaystyle O}{\|}}{C}CH_3$

(*b*)

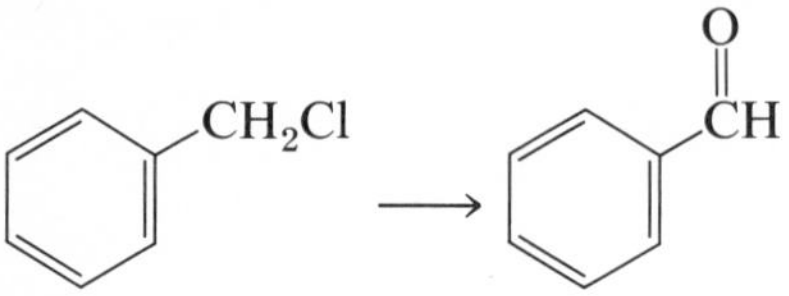

(*c*)

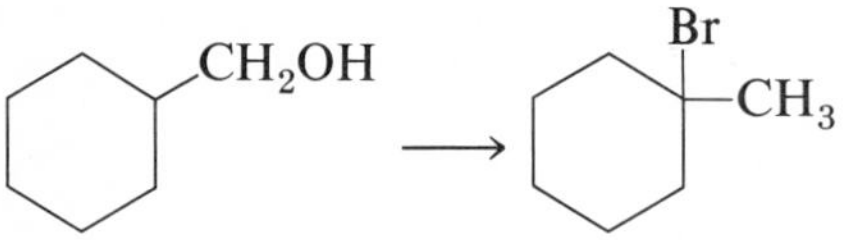

(*d*)

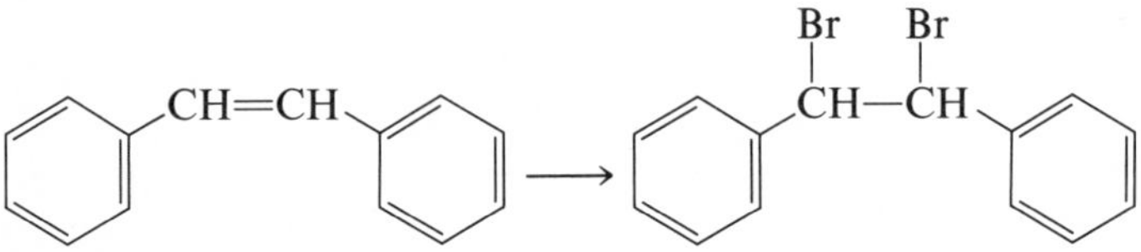

(*e*) $CH_3CH_2CH_2\overset{\overset{\displaystyle Br}{|}}{C}HCH_2CH_2CH_3 \longrightarrow CH_3CH_2\overset{\overset{\displaystyle Br}{|}}{C}HCHCH_2CH_2CH_3$
with lower Br on the second carbon

(*f*)

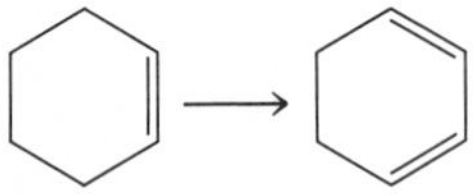

(*g*)

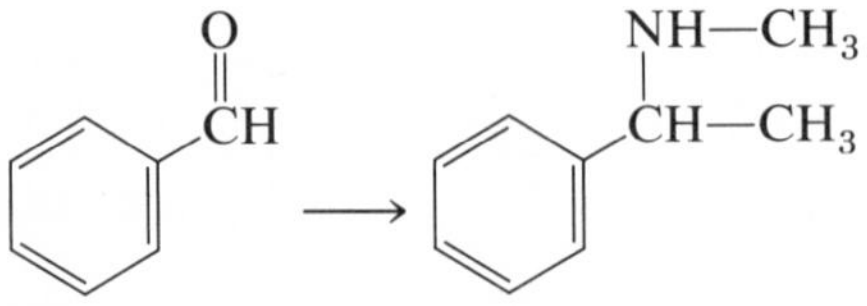

(*h*)

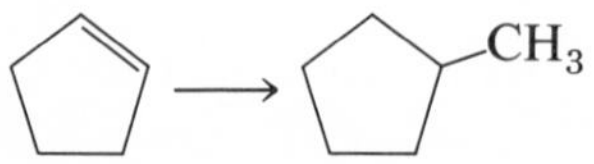

(i)

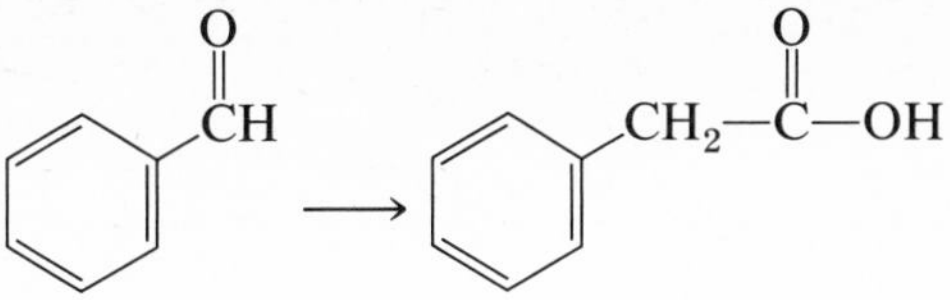

(j)

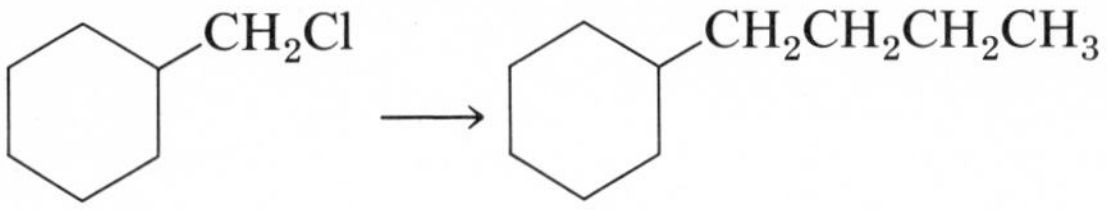

(k)

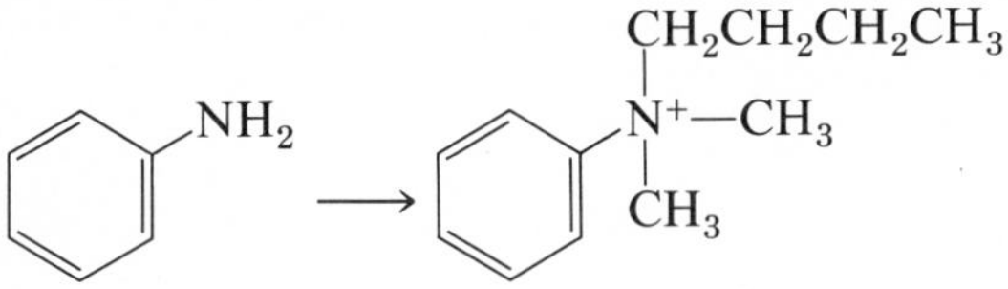

(l)

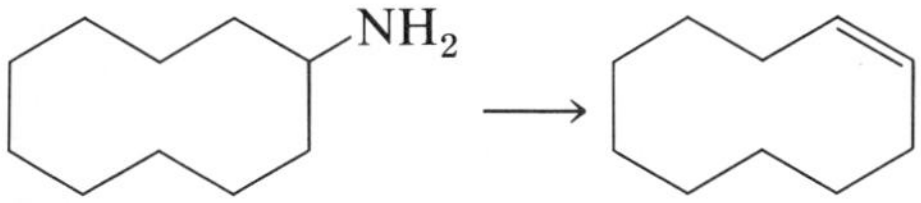

(m)

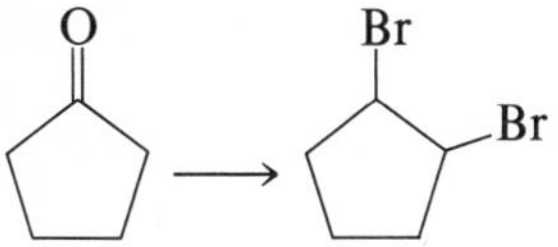

trans isomer

(n)

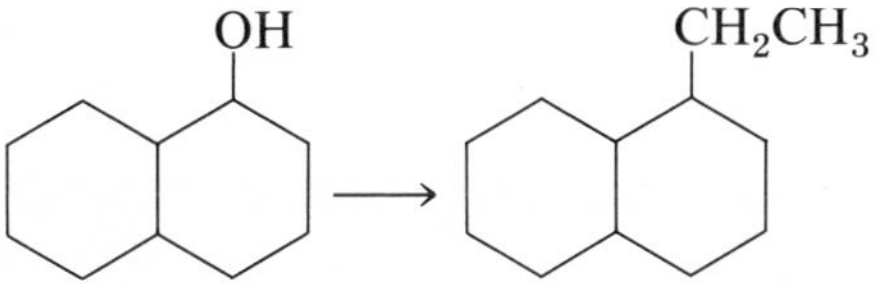

(o)

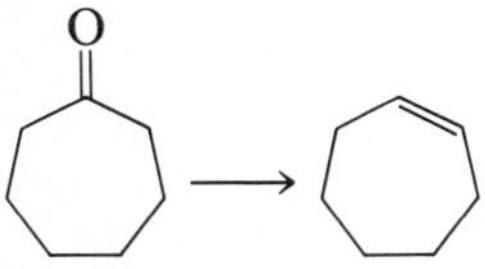

(p)

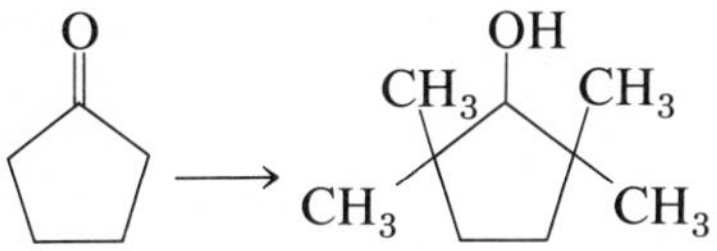

(q)

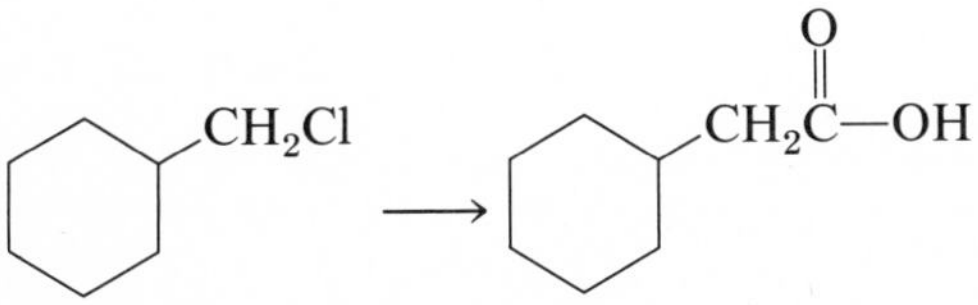

(r) $\underset{\text{OH}}{\underset{|}{\text{CH}_3\text{CH}_2\text{CH}_2\text{CH}_2\overset{|}{\underset{\text{CH}_3}{C}}\text{CH}_3}} \longrightarrow \underset{\text{CH}_3}{\underset{|}{\text{CH}_3\text{CH}_2\text{CH}_2\text{CH}_2\overset{\text{CH}_3}{\underset{|}{C}}\text{—O—CH}_2\text{CH}_3}}$

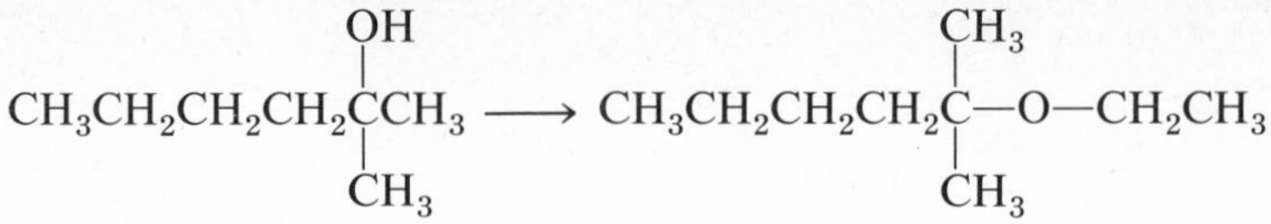

(s) $\text{CH}_2{=}\text{CHCH}_2\text{CH}_2\text{CH}{=}\text{CH}_2 \longrightarrow$

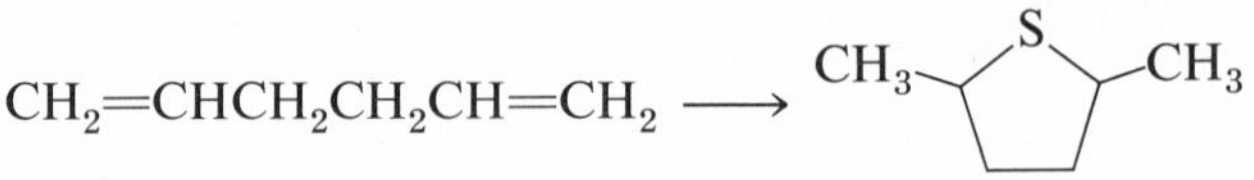

8-4 Propose a synthesis for each of the following compounds starting with one or more of the compounds given in table 8-2.

(a)

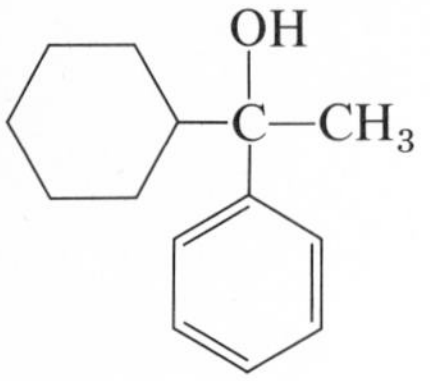

(b)

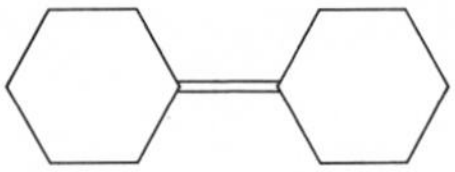

(c)

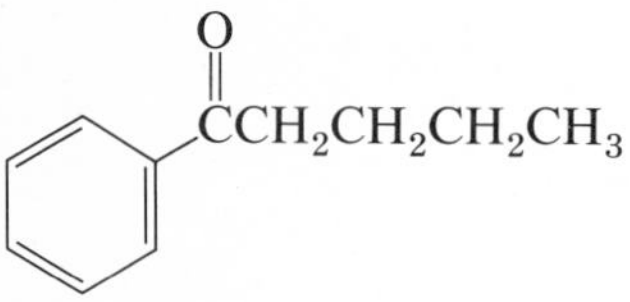

(d) $\underset{\text{CH}_3}{\underset{|}{\text{CH}_3\text{CH}_2\text{CH}_2\text{CHCH}_2\text{CH}_2\text{CH}_2\text{CH}_3}}$

(e)

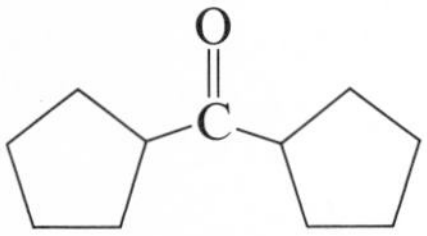

(f)

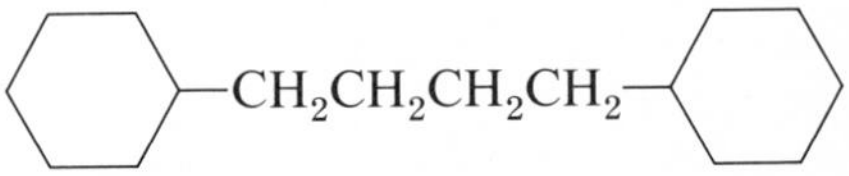

(g) $\text{CH}_2{=}\text{CH—}\overset{\text{O}}{\underset{||}{C}}\text{—CH}{=}\text{CH}_2$

(h)

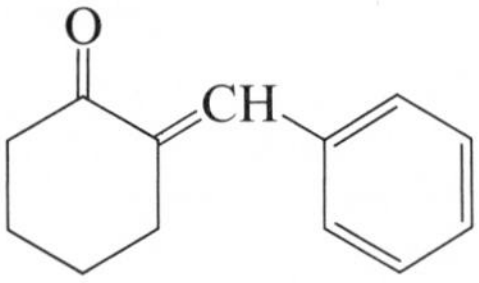

SUGGESTED READING Ireland, R. E.: *Organic Synthesis,* Prentice-Hall, Englewood Cliffs, N.J., 1969. An introduction to the theory and practice of organic synthesis.

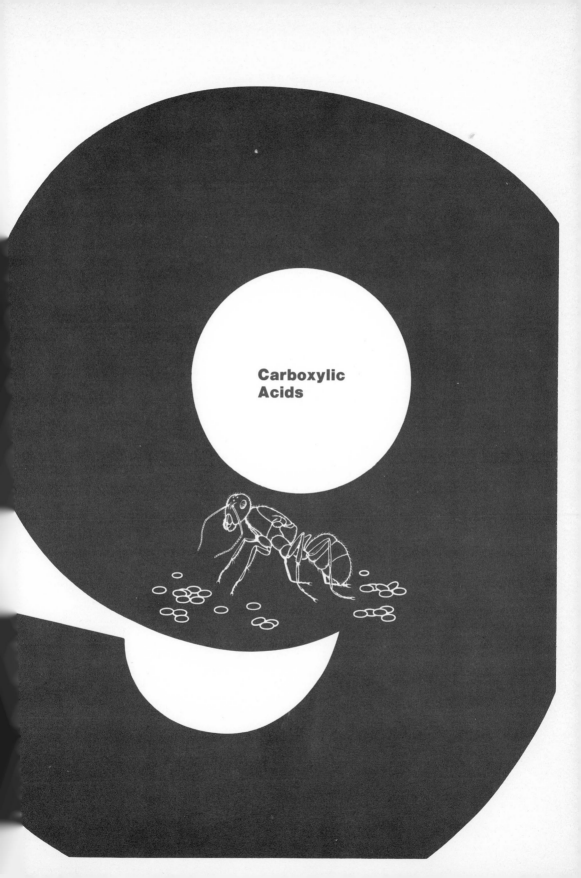

Carboxylic
Acids

Carboxylic acids are compounds having a *carboxyl group,* $-C\overset{\displaystyle O}{\underset{\displaystyle OH}{\diagdown}}$

This group is simultaneously a carbonyl group and a hydroxyl group, and to an extent it has the properties of both. These compounds provide a framework for the study of a number of related compounds in chapter 10.

9-1 THE STRUCTURES OF CARBOXYLIC ACIDS

Acetic acid, a typical carboxylic acid, is shown in figure 9-1. The carbonyl carbon atom in carboxylic acids is sp^2-hybridized, and the carbon atom lies in the same plane as the three atoms attached to it.

Since oxygen is more electronegative than carbon, the oxygen atoms of the carboxyl group are electron-rich and the carbonyl carbon is electron-poor. This bond polarity has an important effect on the properties of carboxylic acids and their derivatives.

PROBLEM 9-1 Considering that the carbonyl carbon atom is sp^2-hybridized, what would you expect the bond angles around this carbon atom to be?

9-2 NOMENCLATURE

Two systems for naming carboxylic acids are in general use, the IUPAC system and the older system of common names. IUPAC

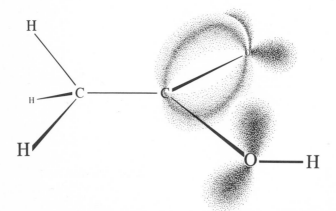

Figure 9-1
Acetic acid.
Lines indicate σ bonds. The carbon-oxygen π bond is shown in red. Each oxygen atom has two pairs of unshared electrons. Because the carbonyl carbon atom is sp²-hybridized, the two carbon atoms and the two oxygen atoms of acetic acid all lie in the same plane.

222

names are based on the same roots as the IUPAC names of other compounds. Common names, on the other hand, are based on an entirely different set of roots, the set used for common names of aldehydes.

Carboxylic acids with five or fewer carbon atoms are almost never called by their correct IUPAC names; the common names were so firmly entrenched in the minds of chemists that the IUPAC names were unable to replace them. The IUPAC has acknowledged this fact and has allowed the common names of the carboxylic acids with up to five carbon atoms to become a part of the IUPAC system.

The common names of all carboxylic acids with five or fewer carbon atoms and of straight-chain carboxylic acids up to twenty carbons are still in general use. Common names of carboxylic acids are derived from the same roots as the common names of aldehydes (table 7-1) and are shown in table 9-1.   *COMMON NAMES*

The straight-chain acids with about ten to twenty carbon atoms are called *fatty acids* because of their occurrence as a part of natural fats. For reasons which will be explained in chapter 10, fatty acids with an odd number of carbon atoms are rare. Of the acids with an even number of carbons, palmitic and stearic acids are the most common.

As with aldehydes and ketones, substituents on the carbon chain of a carboxylic acid are identified by Greek letters, starting with α for the carbon adjacent to the carbonyl carbon. Some examples are

$$Cl-CH_2CH_2CH_2-\overset{\overset{\textstyle O}{\|}}{C}-OH \qquad CH_3CH_2CH_2\overset{\overset{\textstyle OH}{|}}{CH}-\overset{\overset{\textstyle O}{\|}}{C}-OH$$

γ-chlorobutyric acid $\qquad$ α-hydroxyvaleric acid

According to the IUPAC system, carboxylic acids are named using the same roots as other organic compounds.   *IUPAC NAMES*

1 Find the longest continuous chain of carbon atoms that contains the carboxyl group as one end. Name the compound as a derivative of the hydrocarbon having that number of carbons by substituting the ending *-oic acid* for the terminal *-e* of the hydrocarbon name.

2 Number the carbon chain, starting with 1 at the carboxyl carbon.

3 Identify and locate substituents on the carbon chain by name and number (table 3-2). These substituent names become prefixes on the name of the carboxylic acid.

Some examples of the nomenclature of carboxylic acids are shown in figure 9-2.

Table 9-1
Names of Carboxylic Acids

Structure
$\mathrm{H-\overset{O}{\overset{\|}{C}}-OH}$
$\mathrm{CH_3-\overset{O}{\overset{\|}{C}}-OH}$
$\mathrm{CH_3CH_2-\overset{O}{\overset{\|}{C}}-OH}$
$\mathrm{CH_3CH_2CH_2-\overset{O}{\overset{\|}{C}}-OH}$
$\mathrm{CH_3CH_2CH_2CH_2-\overset{O}{\overset{\|}{C}}-OH}$
$\mathrm{CH_3CH_2CH_2CH_2CH_2-\overset{O}{\overset{\|}{C}}-OH}$
$\mathrm{CH_3CH_2CH_2CH_2CH_2CH_2CH_2-\overset{O}{\overset{\|}{C}}-OH}$
$\mathrm{CH_3CH_2CH_2CH_2CH_2CH_2CH_2CH_2CH_2-\overset{O}{\overset{\|}{C}}-OH}$
$\mathrm{CH_3CH_2CH_2CH_2CH_2CH_2CH_2CH_2CH_2CH_2CH_2-\overset{O}{\overset{\|}{C}}-OH}$
$\mathrm{CH_3CH_2CH_2CH_2CH_2CH_2CH_2CH_2CH_2CH_2CH_2CH_2CH_2-\overset{O}{\overset{\|}{C}}-OH}$
$\mathrm{CH_3CH_2CH_2CH_2CH_2CH_2CH_2CH_2CH_2CH_2CH_2CH_2CH_2CH_2CH_2-\overset{O}{\overset{\|}{C}}-OH}$
$\mathrm{CH_3CH_2CH_2CH_2CH_2CH_2CH_2CH_2CH_2CH_2CH_2CH_2CH_2CH_2CH_2CH_2CH_2-\overset{O}{\overset{\|}{C}}-OH}$
$\mathrm{CH_3CH_2CH_2CH_2CH_2CH_2CH_2CH_2CH_2CH_2CH_2CH_2CH_2CH_2CH_2CH_2CH_2CH_2CH_2-\overset{O}{\overset{\|}{C}}-OH}$

*The IUPAC names for the first four compounds are seldom used.

Number of Carbons	Common Name	IUPAC Name
1	formic acid	methanoic acid*
2	acetic acid	ethanoic acid*
3	propionic acid	propanoic acid*
4	butyric acid	butanoic acid*
5	valeric acid	pentanoic acid
6	caproic acid	hexanoic acid
8	caprylic acid	octanoic acid
10	capric acid	decanoic acid
12	lauric acid	dodecanoic acid
14	myristic acid	tetradecanoic acid
16	palmitic acid	hexadecanoic acid
18	stearic acid	octadecanoic acid
20	arachidic acid	eicosanoic acid

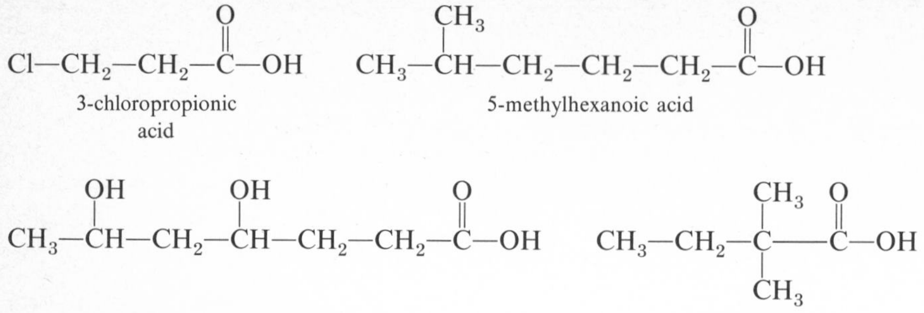

$$Cl-CH_2-CH_2-\overset{\overset{\displaystyle O}{\|}}{C}-OH \qquad CH_3-\overset{\overset{\displaystyle CH_3}{|}}{CH}-CH_2-CH_2-CH_2-\overset{\overset{\displaystyle O}{\|}}{C}-OH$$

3-chloropropionic acid 　　　　　　　　5-methylhexanoic acid

$$CH_3-\overset{\overset{\displaystyle OH}{|}}{CH}-CH_2-\overset{\overset{\displaystyle OH}{|}}{CH}-CH_2-CH_2-\overset{\overset{\displaystyle O}{\|}}{C}-OH \qquad CH_3-CH_2-\overset{\overset{\displaystyle CH_3}{|}}{\underset{\underset{\displaystyle CH_3}{|}}{C}}-\overset{\overset{\displaystyle O}{\|}}{C}-OH$$

4,6-dihydroxyheptanoic acid 　　　　　　　2,2-dimethylbutyric acid

Figure 9-2
The names of some carboxylic acids.

PROBLEM 9-2 Give the structure of (*a*) 3-methylhexanoic acid; (*b*) 2,3-diethyl-octanoic acid; (*c*) 4-bromodecanoic acid.

9-3 ACIDITY OF CARBOXYLIC ACIDS

Acids are, by definition, proton donors. When an inorganic acid is dissolved in water, the acid undergoes ionization, forming hydronium ion (the predominant form of H^+ in water) and an anion

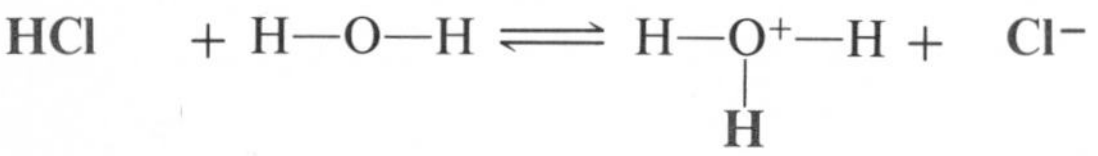

$$HCl \quad + H-O-H \rightleftharpoons H-\overset{\overset{\displaystyle +}{O}}{\underset{\underset{\displaystyle H}{|}}{}}-H + \quad Cl^-$$

acid 　　　　solvent 　　　　hydronium 　　　anion
(hydrochloric 　　　　　　　　　ion 　　　　(chloride)
acid)

When a carboxylic acid is dissolved in water, ionization also occurs, but to a lesser extent than with strong inorganic acids.

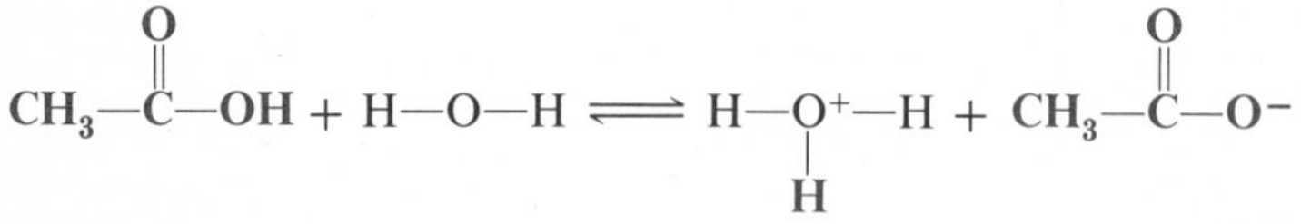

$$CH_3-\overset{\overset{\displaystyle O}{\|}}{C}-OH + H-O-H \rightleftharpoons H-\overset{\overset{\displaystyle +}{O}}{\underset{\underset{\displaystyle H}{|}}{}}-H + CH_3-\overset{\overset{\displaystyle O}{\|}}{C}-O^-$$

acid 　　　　　solvent 　　　　hydronium 　　　　anion
(carboxylic acid) 　　　　　　　　ion 　　　　(carboxylate anion)

ACID STRENGTH　When a strong acid like HCl is dissolved in water, it is completely converted into hydronium ions and chloride ions. Carboxylic acids, however, are weak acids. When acetic acid is dissolved in water, for

Table 9-2
Ionization
Constants of
Some Acids*

Compound	Name	pK_a
H—C(=O)—OH	formic acid	3.8
CH_3—C(=O)—OH	acetic acid	4.8
CH_3CH_2—C(=O)—OH	propionic acid	4.9
$CH_3(CH_2)_6$—C(=O)—OH	caprylic acid	4.9
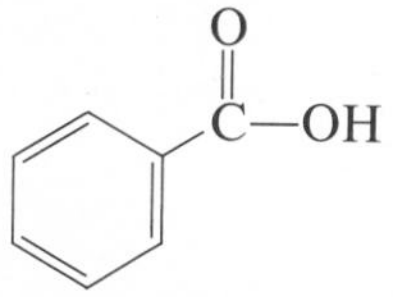 C₆H₅—C(=O)—OH (benzene ring)	benzoic acid	4.2

*Data from H. A. Sober, *Handbook of Biochemistry,* Chemical Rubber Co., Cleveland, Ohio, 1968.

example, it is converted into hydronium ions and acetate ions to only a small extent (about 1 percent when the acid concentration is 1 M).

Put another way, the pK_a values for carboxylic acids are much higher than those for the strong inorganic acids. Some pK_a values are given in table 9-2. Comparison of table 9-2 with table 5-1 and figure 6-1 reveals that carboxylic acids are more acidic than alcohols by a factor of at least 10^{10} and more acidic than ammonium ions by approximately 10^5.

RESONANCE STABILIZATION

The acidity of carboxylic acids is due in large part to the *resonance stabilization* of the carboxylate anion (figure 9-3). As a result of this resonance the negative charge on the carboxylate anion is spread equally over two oxygen atoms. The delocalization of the negative charge is the cause of the greater acidity of carboxylic acids compared with alcohols.

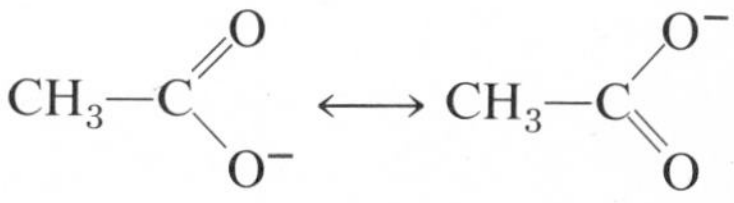

Figure 9-3
Resonance
stabilization of the
carboxylate anion.
The two resonance
forms are of equal
energy and
contribute equally to
the structure of the
carboxylate ion.

Acids have two distinguishing properties: (1) their ability to form hydronium ions on reaction with water and (2) their ability to form *salts* on reaction with bases

$$HCl + NaOH \longrightarrow Na^+ \ Cl^- + HOH$$

acid   base   salt   water

Likewise, for carboxylic acids

$$\underset{\substack{\text{acid} \\ \text{(acetic} \\ \text{acid)}}}{CH_3-\overset{\overset{\displaystyle O}{\|}}{C}-OH} + \underset{\text{base}}{KOH} \longrightarrow \underset{\substack{\text{salt} \\ \text{(potassium} \\ \text{acetate)}}}{CH_3-\overset{\overset{\displaystyle O}{\|}}{C}-O^-} \ K^+ + \underset{\text{water}}{HOH}$$

The acetate anion formed in this reaction is the same resonance-stabilized species as that formed (to a lesser extent) when acetic acid is dissolved in water.

Since amines are bases, ammonium salts of carboxylic acids can be formed by the reaction of ammonia or amines with carboxylic acids

$$\underset{\substack{\text{propionic} \\ \text{acid}}}{CH_3CH_2-\overset{\overset{\displaystyle O}{\|}}{C}-OH} + \underset{n\text{-butylamine}}{CH_3CH_2CH_2CH_2-NH_2} \longrightarrow$$

$$CH_3CH_2-\overset{\overset{\displaystyle O}{\|}}{C}-O^- \quad CH_3CH_2CH_2CH_2-\overset{\overset{\displaystyle H}{|}}{\underset{\underset{\displaystyle H}{|}}{{}^+N}}-H$$

n-butylammonium propionate

Because carboxylic acids are several orders of magnitude weaker than such strong inorganic acids as sulfuric acid, hydrochloric acid, and nitric acid, a carboxylic acid can be made from a carboxylic acid salt by the addition of a strong acid

$$\underset{\text{lithium butyrate}}{2CH_3CH_2CH_2-\overset{\overset{\displaystyle O}{\|}}{C}-O^-} \ Li^+ + \underset{\substack{\text{sulfuric} \\ \text{acid}}}{H_2SO_4} \longrightarrow \underset{\text{butyric acid}}{2CH_3CH_2CH_2-\overset{\overset{\displaystyle O}{\|}}{C}-OH} + \underset{\substack{\text{lithium} \\ \text{sulfate}}}{Li_2SO_4}$$

PROBLEM 9-3 Complete each of the following equilibria and use the data in tables 5-1 and 9-3 to predict the position of each equilibrium:

(a) acetic acid + sodium methoxide; (b) formic acid + methylamine; (c) sodium benzoate + ethanol; (d) sodium acetate + H_3PO_4; (e) sodium formate + acetic acid.

Salts may be named with either IUPAC names or common names; in either case, the rules are the same. The name consists of two words: the first is the name of the cation, and the second is the name of the carboxylic acid from which the anion is derived, with the ending *-ate* substituted for the acid ending *-ic acid:*

$$CH_3CH_2-\overset{\overset{\displaystyle O}{\|}}{C}-O^- \quad Li^+ \qquad CH_3CH_2CH_2CH_2CH_2-\overset{\overset{\displaystyle O}{\|}}{C}-O^- \quad NH_4^+$$

lithium propionate   ammonium hexanoate

The pK_a values for most carboxylic acids lie in the range between 2 and 5 (table 9-2). The pH of the environment of most cells is between 5 and 9; as a result, carboxylic acids exist in living systems as salts rather than as free acids.

This ionization increases the solubilities of carboxylic acids. The carboxyl group itself is polar and confers a degree of water solubility on a molecule, but the carboxylate ion is both polar and charged; it confers a much larger degree of water solubility on a molecule than a carboxyl group does. Indeed, the sole function of carboxyl groups in a number of natural molecules is to make the compound more water-soluble.

9-4 CARBOXYLIC ACIDS IN NATURE

Carboxylic acids are ubiquitous in living systems. They serve almost every conceivable function for living organisms. The naturally occurring carboxylic acids range from the simplest acids, formic acid (first observed by S. Fisher in 1670 as one of the products of the distillation of ants) and acetic acid (sometimes obtained by the destructive distillation of wood) to very complex acids such as the prostaglandins and the amino acids.

The prostaglandins are a family of polyfunctional carboxylic acids containing twenty carbon atoms. They are characterized by the presence of a cyclopentane ring bearing two long side chains. Several prostaglandins occur as natural compounds in mammals, and a large number of synthetic prostaglandin analogs have been made as well. A number of these compounds are shown in figure 9-4.

At the cellular level, prostaglandins seem to perform a large number of functions. They control the secretion of acid in the stomach, the contraction and relaxation of muscle, body temperature, and a number of other things as well. Extensive research is underway to find synthetic prostaglandins which can be used as drugs. Two naturally occurring

Figure 9-4
Structures of some
prostaglandins.

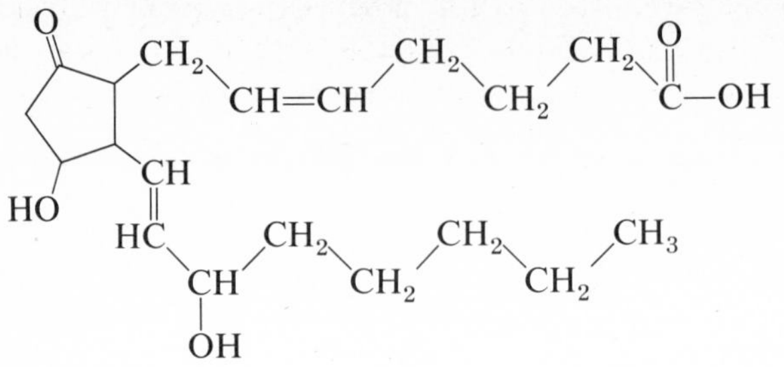

prostaglandin E_2

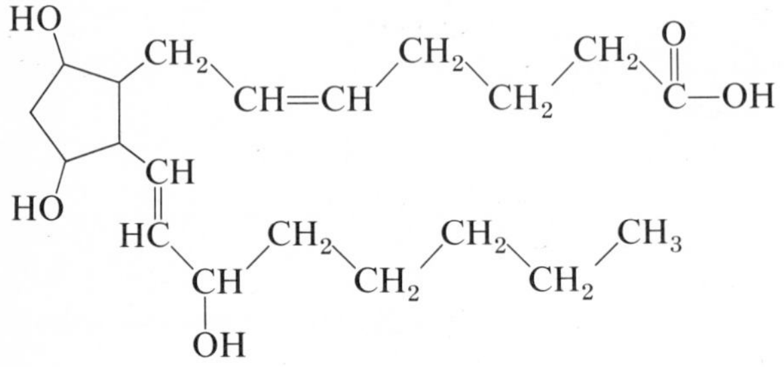

prostaglandin $F_{2\alpha}$

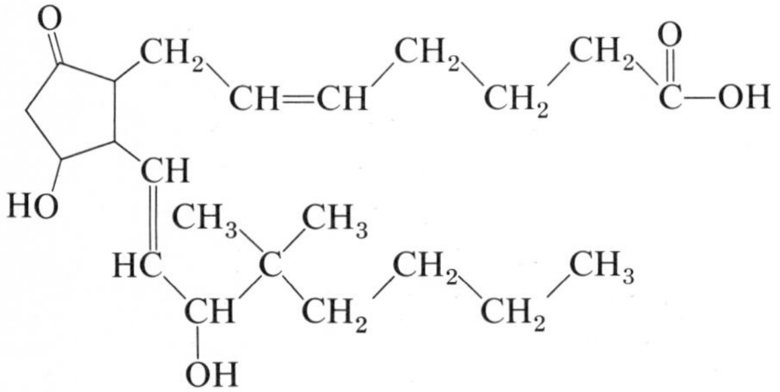

16,16-dimethyl prostaglandin E_2

prostaglandins are already in limited use, prostaglandin E_2 and prostaglandin $F_{2\alpha}$. These compounds can be used for inducing labor and for terminating pregnancy, effects which apparently occur because the compounds cause contraction of smooth muscle.

One of the main parts of metabolism in animals is the controlled oxidation of food to produce energy. Overall, this process can be expressed as

Food + O_2 $\longrightarrow$ CO_2 + H_2O + energy

The last reaction in the long sequence by which food is converted into CO_2 and H_2O is called *decarboxylation* and occurs by the loss of CO_2 from a carboxylic acid or its salt.

Simple carboxylic acids such as acetic acid generally undergo decarboxylation only at very high temperature. However, decarboxylation of β-keto acids takes place readily on gentle heating

$$CH_3-\overset{O}{\underset{\|}{C}}-CH_2-\overset{O}{\underset{\|}{C}}-OH \xrightarrow{heat} CH_3-\overset{O}{\underset{\|}{C}}-CH_3 + CO_2$$

The mechanism of this reaction is shown in mechanism 9-1.

The enzyme-catalyzed decarboxylations of a number of carboxylic acids take place via mechanisms of this type. The oxidative decarboxylation of malic acid by malic enzyme takes place in two steps. Oxidation of enzyme-bound malic acid takes place first

$$HO-\overset{O}{\underset{\|}{C}}-\overset{OH}{\underset{|}{CH}}-CH_2-\overset{O}{\underset{\|}{C}}-OH + NADP^+ \rightleftharpoons$$

malic acid   oxidizing agent*

$$HO-\overset{O}{\underset{\|}{C}}-\overset{O}{\underset{\|}{C}}-CH_2-\overset{O}{\underset{\|}{C}}-OH + NADPH + H^+$$

oxalacetic acid

followed by decarboxylation of the β-keto acid

$$HO-\overset{O}{\underset{\|}{C}}-\overset{O}{\underset{\|}{C}}-CH_2-\overset{O}{\underset{\|}{C}}-OH \rightleftharpoons HO-\overset{O}{\underset{\|}{C}}-\overset{O}{\underset{\|}{C}}-CH_3 + CO_2$$

pyruvic acid

Both these reactions occur while the substrate is attached firmly to the surface of the enzyme.

---

*The oxidizing agent used in this reaction, $NADP^+$, is identical with NAD^+ (figure 7-7), except for the presence of an additional phosphate in the $NADP^+$ molecule. NADPH and NADH differ in a similar way.

232

Mechanism 9-1
Decarboxylation of
Acetoacetic Acid

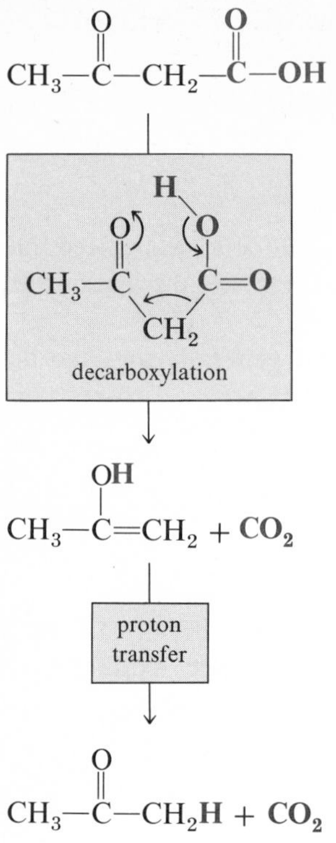

$$CH_3-\overset{\overset{O}{\|}}{C}-CH_2-\overset{\overset{O}{\|}}{C}-OH$$

This decarboxylation takes place most readily when the carboxylic acid, rather than a carboxylate ion, is the reactive species.

decarboxylation

The key step in this reaction is a cyclic step in which the acidic proton from the carboxyl group is transferred to the ketone oxygen and simultaneously carbon dioxide is formed.

$$CH_3-\overset{\overset{OH}{|}}{C}=CH_2 + CO_2$$

The products of this step are carbon dioxide and the enol of acetone.

proton
transfer

$$CH_3-\overset{\overset{O}{\|}}{C}-CH_2H + CO_2$$

Proton transfer occurs quickly, and the final products are acetone and carbon dioxide.

PROBLEMS   **9-4**   Explain why $CH_3\overset{\overset{O}{\|}}{C}CH_2CH_2\overset{\overset{O}{\|}}{C}OH$ does not readily undergo decarboxylation when it is heated.

9-5   Give the structure of the predominant form of propionic acid (*a*) at pH 3; (*b*) at pH 7; (*c*) at pH 10.

9-6   The oxidative decarboxylation of isocitric acid in nature takes

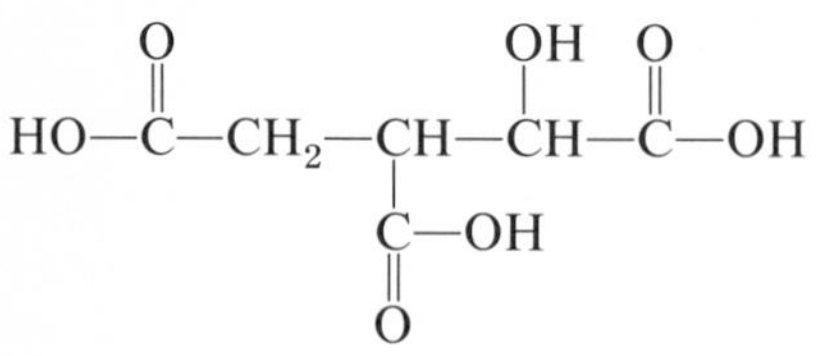

$$HO-\overset{\overset{O}{\|}}{C}-CH_2-\underset{\underset{\underset{\overset{\|}{O}}{C-OH}}{|}}{CH}-\overset{\overset{OH}{|}}{CH}-\overset{\overset{O}{\|}}{C}-OH$$

isocitric acid

place by a mechanism like that described in section 9-5 for malic acid; that is, oxidation takes place first, followed by decarboxylation. Give the structure of the product of the decarboxylation.

9-7 Show the acid-base reaction which occurs when acetic acid is dissolved in methanol. Label pairs of conjugate acids and bases, and indicate approximately where the equilibrium lies.

9-8 Name the following compounds:

(a) $CH_3CHCH_2CH_2\overset{\displaystyle O}{\overset{\|}{C}}$—OH
　　　|
　　CH_3

(b) $CH_3CH_2CH_2\overset{\displaystyle O}{CH\overset{\|}{C}}$—OH
　　　　　　|
　　　　CH_3CH
　　　　　　|
　　　　　CH_3

(c) $CH_3CH_2\overset{\displaystyle CH_3}{\overset{|}{CH}}CH_2CH_2\overset{\displaystyle O}{\overset{\|}{C}}$—O⁻ $CH_3NH_3{}^+$

(d) $CH_3\overset{\displaystyle O}{\overset{\|}{C}}$—O⁻ Li⁺

9-9 Arrange the following compounds in order of increasing acidity:

$CH_3\overset{\displaystyle O}{\overset{\|}{CH}}$, $CH_3NH_3{}^+$, $CH_3\overset{\displaystyle O}{\overset{\|}{C}}$—OH, CH_3OH.

9-10 Give the reagents needed to carry out each of the following transformations. More than one step may be required.

(a) $CH_3CH_2CH_2Br \longrightarrow CH_3CH_2CH_2\overset{\displaystyle O}{\overset{\|}{C}}$—OH

(b) $CH_3CH_2CH_2Br \longrightarrow CH_3CH_2\overset{\displaystyle O}{\overset{\|}{C}}$—OH

(c) ⬠—$CH_2OH \longrightarrow$ ⬠—$\overset{\displaystyle O}{\overset{\|}{C}}$—OH

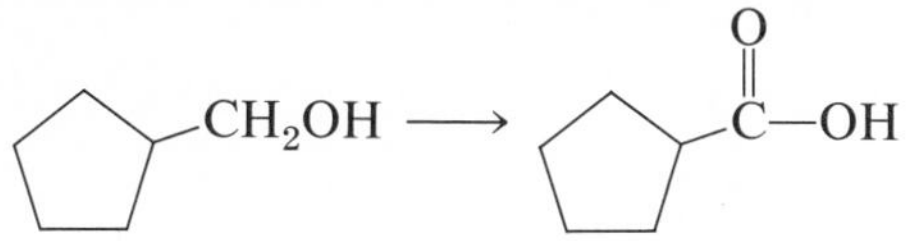

(d) $CH_3CH_2\overset{\displaystyle O}{\overset{\|}{CH}} \longrightarrow CH_3CH_2\overset{\displaystyle O}{\overset{\|}{C}}$—OH

9-11 Give the structure of the product of each of the following reactions:

(a) $CH_3CH_2CH_2\overset{\displaystyle O}{\overset{\|}{C}}$—OH + KOH $\longrightarrow$

(b) $CH_3CH_2\overset{\displaystyle O}{\overset{\|}{C}}$—O⁻ NH₄⁺ + HI $\longrightarrow$

(c) + CH₃CH₂NH₂ $\longrightarrow$

(d) $\xrightarrow{\text{heat}}$

SUGGESTED READING   Pike, J. E.: Prostaglandins, p. 66 in *Organic Chemistry of Life: Readings from Scientific American,* Freeman, San Francisco, 1973.

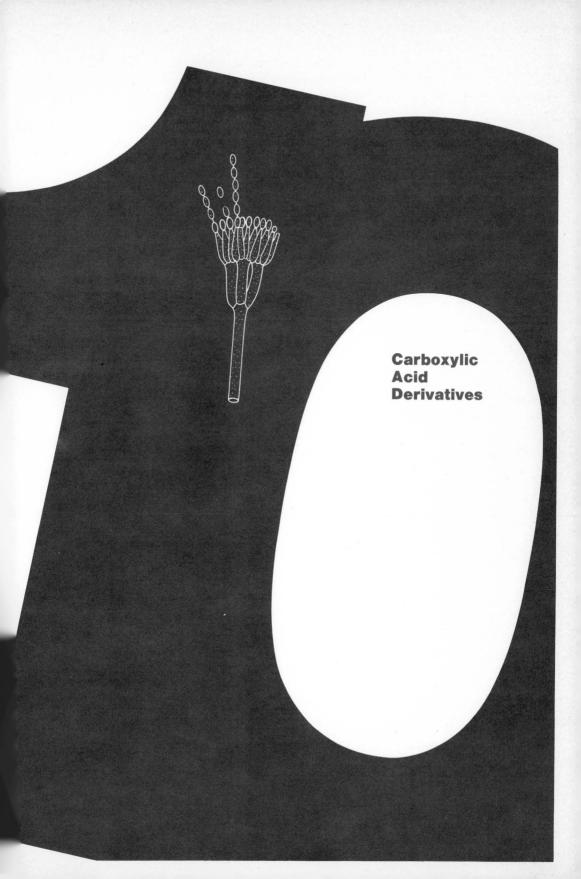

**Carboxylic
Acid
Derivatives**

The carboxylic acids, considered in chapter 9, are the parents of a large family of compounds, all of which have the —OH of the carboxyl group replaced by some other group. The important members of this family are shown in figure 10-1. In spite of their structural diversity, all these compounds undergo similar reactions. Like aldehydes and ketones, they undergo reactions characteristic of the carbonyl group:

1 Reactions in which the first step is nucleophilic attack on the carbonyl carbon atom
2 Reactions in which protonation of the carbonyl oxygen is followed by nucleophilic attack on the carbonyl carbon
3 Reactions in which the first step is removal of an α hydrogen by base

10-1 ESTERS

Esters can be formed by the combination of a carboxylic acid with an alcohol

$$CH_3-\overset{\overset{\textstyle O}{\|}}{C}-OH + HO-CH_3 \xrightarrow{H^+} CH_3-\overset{\overset{\textstyle O}{\|}}{C}-O-CH_3 + HOH$$

acid   alcohol   ester

The synthesis of esters by this method requires an acid catalyst and a large excess of alcohol (section 10-8). Other reactions which can be used to form esters are discussed in section 10-5.

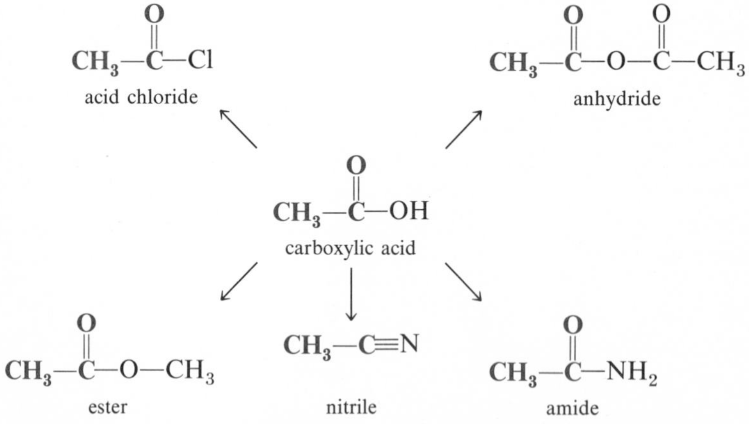

$$CH_3-\overset{\overset{\textstyle O}{\|}}{C}-Cl$$
acid chloride

$$CH_3-\overset{\overset{\textstyle O}{\|}}{C}-O-\overset{\overset{\textstyle O}{\|}}{C}-CH_3$$
anhydride

$$CH_3-\overset{\overset{\textstyle O}{\|}}{C}-OH$$
carboxylic acid

$$CH_3-\overset{\overset{\textstyle O}{\|}}{C}-O-CH_3$$
ester

$$CH_3-C\equiv N$$
nitrile

$$CH_3-\overset{\overset{\textstyle O}{\|}}{C}-NH_2$$
amide

Figure 10-1
Carboxylic acid
derivatives.
All these compounds except nitriles are derived by the combination of a carboxylic acid with another compound, with the elimination of water. Acid chlorides are derived by combination with HCl; esters, by combination with an alcohol; amides, with an amine; anhydrides, with another molecule of the carboxylic acid.

Esters are named by giving the name of the *alkyl group* attached to oxygen, followed by the name of the *carboxylate group*. Either IUPAC names or common names can be used:

$$CH_3-\overset{\overset{\textstyle O}{\|}}{C}-O-CH_2CH_2CH_2CH_3 \qquad CH_3\overset{\overset{\textstyle CH_3}{|}}{C}HCH_2CH_2CH_2-\overset{\overset{\textstyle O}{\|}}{C}-O-CH_2CH_3$$

n-butyl acetate   ethyl 5-methylhexanoate

Esters in which the ester group is part of a ring are derived from hydroxy acids and are called *lactones*

derived from $HO-CH_2CH_2CH_2-\overset{\overset{\textstyle O}{\|}}{C}-OH$

γ-butyrolactone   γ-hydroxybutyric
                                   acid

A vast number of esters occur in nature. In many cases they serve as precursors for carboxylic acids. A number of examples will be considered in this chapter, including glycerides, acetyl choline, and acetyl coenzyme A (section 10-5).

Give the structure of an ester formed by the combination of octanoic acid with 2-butanol. Name the compound.

Although amides can be viewed as having been constructed from a carboxylic acid and an amine, amides cannot be synthesized by this method. Instead, amides are formed by the reaction of ammonia or amines with acid chlorides, esters, or anhydrides (section 10-5).

Tertiary amines do not differ appreciably in reactivity from corresponding primary or secondary amines, but tertiary amines do not form stable amides. Such amides would have a positive charge on the amide nitrogen and would be very susceptible to reaction with nucleophiles.

The nitrogen atom in amides is sp^2-hybridized and is involved in resonance with the carbonyl group. The three resonance forms shown in figure 10-2 do not contribute equally to the structure of amides; the first resonance form is the most important, but the second is important for understanding the chemistry of amides and the third for understanding the geometry of amides.

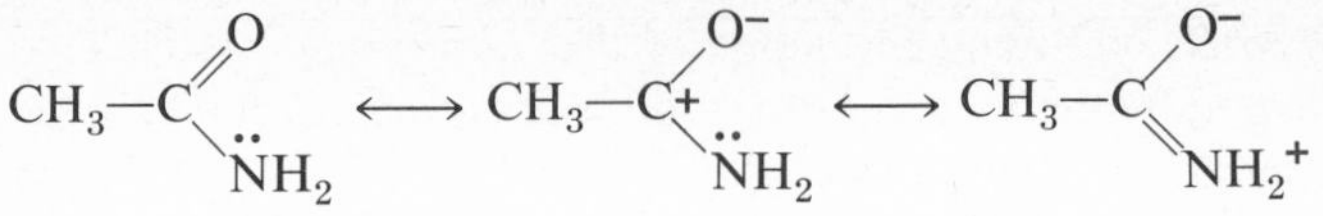

Figure 10-2
Resonance in
amides. The first of
the three forms is
the most important.
The third is
responsible for the
geometry and the
reduced basicity of
amides.

$$CH_3-C\overset{O}{\underset{\ddot{N}H_2}{\Big\backslash}} \longleftrightarrow CH_3-\overset{O^-}{\underset{\ddot{N}H_2}{C+}} \longleftrightarrow CH_3-C\overset{O^-}{\underset{NH_2^+}{\Big\backslash}}$$

The last resonance form in figure 10-2 can be of importance only if the amide nitrogen is situated so that the unshared pair of electrons in the *p* orbital on nitrogen can overlap efficiently with the adjacent π molecular orbital. As a result, amides exist in a conformation which places all the atoms bonded to the carbonyl carbon and the amide nitrogen in the same plane (figure 10-3). Two different conformations are possible which have the kind of extensive resonance interactions shown in figure 10-2. These two structures represent conformational isomers, rather than geometrical isomers, because their interchange does not break a full double bond. The barrier to interconversion of these two conformations is approximately 20 kilocalories/mole. This figure is about six times greater than the barrier to internal rotation in ethane but five times less than the energy needed to break a carbon-carbon single bond. Rotation of the carbon-nitrogen bond in amides is relatively slow at room temperature.

Although this 20 kilocalorie/mole barrier is not sufficient to prevent the rotation of the carbon-nitrogen bond in amides, it is an important factor in defining the structures of proteins, which are long chains of amino acids held together by amide linkages (chapter 12).

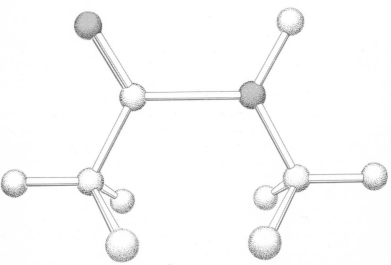

Figure 10-3
Geometry of
N-methylacetamide.

Because of resonance the carbonyl group, the nitrogen, the two atoms attached to nitrogen, and the atom attached to the carbonyl carbon all lie in the same plane.

Unlike amines, amides are not appreciably basic. There are two reasons for this: (1) the resonance interactions shown in figure 10-2 involve the unshared pair of electrons on nitrogen, and protonation of the nitrogen would result in destruction of those resonance interactions; (2) the nitrogen atom in amides is sp^2-hybridized, and such nitrogen atoms are usually less basic than sp^3-hybridized nitrogen atoms.

BASICITY OF AMIDES

Simple amides which have only hydrogen on the amide nitrogen are named by substituting the ending *-amide* for the ending *-ic acid* or *-oic acid* of the parent acid

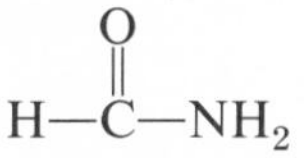

NAMES OF AMIDES

formamide 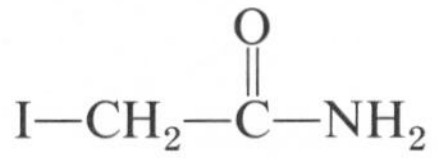 iodoacetamide

Amides which have substituents on nitrogen are named by giving the names of the substituents (preceded by *N-* to show that they are attached to nitrogen) as the first prefixes on the name of the parent amide

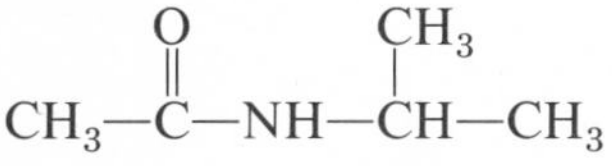

N-isopropylacetamide 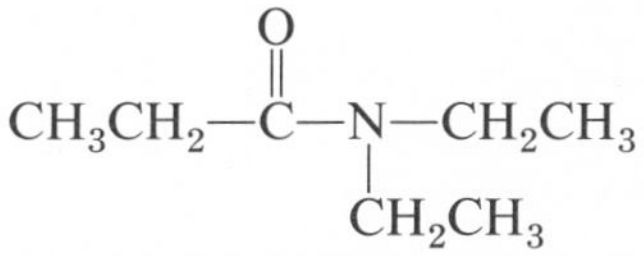 *N*,*N*-diethylpropionamide

Cyclic amides derived from amino acids are called *lactams:*

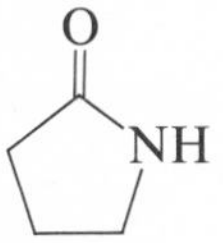

γ-butyrolactam derived from γ-aminobutyric acid

The most widespread occurrence of amides in nature is in the form of *proteins,* which are long chains of amino acids held together by amide groups (chapter 12).

AMIDES IN NATURE

The metabolism of ammonia involves a number of important amides, including *glutamine* and *carbamyl phosphate,* which serve to store ammonia and provide readily available ammonia when needed

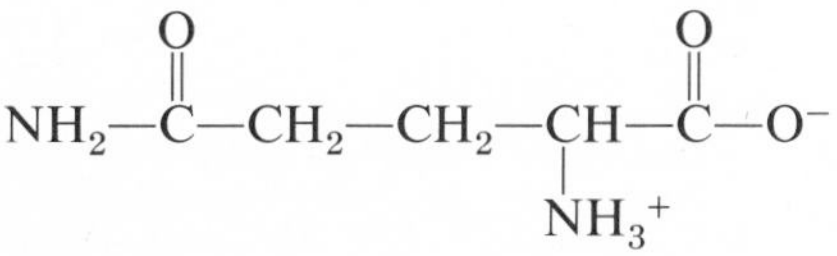

glutamine 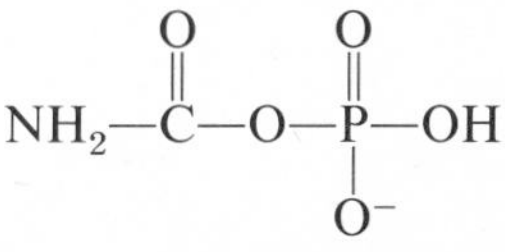 carbamyl phosphate

Urea, which is an amide derived from carbonic acid (section 10-10), is the form in which ammonia is excreted by terrestrial vertebrates

$$NH_2-\overset{\overset{\displaystyle O}{\|}}{C}-NH_2$$

urea

PROBLEM 10-2 Give the structure and name of the amide formed by combination of methylethylamine with 2-methylbutyric acid. Draw the two principal conformations of the amide, and show three resonance structures for each conformation.

PENICILLIN *Antibiotics* are compounds which inhibit the growth of microorganisms in the body. Because many diseases are caused by invading microorganisms, the study of antibiotics is an important part of modern chemistry.

One of the most important antibiotics in current use is penicillin (figure 10-4), which is at the same time a lactam, an amide, a carboxylic acid, and a sulfide.

Penicillin was discovered in 1928 by Sir Alexander Fleming, but as is so often the case in biochemistry, early attempts to isolate and study this compound were unsuccessful because of the very small amounts present in Fleming's preparations. The serious study of penicillin began in 1939, and by 1949 penicillin had become a widely used antibiotic.

The naturally occurring form of penicillin is penicillin G (figure 10-4), and this is still the best for the majority of uses. However, penicillin is rapidly degraded by an enzyme called *penicillinase,* which is secreted by a number of the microorganisms that penicillin is intended to control. In addition, penicillin is hydrolyzed by the acid in the stomach, so that oral administration of penicillin must be carefully regulated. To alleviate these problems, a search soon began for structural analogs of penicillin which would not be subject to such rapid degradation. The search centered on compounds which have the same ring structure as penicillin G but different acid groups in the amide. Several hundred such analogs were synthesized and tested, and a few of them have proved effective and have become useful drugs (figure 10-4).

10-3 *ANHYDRIDES AND ACID CHLORIDES*

Anhydrides appear to result from the combination of two molecules of a carboxylic acid

$$CH_3-\overset{\overset{\displaystyle O}{\|}}{C}-OH + HO-\overset{\overset{\displaystyle O}{\|}}{C}-CH_3 \longrightarrow CH_3-\overset{\overset{\displaystyle O}{\|}}{C}-O-\overset{\overset{\displaystyle O}{\|}}{C}-CH_3 + H_2O$$

acid   acid   anhydride

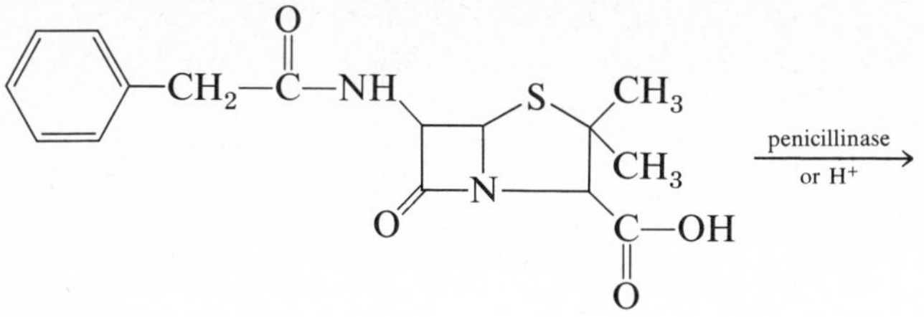

penicillin G

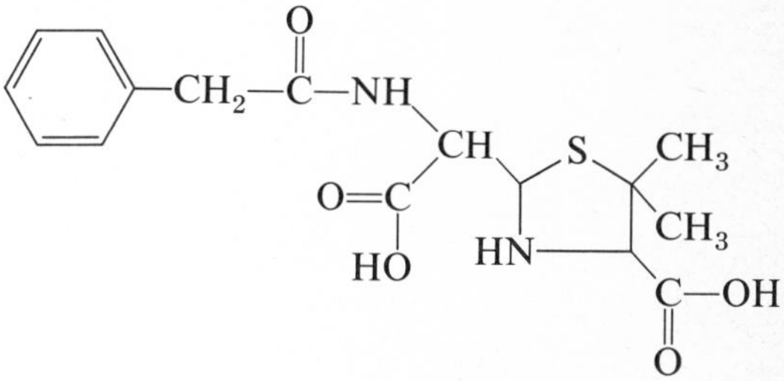

penicilloic acid

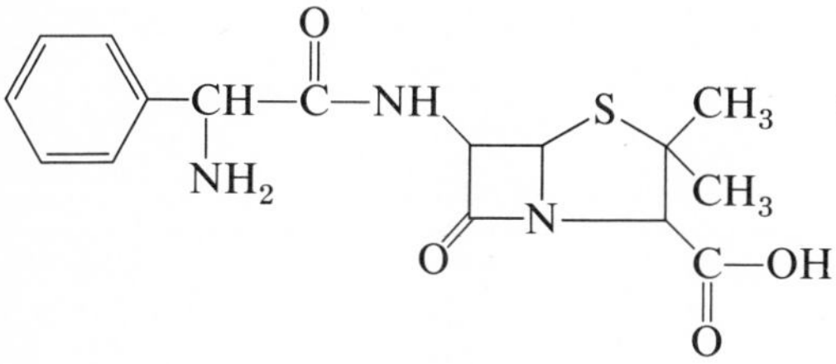

ampicillin

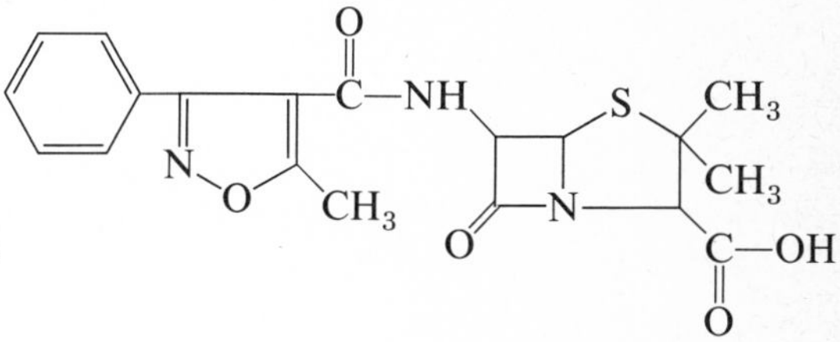

oxacillin

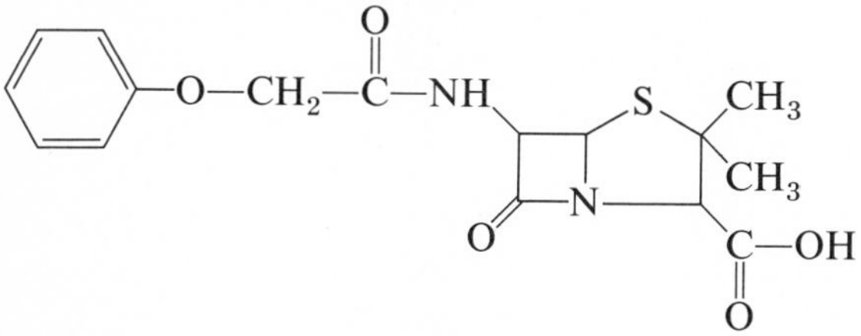

phenoxymethyl
penicillin

*This naturally occurring antibiotic is both an amide and a lactam. Decomposition in the stomach or in the presence of the enzyme **penicillinase** produces penicilloic acid. The synthetic penicillins have the same ring structure as penicillin G, but differ in the nature of the side-chain amide.*

Figure 10-4
Penicillin.

However, this method is not usually useful for the synthesis of anhydrides; instead, they must be prepared from acid chlorides (section 10-5).

Acid chlorides are prepared by the reaction of carboxylic acids with thionyl chloride, $SOCl_2$,

$$CH_3-\overset{\overset{\displaystyle O}{\|}}{C}-OH + SOCl_2 \longrightarrow CH_3-\overset{\overset{\displaystyle O}{\|}}{C}-Cl + SO_2 + HCl$$

acid   thionyl   acid
                chloride   chloride

NOMENCLATURE The names of anhydrides and acid chlorides are derived from either set of names of carboxylic acids. Anhydrides are named by placing the word *anhydride* in place of the word *acid* in the name of the acid. Acid chlorides are named by replacing the acid ending *-ic acid* by *-yl chloride*.

$$CH_3CH_2-\overset{\overset{\displaystyle O}{\|}}{C}-O-\overset{\overset{\displaystyle O}{\|}}{C}-CH_2CH_3 \qquad CH_3CH_2CH_2CH_2CH_2-\overset{\overset{\displaystyle O}{\|}}{C}-Cl$$

propionic anhydride   hexanoyl chloride

PROBLEM 10-3 Give the structures and names of the anhydride and acid chloride of butyric acid.

ANHYDRIDES AND ACID CHLORIDES IN NATURE Carboxylic acid anhydrides and acid chlorides seldom occur in nature. Both types of compounds are so reactive that they would rapidly be destroyed by the aqueous environment of the living cell. A number of mixed anhydrides of carboxylic and phosphoric acids are important natural compounds (section 10-11).

10-4 NITRILES

Nitriles, or cyano compounds, do not contain a carbonyl group. Nonetheless, the carbon atom in the cyano group is at oxidation level 3, and hydrolysis of a nitrile produces an amide or a carboxylic acid (section 10-5).

Nitriles are often produced by the reaction of an alkyl halide with cyanide ion (section 5-6)

$$CH_3CH_2-Br + NaCN \longrightarrow CH_3CH_2-C{\equiv}N + NaBr$$

alkyl halide   nitrile

Nitriles can also be produced by heating an amide with a strong dehydrating agent like P_2O_5

$$3 \quad \underset{\text{amide}}{\overset{\overset{\displaystyle O}{\|}}{\text{C}-\text{NH}_2}} \quad + \quad P_2O_5 \quad \xrightarrow{\text{heat}} \quad 3 \quad \underset{\text{nitrile}}{\text{C}\equiv\text{N}} \quad + \quad 2H_3PO_4$$

The names of nitriles are constructed by removing the final *-oic acid* or *-ic acid* of the name of the carboxylic acid having the same number of carbon atoms and replacing it with the ending *-onitrile*.

Give the structure and name of the nitrile of pentanoic acid.   **PROBLEM 10-4**

10-5   REACTIONS OF CARBOXYLIC ACID DERIVATIVES WITH NUCLEOPHILES

The common denominator in carboxylic acid chemistry is the polarization of the carbonyl group; carboxylic acid derivatives react with nucleophiles because the carbonyl carbon atom is electron-poor, and they react with electrophiles because the carbonyl oxygen atom is electron-rich. In that respect, the chemistry of these compounds is much like that of aldehydes and ketones. However, the reactions of carboxylic acid derivatives frequently involve cleavage of a bond between the carbonyl carbon atom and an atom to which it is attached, a reaction seldom observed with aldehydes and ketones.

The names of these reactions reflect the nature of the attacking group: *hydrolysis* is replacement of some group by water; *alcoholysis* is replacement by an alcohol; *aminolysis* is replacement by ammonia or an amine.

The mechanisms of all these reactions are quite similar, except for the difference in the nature of the attacking and leaving groups. As with aldehydes and ketones, reactions may occur either by attack of the nucleophile on the unprotonated substrate or by attack on the protonated substrate, depending on the reactivity of the nucleophile and the reactivity of the acid derivative.

The key intermediate in all these reactions is a *tetrahedral intermediate;* it is formed in virtually all the reactions we will consider, and an understanding of its properties will be essential.*

---

*There are a few hydrolysis reactions and related reactions of acid derivatives which do not occur via tetrahedral intermediates, but we will have little occasion to consider them.

The reaction of an ester with hydroxide is called *ester hydrolysis* or *saponification*. The second name derives from the fact that this reaction was once used for making soap from animal fat. By this reaction an ester is converted into its constituent alcohol and acid

$$CH_3CH_2CH_2-\overset{\overset{\displaystyle O}{\|}}{C}-O-CH_2CH_3 + Na^+ \ ^-OH \longrightarrow$$

ethyl butyrate

$$CH_3CH_2CH_2-\overset{\overset{\displaystyle O}{\|}}{C}-O^- \ Na^+ + HO-CH_2CH_3$$

sodium butyrate   ethanol

The mechanism of this reaction involves three steps: nucleophilic attack by hydroxide on the carbonyl carbon, loss of ethoxide, and proton transfer (mechanism 10-1).

Phenyl benzoate also undergoes hydrolysis by this same mechanism

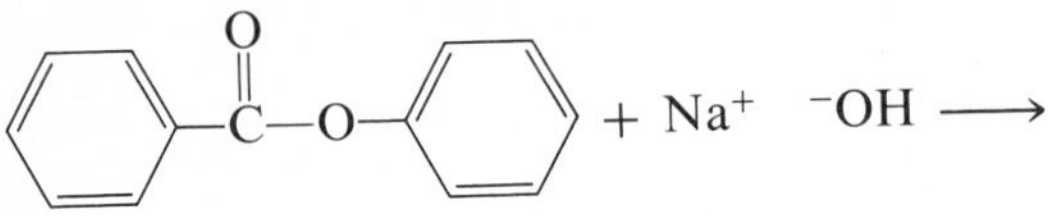

phenyl benzoate

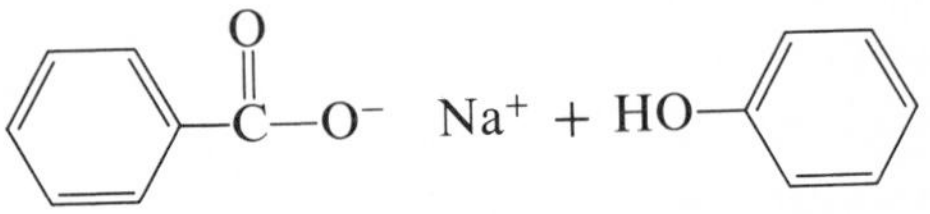

sodium benzoate   phenol

PROBLEM 10-5   Give the mechanism of the above reaction.

SOAP   The principal constituents of animal fat are compounds called *glycerides,* esters of long-chain carboxylic acids with the triol *glycerol* (also called glycerin)

$$HO-CH_2-\underset{\underset{\displaystyle OH}{|}}{CH}-CH_2-OH$$

glycerol

Such compounds are called *monoglycerides, diglycerides,* or *triglycerides,* according to whether glycerol is involved in one, two, or three ester linkages. The most common glycerides are the triglycerides.

$$CH_3-\overset{\overset{\displaystyle O}{\|}}{C}-O-CH_3 + OH^-$$

Polarization of the carbonyl group makes the carbonyl carbon atom susceptible to nucleophilic attack by hydroxide and other nucleophiles.

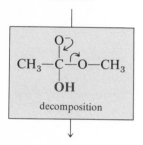

nucleophilic
attack

In the first step, the nucleophile attacks the carbonyl carbon, and the carbon-oxygen double bond becomes a single bond.

$$\left[CH_3-\overset{\overset{\displaystyle O^-}{\|}}{\underset{\underset{\displaystyle OH}{\|}}{C}}-O-CH_3 \right]$$

tetrahedral
intermediate

The product of this first step is a short-lived **tetrahedral intermediate.** The carbonyl group has temporarily been destroyed, and the carbonyl carbon is sp³-hybridized. In this case, the carbonyl oxygen bears a negative charge.

decomposition

Decomposition of the tetrahedral intermediate takes place by a reaction similar to that by which it was formed but with the roles of hydroxide and methoxide reversed. In this step, a carbon-oxygen bond breaks, and the carbonyl carbon-oxygen double bond is reformed.

$$CH_3-\overset{\overset{\displaystyle O}{\|}}{C}-OH + {}^-O-CH_3$$

The products of this step are a carboxylic acid and an alkoxide ion. These are stable molecules, except that the alkoxide is a strong base and very rapidly removes the acidic proton from the carboxylic acid (section 9-3).

proton
transfer

$$CH_3-\overset{\overset{\displaystyle O}{\|}}{C}-O^- + HO-CH_3$$

The final products of the reaction are an alcohol and a carboxylate ion.

Soap is made by the hydrolysis of natural fats with sodium hydroxide:

$$CH_2-O-\overset{\overset{\displaystyle O}{\|}}{C}-(CH_2)_{16}CH_3$$

$$CH-O-\overset{\overset{\displaystyle O}{\|}}{C}-(CH_2)_{16}CH_3 \;+\; 3NaOH \longrightarrow$$

$$CH_2-O-\overset{\overset{\displaystyle O}{\|}}{C}-(CH_2)_{16}CH_3$$

tristearin
(a triglyceride)

$$HO-CH_2-\overset{\overset{\displaystyle OH}{|}}{CH}-CH_2-OH \;+\; 3CH_3(CH_2)_{16}-\overset{\overset{\displaystyle O}{\|}}{C}-O^- \;\; Na^+$$

glycerol   sodium stearate

This reaction involves stepwise hydrolysis of three esters, each proceeding via a tetrahedral intermediate. It is not known which of the three esters groups hydrolyzes first.

PROBLEM 10-6   Draw the structures of all mono-, di-, and triglycerides of palmitic acid.

TRANSESTER-   The mechanism described in the preceding section for reaction of esters
IFICATION   with hydroxide ion can also be applied to the reaction of esters with alkoxide ion. By this process an ester is converted into another ester having the same acid fragment but a different alcohol fragment

$$CH_3CH_2CH_2-\overset{\overset{\displaystyle O}{\|}}{C}-O-CH_2CH_2CH_3 \;+\; Na^+ \;\; {}^-O-CH_2CH_3 \rightleftharpoons$$

n-propyl butyrate   sodium ethoxide
(excess)

$$CH_3CH_2CH_2-\overset{\overset{\displaystyle O}{\|}}{C}-O-CH_2CH_3 \;+\; Na^+ \;\; {}^-O-CH_2CH_2CH_3$$

ethyl butyrate   sodium n-propoxide

The reaction is reversible, but it can be carried to completion by use of a large excess of sodium ethoxide. This *transesterification* can also occur by an acid-catalyzed pathway.

Show how to make *n*-hexyl acetate from *n*-hexanol and ethyl acetate   ***PROBLEM 10-7***
by transesterification.

Acetylcholine is involved in the transmission of nerve impulses in   *ACETYLCHOLINE*
animals

$$CH_3-\overset{\overset{\displaystyle O}{\|}}{C}-O-CH_2-CH_2-\overset{\overset{\displaystyle CH_3}{|}}{\underset{\underset{\displaystyle CH_3}{|}}{\overset{+}{N}}}-CH_3$$

acetylcholine

Communication from a nerve cell to an adjacent muscle is accomplished by the release of acetylcholine by the nerve cell. Receptors in muscle detect the presence of acetylcholine and trigger a sequence of events leading to muscle contraction. Before this sequence can be repeated, the liberated acetylcholine must be destroyed. This destruction is the role of the enzyme *acetylcholinesterase,* which hydrolyzes acetylcholine to choline and acetate. The mechanism of action of this enzyme involves first a transesterification and then an ester hydrolysis (figure 10-5).

Like esters, amides react with hydroxide ion to form acids   *AMIDE*
                                                               HYDROLYSIS

$$CH_3CH_2-\overset{\overset{\displaystyle O}{\|}}{C}-NH_2 + OH^- \longrightarrow CH_3CH_2-\overset{\overset{\displaystyle O}{\|}}{C}-O^- + NH_3$$

propionamide   propionate

As with esters, the mechanism involves nucleophilic attack of hydroxide ion on the carbonyl carbon atom and formation of a tetrahedral intermediate

$$\left[CH_3CH_2-\overset{\overset{\displaystyle O^-}{|}}{\underset{\underset{\displaystyle OH}{|}}{C}}-NH_2 \right]$$

tetrahedral
intermediate

After a proton transfer, ammonia is lost, and a carboxylate ion is formed.

This reaction is very common. For example, hippuric acid, which is found in urine following ingestion of sodium benzoate, hydrolyzes

248

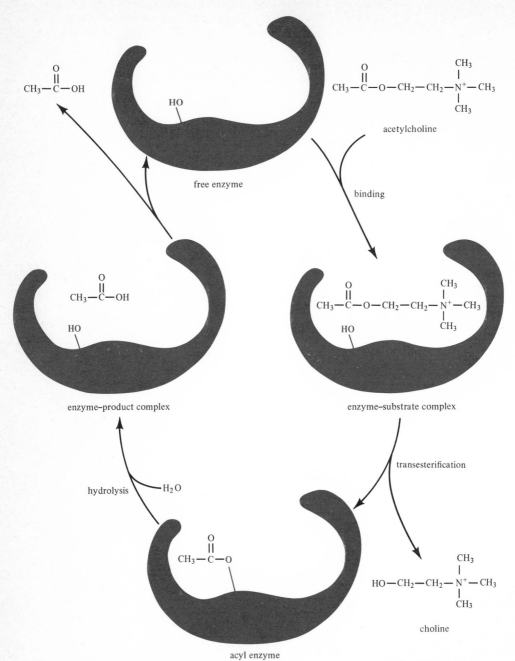

Figure 10-5
Action of the enzyme acetylcholinesterase.

Binding of acetylcholine to the enzyme involves recognition by the enzyme of the quaternary ammonium group and the ester group. Following binding, a transesterification occurs; a hydroxyl group of the enzyme acts as a nucleophile and attacks the carbonyl carbon of acetylcholine. Following release of choline, hydrolysis of the acyl enzyme ester occurs. Tetrahedral intermediates are presumably involved in both these reactions, but this has not been demonstrated.

readily, forming sodium benzoate and glycine

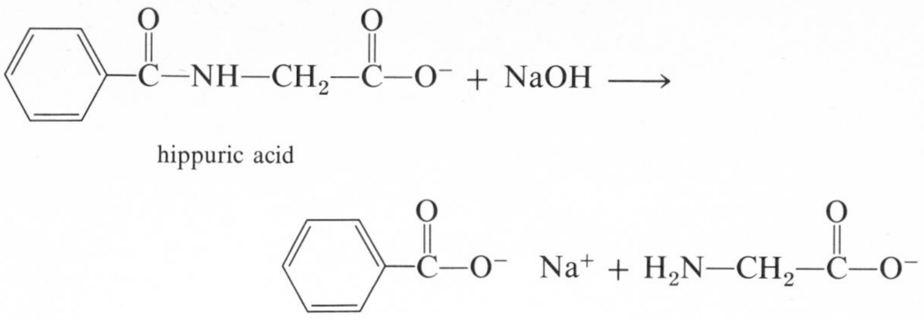

hippuric acid

sodium benzoate   glycine

The mechanism of amide hydrolysis differs from that of ester hydrolysis in one important respect: proton transfer to nitrogen occurs *before* decomposition of the tetrahedral intermediate. As a result, the leaving group is NH_3, rather than NH_2^-. Write the mechanism of the reaction of propionamide with hydroxide. **PROBLEM 10-8**

Hydroxide ion is not the only nucleophile capable of attacking the carbonyl carbon atom of esters and other carboxylic acid derivatives. Ammonia and primary and secondary amines react readily with esters, anhydrides, and acid chlorides **REACTION OF ESTERS WITH AMINES**

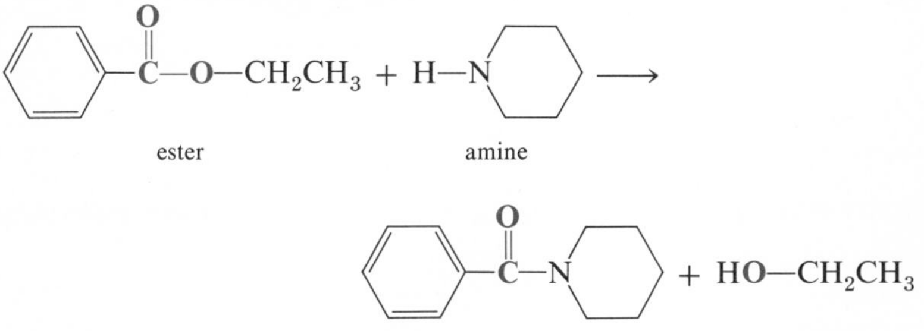

ester   amine

amide   alcohol

The mechanism of this reaction is shown in mechanism 10-2. The similarity to mechanism 10-1 is obvious.

Ammonia also reacts readily with esters

$$H-\overset{O}{\overset{\|}{C}}-O-CH_3 + NH_3 \longrightarrow H-\overset{O}{\overset{\|}{C}}-NH_2 + HO-CH_3$$

methyl   formamide   methanol
formate

The acyl enzyme intermediate formed during the hydrolysis of acetylcholine (figure 10-5) reacts readily with hydroxylamine, NH_2OH. Give the product of that reaction. **PROBLEM 10-9**

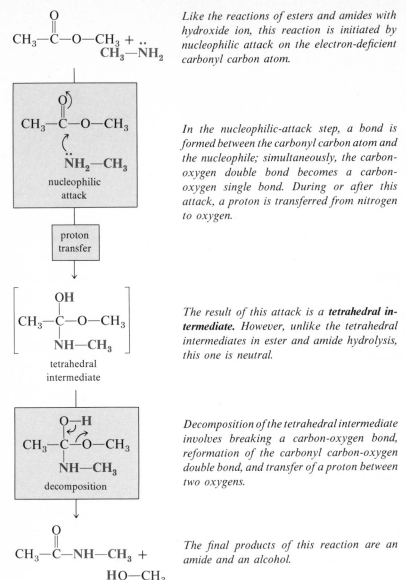

Mechanism 10-2
Ester Aminolysis

Like the reactions of esters and amides with hydroxide ion, this reaction is initiated by nucleophilic attack on the electron-deficient carbonyl carbon atom.

In the nucleophilic-attack step, a bond is formed between the carbonyl carbon atom and the nucleophile; simultaneously, the carbon-oxygen double bond becomes a carbon-oxygen single bond. During or after this attack, a proton is transferred from nitrogen to oxygen.

The result of this attack is a **tetrahedral intermediate.** However, unlike the tetrahedral intermediates in ester and amide hydrolysis, this one is neutral.

Decomposition of the tetrahedral intermediate involves breaking a carbon-oxygen bond, reformation of the carbonyl carbon-oxygen double bond, and transfer of a proton between two oxygens.

The final products of this reaction are an amide and an alcohol.

THIOL ESTERS   A thiol ester is an ester which has a sulfur atom in place of the singly bonded oxygen. Thus, thiol esters are derived from carboxylic acids and thiols. The reactivity of thiol esters is similar to that of oxygen esters

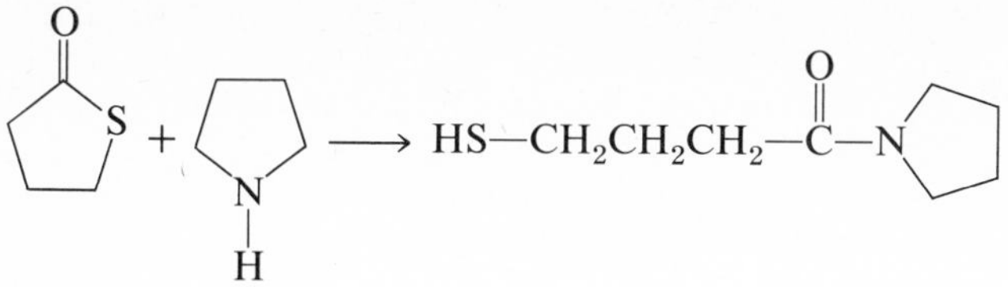

Presumably these reactions, too, go by way of tetrahedral intermediates.

Give the mechanism of the last reaction above.   **PROBLEM 10-10**

The reactions discussed in the preceding sections and many of those to be discussed in subsequent sections are frequently called *acyl transfer reactions,* reactions in which an acyl group (the carbonyl group and the carbon substituent attached to it) is transferred from one group to another

ACYL-GROUP TRANSFER FROM ACETYL COENZYME A

$$R-\overset{O}{\underset{\|}{C}}-A + B \longrightarrow R-\overset{O}{\underset{\|}{C}}-B + A$$

Coenzyme A, a very complex thiol which occurs in all living organisms, forms a number of important thiol esters

coenzyme A

A particularly important class of acyl transfer reactions in biochemistry is the group of transfers from these esters. Some examples involving the most common of these esters, acetyl coenzyme A, are shown in figure 10-6. The mechanisms of these acetyl transfer reactions have not been investigated in detail, but it seems safe to assume that these reactions, too, involve nucleophilic attack on the carbonyl carbon of the ester, followed by formation and decomposition of a tetrahedral intermediate.

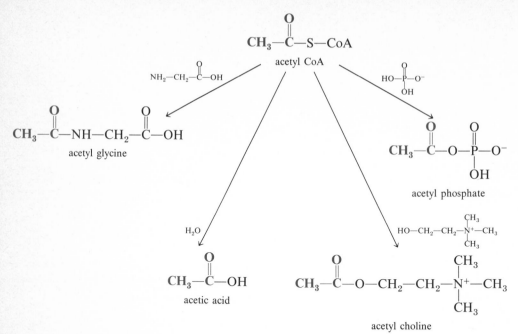

Figure 10-6
Acyl transfer
reactions involving
acetyl coenzyme A.

All these reactions presumably involve nucleophilic attack on the carbonyl carbon atom of acetyl coenzyme A. Each reaction is catalyzed by a different enzyme.

REACTION OF ACID CHLORIDES AND ANHYDRIDES WITH NUCLEOPHILES

Acid chlorides and anhydrides are like esters and amides in that they react with nucleophiles via a mechanism involving a tetrahedral intermediate. However, they are unlike esters and amides in that they react readily with even such poor nucleophiles as water and alcohols, which ordinarily react with amides or esters only in the presence of an acid catalyst (section 10-8).

Benzoyl chloride, for example, reacts rapidly with water in neutral solution,

benzoyl chloride   benzoic acid

presumably via a tetrahedral intermediate*

*The evidence for the occurrence of this intermediate is more equivocal in this case than with esters and amides. In addition, it is clear that in polar solvents (pure water, for example) a completely different mechanism occurs, one involving ionization of the acid chloride to $R—C^+{=}O + Cl^-$, analogous to the carbonium-ion-forming reactions of alkyl halides.

$$\left[\begin{array}{c} \overset{O^-}{\underset{OH}{\overset{|}{\underset{|}{C}}}}-Cl \end{array} \right]$$

tetrahedral
intermediate

The hydrolysis of anhydrides is similar to that of acid chlorides, except that two molecules of carboxylic acid are produced

$$CH_3CH_2-\overset{O}{\overset{\|}{C}}-O-\overset{O}{\overset{\|}{C}}-CH_2CH_3 + H_2O \longrightarrow CH_3CH_2-\overset{O}{\overset{\|}{C}}-OH + HO-\overset{O}{\overset{\|}{C}}-CH_2CH_3$$

propionic anhydride   propionic acid   propionic acid

Acid chlorides and anhydrides react with alcohols to form esters and with amines to form amides

$$Cl-\overset{O}{\overset{\|}{C}}-CH_2-\overset{O}{\overset{\|}{C}}-Cl + 2CH_3-\overset{CH_3}{\underset{CH_3}{\overset{|}{\underset{|}{C}}}}-OH \longrightarrow$$

$$CH_3-\overset{CH_3}{\underset{CH_3}{\overset{|}{\underset{|}{C}}}}-O-\overset{O}{\overset{\|}{C}}-CH_2-\overset{O}{\overset{\|}{C}}-O-\overset{CH_3}{\underset{CH_3}{\overset{|}{\underset{|}{C}}}}-CH_3 + 2HCl$$

$$\text{⬡}-CH_2-\overset{O}{\overset{\|}{C}}-Cl + CH_3CH_2CH_2CH_2-NH_2 \longrightarrow$$

$$\text{⬡}-CH_2-\overset{O}{\overset{\|}{C}}-NH-CH_2CH_2CH_2CH_3 + HCl$$

$$CH_3-\overset{O}{\overset{\|}{C}}-O-\overset{O}{\overset{\|}{C}}-CH_3 + CH_3-\underset{NH_2}{\overset{|}{CH}}-\overset{O}{\overset{\|}{C}}-OH \longrightarrow$$

$$CH_3-\overset{O}{\overset{\|}{C}}-NH-\underset{}{\overset{CH_3}{\overset{|}{CH}}}-\overset{O}{\overset{\|}{C}}-OH + CH_3-\overset{O}{\overset{\|}{C}}-OH$$

$$\underset{CH_2}{\overset{CH_2}{}} \begin{array}{c} \overset{O}{\overset{\|}{C}} \\ \diagdown \\ O \\ \diagup \\ \overset{}{\underset{\|}{C}} \\ O \end{array} + CH_3CH_2-OH \longrightarrow HO-\overset{O}{\overset{\|}{C}}-CH_2CH_2-\overset{O}{\overset{\|}{C}}-O-CH_2CH_3$$

PROBLEM 10-11   If a particularly expensive carboxylic acid is to be converted into an amide, it is advantageous to prepare the acid chloride and react this compound with the appropriate amine, rather than prepare the anhydride and react it with the amine. Why?

HYDROLYSIS OF NITRILES   Although nitriles do not have a carbonyl group, they are considered to be derivatives of carboxylic acids because hydrolysis produces a carboxylic acid. However, in this case, the first hydrolysis product is an amide

$$C \equiv N \quad + H_2O \xrightarrow{\text{OH}^-} \quad \overset{\displaystyle O}{\overset{\|}{C}}-NH_2$$

benzonitrile   benzamide

The amide can be isolated if the reaction is conducted carefully, but in the presence of excess hydroxide or high temperature, the amide is hydrolyzed to a carboxylic acid*

$$\overset{\displaystyle O}{\overset{\|}{C}}-NH_2 \quad + OH^- \longrightarrow \quad \overset{\displaystyle O}{\overset{\|}{C}}-O^- \quad + NH_3$$

benzamide   benzoate

The mechanism of nitrile hydrolysis is given in mechanism 10-3. The first step in the reaction is nucleophilic attack by hydroxide ion on the carbon atom of the nitrile. Following a series of proton transfers, an amide is produced.

RELATIVE REACTIVITIES   All types of carboxylic acid derivatives react with nucleophiles, but the reaction rates vary over a wide range. Acid chlorides and anhydrides are the most reactive; they react rapidly at room temperature with even such poor nucleophiles as water and alcohols. Amides, on the other hand, are the least reactive; they do not react with water or alcohols even on long standing at 100°C. In the presence of a good nucleophile such as hydroxide ion or an alkoxide ion, an amide will react at a moderate rate.

---

*Note that hydroxide serves as a catalyst for hydrolysis of the nitrile, but is consumed in the hydrolysis of the amide.

$CH_3-C\equiv N\ +$

$OH^-\ +\ H_2O$

The carbon-nitrogen triple bond is polarized much as carbon-oxygen double bonds are polarized: the carbon is electron-poor, and the nitrogen is electron-rich. As a result, nucleophilic attack on carbon can occur.

$CH_3-C\equiv N$

^-OH

nucleophilic
attack

In the first step, the attacking nucleophile forms a bond to carbon, and the triple bond is converted into a double bond.

$$\left[CH_3-\underset{OH}{C}=N^- \right]$$

intermediate

The intermediate formed as a result of this step is very unstable because it has a negative charge on nitrogen. It immediately decomposes by transfer of two protons, one from the oxygen in the intermediate and one from the water present.

two
proton
transfers

$$CH_3-\overset{O}{\underset{H}{\underset{|}{\overset{||}{C}}}}-N-H\ +\ OH^-$$

The product formed after two proton transfers is an amide. This amide is itself susceptible to nucleophilic attack by the hydroxide present, but reaction of the amide with hydroxide is slower than reaction of the original nitrile, and the amide can frequently be isolated.

The relative reactivities of important carboxylic acid derivatives and of important nucleophiles are summarized in figure 10-7. The rate of reaction of a particular type of compound with a particular nucleophile is a function of how near the top of the chart the two are: the nearer the top either component is, the faster the reaction. Reactions of compounds near the bottom with nucleophiles near the bottom are slow or nonexistent.

10-6 REACTIONS OF GRIGNARD REAGENTS WITH ESTERS

In chapter 7 we discussed the reaction of Grignard reagents with aldehydes and ketones. Esters also react with Grignard reagents, but

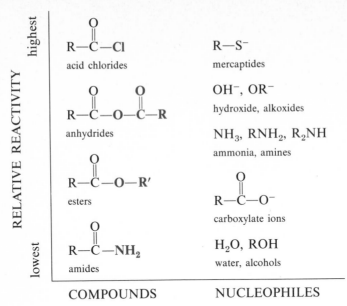

Figure 10-7
Relative reactivities.
The higher a
compound or ion is
on this chart, the
more reactive it is.

RELATIVE REACTIVITY

highest

lowest

O
‖
R—C—Cl

acid chlorides

R—S⁻

mercaptides

O   O
‖   ‖
R—C—O—C—R

anhydrides

OH⁻, OR⁻

hydroxide, alkoxides

NH₃, RNH₂, R₂NH

ammonia, amines

O
‖
R—C—O—R′

esters

O
‖
R—C—O⁻

carboxylate ions

O
‖
R—C—NH₂

amides

H₂O, ROH

water, alcohols

COMPOUNDS   NUCLEOPHILES

each molecule of ester consumes two molecules of Grignard reagent

The mechanism of this reaction is shown in mechanism 10-4. The reaction involves nucleophilic attack on the carbonyl carbon atom of the ester by the Grignard reagent, expulsion of alkoxide to form a ketone, and reaction of the ketone with a second molecule of Grignard reagent to form an alcohol.

When esters of formic acid are used, the product is a secondary alcohol

$$\text{C}_6\text{H}_{11}-\text{Cl} + \text{Mg} \xrightarrow{\text{ether}} \text{C}_6\text{H}_{11}-\text{MgCl}$$

$$2\ \text{C}_6\text{H}_{11}-\text{MgCl} + \overset{\displaystyle O}{\overset{\|}{\text{HC}}}-\text{O}-\text{CH}_2\text{CH}_3 \xrightarrow[\text{2. add } H_2O]{\text{1. reaction}}$$

$$\text{C}_6\text{H}_{11}-\overset{\displaystyle \text{OH}}{\overset{|}{\text{CH}}}-\text{C}_6\text{H}_{11} + 2\text{MgClOH} + \text{HO}-\text{CH}_2\text{CH}_3$$

Show how you would use a Grignard reaction with an ester to form   **PROBLEM 10-12**
4-methyl-4-heptanol.

10-7 REDUCTION

Lithium aluminum hydride readily reduces esters and other carboxylic acid derivatives

$$\text{CH}_3\text{CH}_2\text{CH}_2\text{CH}_2-\overset{\displaystyle O}{\overset{\|}{\text{C}}}-\text{OCH}_2\text{CH}_3 + \text{LiAlH}_4 \xrightarrow[\text{solvent}]{\text{ether}} \xrightarrow[H_2O]{\text{then add}}$$

$$\text{CH}_3\text{CH}_2\text{CH}_2\text{CH}_2-\overset{\displaystyle \text{H}}{\underset{\displaystyle \text{H}}{\overset{|}{\underset{|}{\text{C}}}}}-\text{OH} + \text{CH}_3\text{CH}_2\text{OH} + \text{LiOH} + \text{Al(OH)}_3$$

This reaction occurs because of the propensity of the hydrogens of lithium aluminum hydride to act as nucleophiles (section 7-7). The reaction mechanism is similar to that of the Grignard reaction (mechanism 10-4), except that hydride, H⁻, rather than carbon, is the nucleophile. Attack by hydride presumably forms a tetrahedral intermediate, which decomposes to form an aldehyde. This aldehyde then reacts with

$$\text{CH}_3\text{CH}_2\text{CH}_2\text{CH}_2-\overset{\displaystyle O}{\overset{\|}{\text{C}}}-\text{H}$$
    aldehyde intermediate

a second hydride (section 7-7) and (after hydrolysis) forms a primary alcohol. Sodium borohydride will not reduce carboxylic acid derivatives.

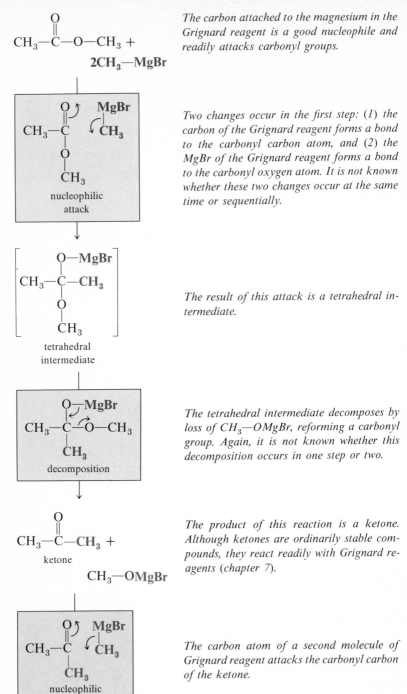

Mechanism 10-4
Reaction of
Grignard Reagents
with Esters

$$CH_3-\overset{\overset{\displaystyle O}{\|}}{C}-O-CH_3 +$$

$$2CH_3-MgBr$$

The carbon attached to the magnesium in the Grignard reagent is a good nucleophile and readily attacks carbonyl groups.

nucleophilic
attack

Two changes occur in the first step: (1) the carbon of the Grignard reagent forms a bond to the carbonyl carbon atom, and (2) the MgBr of the Grignard reagent forms a bond to the carbonyl oxygen atom. It is not known whether these two changes occur at the same time or sequentially.

tetrahedral
intermediate

The result of this attack is a tetrahedral intermediate.

decomposition

The tetrahedral intermediate decomposes by loss of CH_3—$OMgBr$, reforming a carbonyl group. Again, it is not known whether this decomposition occurs in one step or two.

$$CH_3-\overset{\overset{\displaystyle O}{\|}}{C}-CH_3 +$$

ketone

$$CH_3-OMgBr$$

The product of this reaction is a ketone. Although ketones are ordinarily stable compounds, they react readily with Grignard reagents (*chapter 7*).

nucleophilic
attack

The carbon atom of a second molecule of Grignard reagent attacks the carbonyl carbon of the ketone.

$$\left[\begin{array}{c} \text{O--MgBr} \\ | \\ \text{CH}_3\text{--C--CH}_3 \\ | \\ \text{CH}_3 \end{array} \right]$$

The intermediate formed as a result of this attack is stable and remains in solution until water is added.

↓

$\boxed{\text{hydrolysis}}$

Water is not added until all preceding steps are completed.

↓

$$\begin{array}{c} \text{OH} \\ | \\ \text{CH}_3\text{--C--CH}_3 \; + \\ | \\ \text{CH}_3 \end{array}$$

$2\text{MgBr}\text{OH} + \text{CH}_3\text{--OH}$

Addition of water produces a tertiary alcohol.

Amides are reduced to amines by a slightly different mechanism

$$\text{CH}_3\text{CH}_2\text{CH}_2\text{CH}_2\text{CH}_2\text{CH}_2\text{CH}_2\overset{\displaystyle O}{\overset{\|}{\text{--C}}}\text{--NHCH}_3 + \text{LiAlH}_4 \xrightarrow[\text{solvent}]{\text{ether}} \xrightarrow{\text{then add} \atop \text{H}_2\text{O}}$$

$$\text{CH}_3\text{CH}_2\text{CH}_2\text{CH}_2\text{CH}_2\text{CH}_2\text{CH}_2\overset{\displaystyle H}{\underset{\displaystyle H}{\overset{|}{\underset{|}{\text{--C}}}}}\text{--NHCH}_3 + \text{LiOH} + \text{Al(OH)}_3$$

Nitriles can also be reduced with lithium aluminum hydride. Give the product of reduction of butyronitrile with lithium aluminum hydride.

PROBLEM 10-13

10-8 ACID-CATALYZED NUCLEOPHILIC ATTACK

In chapter 7 we showed that the ability of poor nucleophiles such as water and alcohols to attack the carbonyl carbon atom of aldehydes and ketones is very much enhanced by protonation of the carbonyl oxygen atom. The same is true of esters and amides (this additional help is usually unnecessary for acid chlorides and anhydrides).

In aqueous acid, esters are hydrolyzed to alcohols and carboxylic acids

ACID-CATALYZED ESTER HYDROLYSIS

$$\overset{\displaystyle O}{\overset{\|}{\text{CH}_3\text{--C}}}\text{--O--CH}_2\text{CH}_2\text{CH}_3 + \text{H}_2\text{O} \xrightarrow{\text{H}^+} \overset{\displaystyle O}{\overset{\|}{\text{CH}_3\text{--C}}}\text{--OH} + \text{HO--CH}_2\text{CH}_2\text{CH}_3$$

ester   (excess)   acid   alcohol

Mechanism 10-5
Acid-catalyzed
Ester Hydrolysis

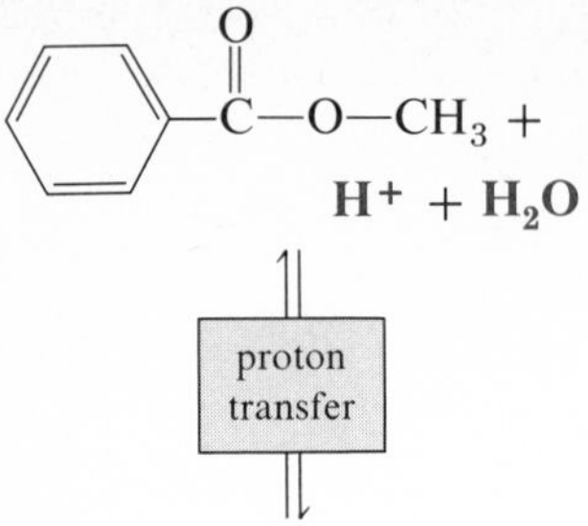

The acid-catalyzed reaction is initiated by protonation of the carbonyl oxygen atom, as with aldehydes and ketones. The protonation is readily reversible, and the equilibrium lies very far on the side of ester plus proton, even though the protonated ester is resonance-stabilized. However, the protonated ester is sufficiently reactive to make up for its low concentration.

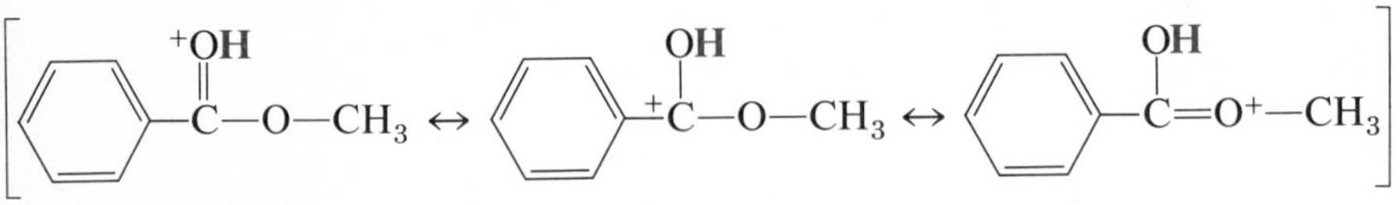

resonance-stabilized protonated ester

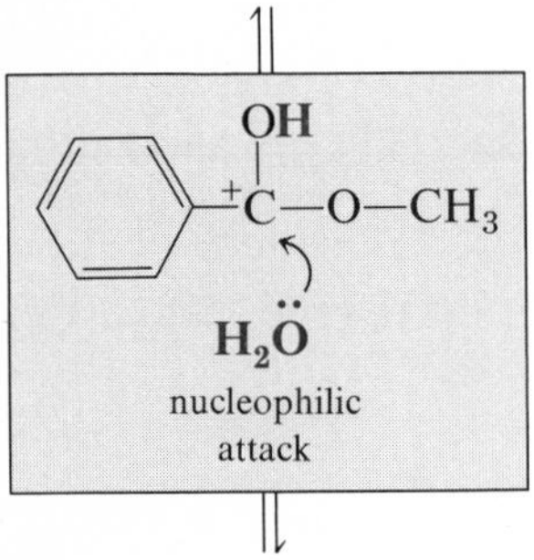

nucleophilic attack

Protonation of the carbonyl group greatly increases its reactivity toward nucleophiles. In this case water attacks the carbonyl carbon.

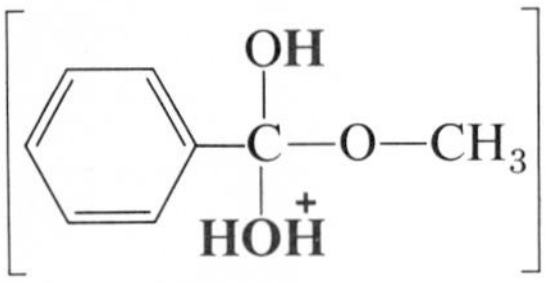

tetrahedral intermediate

The tetrahedral intermediate produced is like the one in the reaction of esters with hydroxide, except that it has two additional protons.

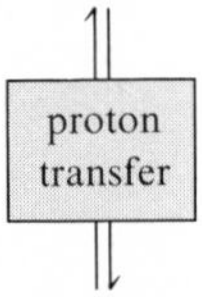

proton transfer

Proton transfer among the three oxygen atoms of the tetrahedral intermediate occurs rapidly.

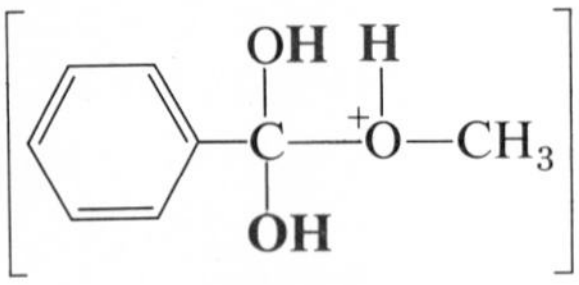

tetrahedral intermediate

This form of the tetrahedral intermediate can decompose to form acid and alcohol.

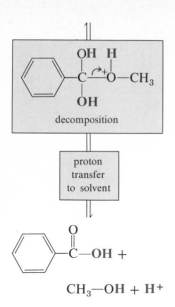

Decomposition of the tetrahedral intermediate occurs in two steps: (1) the carbon-oxygen bond is broken, and an alcohol and the conjugate acid of the carboxylic acid product are produced; (2) the extra proton is lost from the carboxylic acid, and the final products are produced.

The acid, H^+, is written above the arrow because, strictly speaking, it acts as a *catalyst* and is not consumed; nonetheless, it plays an essential role in the reaction, as shown in mechanism 10-5.

This reaction, like the corresponding reaction in alkaline solution (section 8-5), involves a tetrahedral intermediate.

Thiol esters can be hydrolyzed by this same mechanism

$$HC—S—CH_2CH_3 + H_2O \xrightarrow{H^+} HC—OH + HS—CH_2CH_3$$

(excess)

Give the mechanism of the above reaction.   ***PROBLEM 10-14***

Unlike the hydrolysis of esters in alkaline solution, the hydrolysis of esters in acidic solution is *reversible;* that is, the carboxylic acid and alcohol are in equilibrium with ester and water. The equilibrium can be shifted to one side or the other by appropriate choice of conditions: reaction of an ester with a large excess of water in the presence of acid results in the formation of a carboxylic acid; reaction of a carboxylic acid with a large excess of alcohol in the presence of acid results in formation of an ester. This is a useful method of *esterification* (ester formation), particularly for compounds which contain hydroxyl, amino, or other nucleophilic groups and are therefore incapable of forming

ACID-CATALYZED ESTER FORMATION

stable acid chlorides or anhydrides

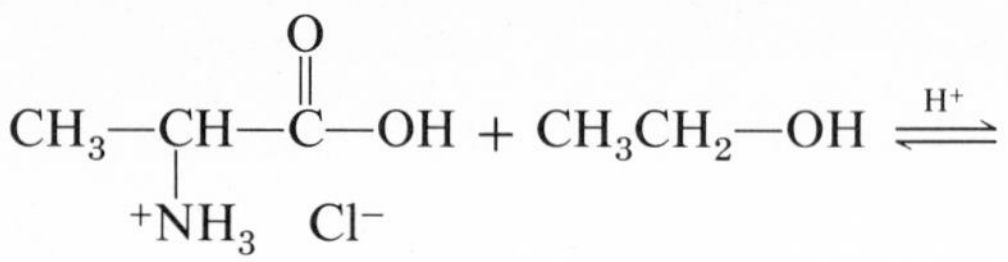

$$
\underset{\substack{\text{alanine} \\ \text{hydrochloride}}}{CH_3-\underset{\substack{| \\ +NH_3 \quad Cl^-}}{CH}-\overset{\overset{O}{\|}}{C}-OH} + \underset{\substack{\text{ethanol} \\ \text{(excess)}}}{CH_3CH_2-OH} \overset{H^+}{\rightleftharpoons}
$$

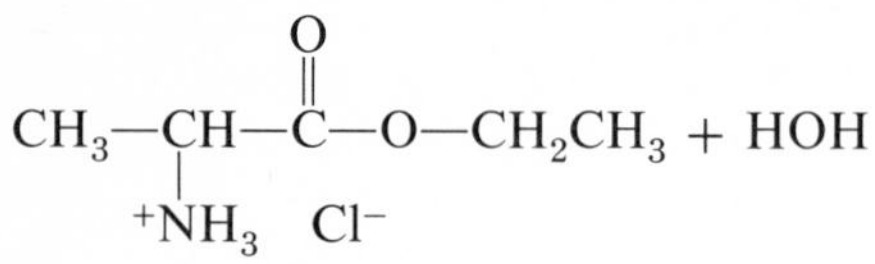

$$
\underset{\substack{\text{alanine ethyl ester} \\ \text{hydrochloride}}}{CH_3-\underset{\substack{| \\ +NH_3 \quad Cl^-}}{CH}-\overset{\overset{O}{\|}}{C}-O-CH_2CH_3} + HOH
$$

This method of esterification is most successful when the water produced is removed as it is formed.

The mechanism of this reaction is precisely the same as mechanism 10-5 taken in reverse.

PROBLEM 10-15   The above reaction can also be used for transesterification, the interconversion of the alcohol groups of esters. Give a mechanism for the reaction

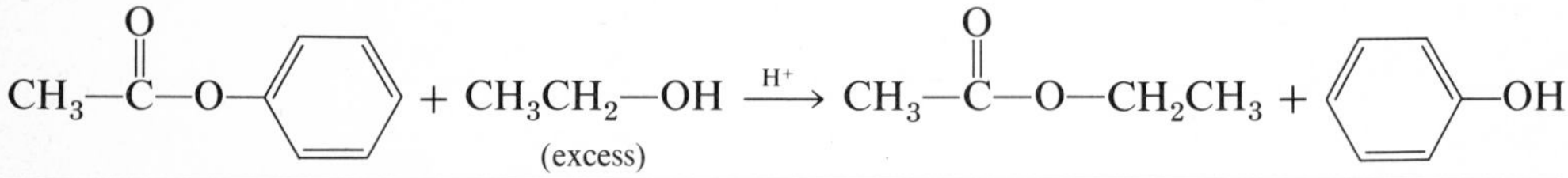

$$
CH_3-\overset{\overset{O}{\|}}{C}-O-\bigcirc + \underset{\text{(excess)}}{CH_3CH_2-OH} \overset{H^+}{\longrightarrow} CH_3-\overset{\overset{O}{\|}}{C}-O-CH_2CH_3 + \bigcirc-OH
$$

ACID-CATALYZED AMIDE HYDROLYSIS   The acid-catalyzed hydrolysis of amides also takes place readily:

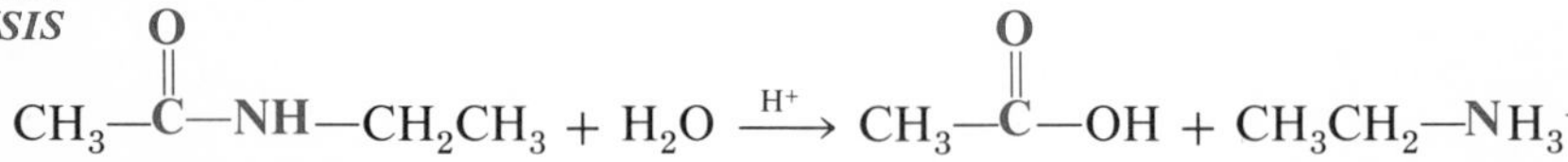

$$
\underset{\text{N-ethyl acetamide}}{CH_3-\overset{\overset{O}{\|}}{C}-NH-CH_2CH_3} + H_2O \overset{H^+}{\longrightarrow} \underset{\text{acetic acid}}{CH_3-\overset{\overset{O}{\|}}{C}-OH} + \underset{\substack{\text{ethylammonium} \\ \text{ion}}}{CH_3CH_2-NH_3^+}
$$

The mechanism of this reaction is the same as that of acid-catalyzed ester hydrolysis. The tetrahedral intermediate which is formed is similar to previous ones:

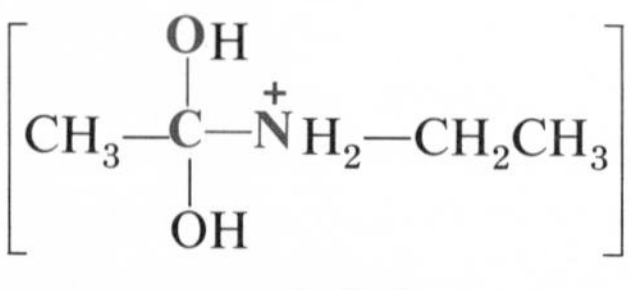

$$
\left[CH_3-\underset{\underset{OH}{|}}{\overset{\overset{OH}{|}}{C}}-\overset{+}{N}H_2-CH_2CH_3 \right]
$$

tetrahedral
intermediate

However, this tetrahedral intermediate exists primarily in the form shown, in which the extra proton is on the amide nitrogen, because the nitrogen atom is more basic than either of the oxygen atoms.

Unlike the acid-catalyzed hydrolysis of esters, the acid-catalyzed hydrolysis of amides is not reversible. Under the conditions of high acid concentration used in this reaction, the product amine is protonated and thus rendered unreactive as soon as it is formed.

Give the mechanism of the following reaction: **PROBLEM 10-16**

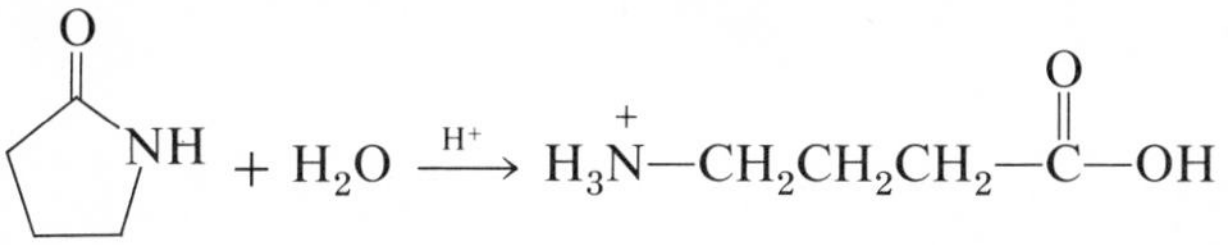

$$\underset{\text{(cyclopentanone-NH)}}{\bigcirc\!\!\!\!=\!\!O\ \ \text{NH}} + H_2O \xrightarrow{H^+} H_3\overset{+}{N}-CH_2CH_2CH_2-\overset{O}{\overset{\|}{C}}-OH$$

Suggest a reason why acid-catalyzed aminolysis of esters does not occur. **PROBLEM 10-17**

10-9 ENOLIZATION

Like aldehydes and ketones, esters undergo *enolization,* the removal of an α hydrogen by base and formation of a resonance-stabilized enolate anion

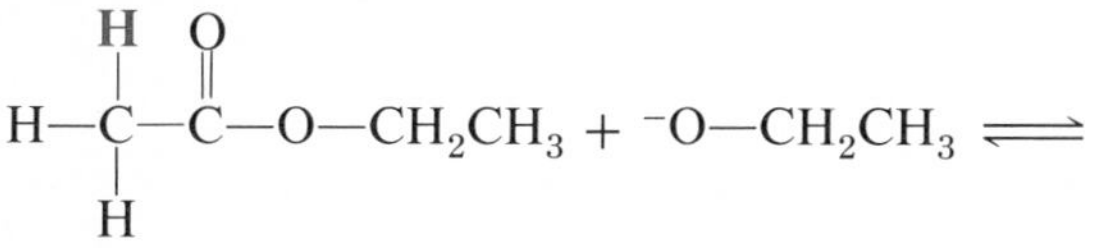

$$\underset{\text{ester}}{H-\overset{H}{\underset{H}{\overset{|}{\underset{|}{C}}}}-\overset{O}{\overset{\|}{C}}-O-CH_2CH_3} + \underset{\text{base}}{^-O-CH_2CH_3} \rightleftharpoons$$

$$\left[\underset{}{H_2^-C-\overset{O}{\overset{\|}{C}}-O-CH_2CH_3} \longleftrightarrow H_2C=\overset{O^-}{\overset{|}{C}}-O-CH_2CH_3\right] + HO-CH_2CH_3$$

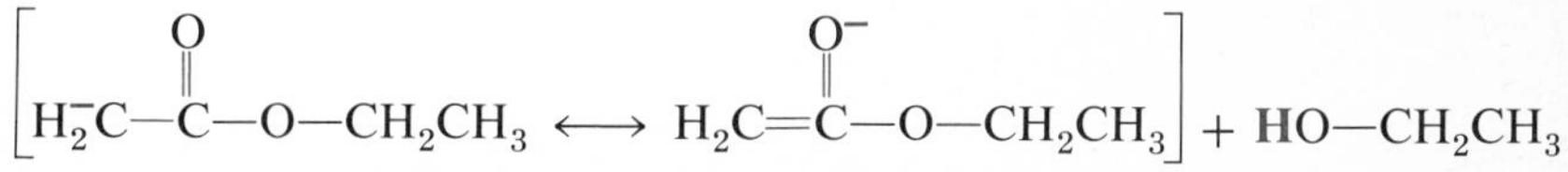

resonance-stabilized enolate anion

In order to prevent hydrolysis or transesterification it is necessary to use as the base in this reaction the conjugate anion of the alcohol portion of the ester. If any other base is used, it will in most cases act as a nucleophile and attack the carbonyl carbon atom. If the conjugate anion of the alcohol is used, then the product of nucleophilic attack is identical with the starting material.

As with aldehydes and ketones, the enolate ion is stabilized by *resonance.* The equilibrium between ester and enolate strongly favors the ester; only a fraction of a percent of enolate is present at equilibrium. However, that small amount of enolate is very reactive and may act as a nucleophile in a variety of reactions.

The carbon atom of the enolate can attack the carbonyl carbon of another molecule of ester. The net result of this reaction is formation of a β-ketoester

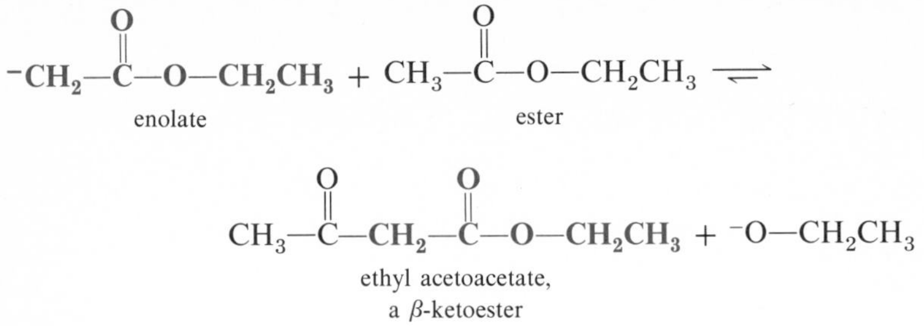

$$^-CH_2-\overset{\overset{\displaystyle O}{\|}}{C}-O-CH_2CH_3 + CH_3-\overset{\overset{\displaystyle O}{\|}}{C}-O-CH_2CH_3 \rightleftharpoons$$

enolate   ester

$$CH_3-\overset{\overset{\displaystyle O}{\|}}{C}-CH_2-\overset{\overset{\displaystyle O}{\|}}{C}-O-CH_2CH_3 + {}^-O-CH_2CH_3$$

ethyl acetoacetate,
a β-ketoester

The mechanism of this reaction is shown in mechanism 10-6 (page 266).

Cyclic ketoesters are readily formed from diesters in which the two ester groups are separated by an appropriate number of carbon atoms

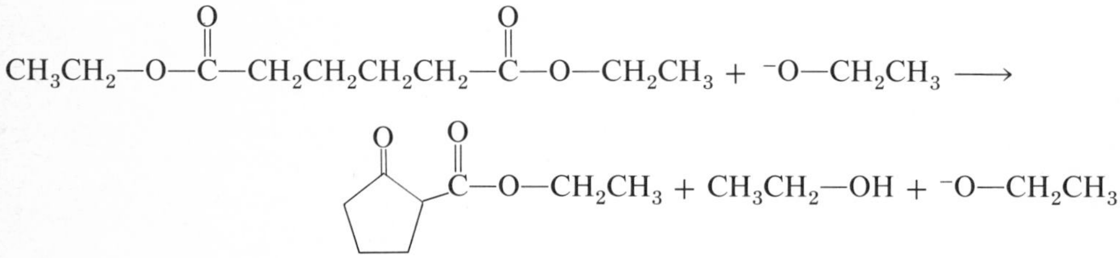

$$CH_3CH_2-O-\overset{\overset{\displaystyle O}{\|}}{C}-CH_2CH_2CH_2CH_2-\overset{\overset{\displaystyle O}{\|}}{C}-O-CH_2CH_3 + {}^-O-CH_2CH_3 \longrightarrow$$

$$\overset{\overset{\displaystyle O}{\|}}{C}-O-CH_2CH_3 + CH_3CH_2-OH + {}^-O-CH_2CH_3$$

PROBLEM 10-18 Suggest a reason why carboxylic acids themselves do not undergo enolate formation.

PROBLEM 10-19 Give the mechanism of the above reaction.

FATTY ACID SYNTHESIS The key step in the synthesis of long-chain fatty acids from two-carbon precursors in living systems is a condensation reaction between an ester and an enolate like those shown above. The alcohol portion of both the ester and the enolate is *acyl carrier protein* (ACP), a sulfhydryl-containing protein with a molecular weight of about 10,000. After the condensation step, subsequent steps reduce the ketoester product to a new longer-chain ester, which then participates in another condensation reaction. The entire sequence of steps (figure 10-8) is repeated several times until a fatty acid ester of the desired length is produced, at which point the thiol ester is hydrolyzed.

The lengths of chains synthesized in this manner are usually not precisely controlled, and most lipids contain a variety of fatty acids of different chain lengths. However, fatty acids produced by this scheme always have an even number of carbon atoms except in the

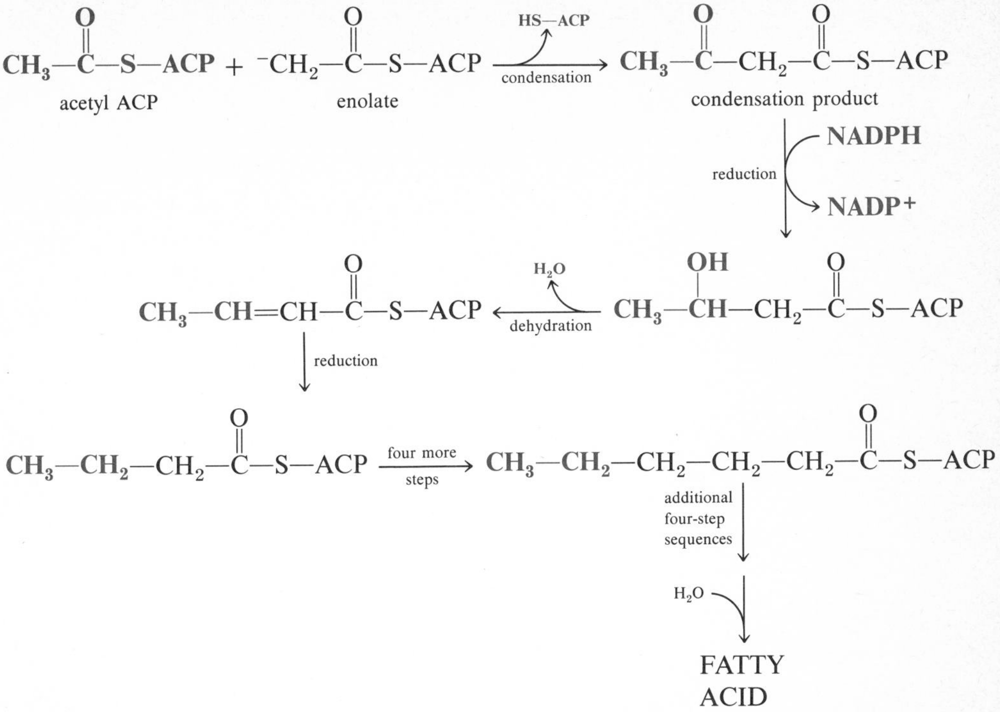

$$CH_3-\overset{\overset{\displaystyle O}{\|}}{C}-S-ACP + {}^-CH_2-\overset{\overset{\displaystyle O}{\|}}{C}-S-ACP \xrightarrow[\text{condensation}]{\text{HS}-\text{ACP}} CH_3-\overset{\overset{\displaystyle O}{\|}}{C}-CH_2-\overset{\overset{\displaystyle O}{\|}}{C}-S-ACP$$

acetyl ACP　　　　　　　　enolate　　　　　　　　　　　　condensation product

reduction $\begin{array}{l} \text{NADPH} \\ \text{NADP}^+ \end{array}$

$$CH_3-CH=CH-\overset{\overset{\displaystyle O}{\|}}{C}-S-ACP \xleftarrow[\text{dehydration}]{H_2O} CH_3-\overset{\overset{\displaystyle OH}{|}}{CH}-CH_2-\overset{\overset{\displaystyle O}{\|}}{C}-S-ACP$$

reduction

$$CH_3-CH_2-CH_2-\overset{\overset{\displaystyle O}{\|}}{C}-S-ACP \xrightarrow[\text{steps}]{\text{four more}} CH_3-CH_2-CH_2-CH_2-CH_2-\overset{\overset{\displaystyle O}{\|}}{C}-S-ACP$$

additional
four-step
sequences

H_2O

FATTY
ACID

The starting compound is the thiol ester acetyl-ACP (acyl carrier protein). Condensation of this ester with the enolate of another acetyl-ACP produces a β-ketoester, which is reduced, dehydrated, and then reduced again. The product is a new, longer-chain acyl-ACP. Condensation of this material with

another enolate of acetyl-ACP and reduction, dehydration, and reduction extends the carbon chain by another two carbon atoms. This sequence continues until an ester of the desired size is obtained, at which point, hydrolysis occurs, producing a fatty acid.

*Figure 10-8
Fatty acid synthesis
in nature.*

unusual case where a propionyl-ACP is used instead of acetyl-ACP in the first step.

10-10　CARBONIC ACID AND ITS CONGENERS

Carbon dioxide and urea are among the compounds at oxidation level 4 which are terminal products of animal metabolism. The structures of carbonic acid and some of its derivatives are shown in table 10-1.

Carbon dioxide and carbonic acid differ simply by gain or loss of a molecule of water:

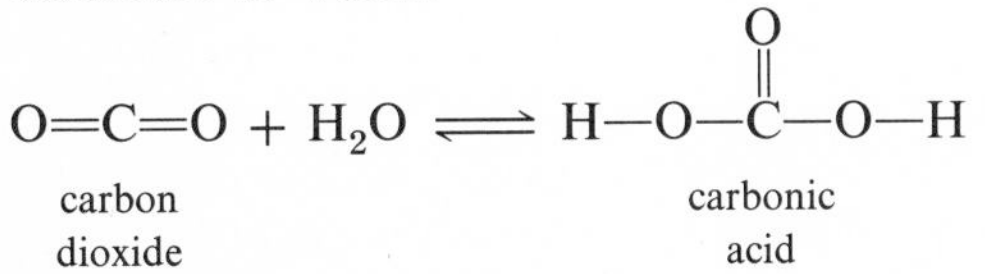

*HYDRATION OF
CARBON DIOXIDE*

$$O=C=O + H_2O \rightleftharpoons H-O-\overset{\overset{\displaystyle O}{\|}}{C}-O-H$$

carbon　　　　　　carbonic
dioxide　　　　　　　acid

Mechanism 10-6
Ester Condensation

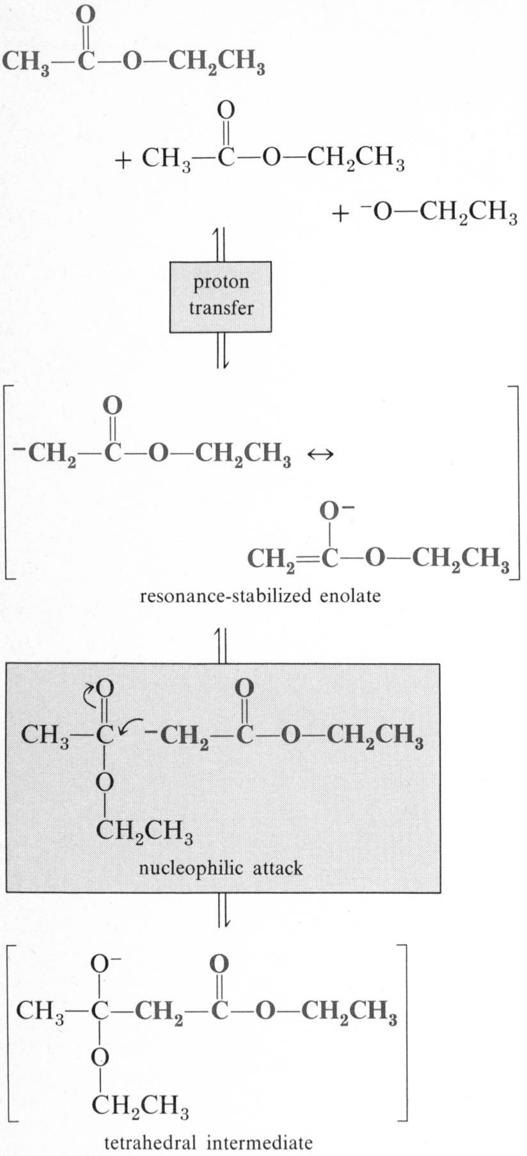

$$CH_3-\overset{\overset{\displaystyle O}{\|}}{C}-O-CH_2CH_3$$

$$+ \ CH_3-\overset{\overset{\displaystyle O}{\|}}{C}-O-CH_2CH_3$$

$$+ \ ^-O-CH_2CH_3$$

The starting materials for this reaction are two molecules of ester and one molecule of alkoxide.

proton
transfer

$$\left[\ ^-CH_2-\overset{\overset{\displaystyle O}{\|}}{C}-O-CH_2CH_3 \ \leftrightarrow \right.$$

$$\left. CH_2=\overset{\overset{\displaystyle O^-}{|}}{C}-O-CH_2CH_3 \right]$$

resonance-stabilized enolate

The first intermediate is the enolate, which is formed by removal of an α proton from the ester.

$$CH_3-\overset{\overset{\displaystyle O}{\|}}{\underset{\underset{\displaystyle CH_2CH_3}{\overset{\displaystyle |}{O}}}{C}}-^-CH_2-\overset{\overset{\displaystyle O}{\|}}{C}-O-CH_2CH_3$$

nucleophilic attack

This enolate is quite nucleophilic and readily attacks the carbonyl carbon atom of the second molecule of ester.

$$\left[CH_3-\overset{\overset{\displaystyle O^-}{|}}{\underset{\underset{\displaystyle CH_2CH_3}{\overset{\displaystyle |}{O}}}{C}}-CH_2-\overset{\overset{\displaystyle O}{\|}}{C}-O-CH_2CH_3 \right]$$

tetrahedral intermediate

The product of this nucleophilic attack is a **tetrahedral intermediate.**

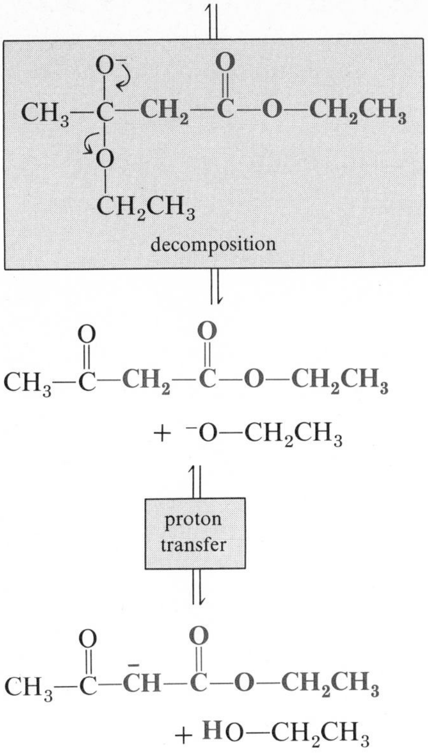

$$CH_3-\overset{\overset{\displaystyle O}{\|}}{\underset{\underset{\displaystyle CH_2CH_3}{|}}{\underset{|}{\overset{|}{O}}}}-CH_2-\overset{\overset{\displaystyle O}{\|}}{C}-O-CH_2CH_3$$

decomposition

The tetrahedral intermediate decomposes by loss of ethoxide.

$$CH_3-\overset{\overset{\displaystyle O}{\|}}{C}-CH_2-\overset{\overset{\displaystyle O}{\|}}{C}-O-CH_2CH_3$$

$$+ \ ^-O-CH_2CH_3$$

The products are a β-ketoester and a molecule of alkoxide.

proton transfer

$$CH_3-\overset{\overset{\displaystyle O}{\|}}{C}-\overset{-}{C}H-\overset{\overset{\displaystyle O}{\|}}{C}-O-CH_2CH_3$$

$$+ \ HO-CH_2CH_3$$

The condensation reaction is reversible, but the equilibrium favors the condensation product because of the formation of this very stable enolate.

Table 10-1
Carbonic Acid and Its Derivatives

Type	Structure	Name
acid	$HO-\overset{\overset{\displaystyle O}{\|}}{C}-OH$	carbonic acid
salt	$2Na^+ \ ^-O-\overset{\overset{\displaystyle O}{\|}}{C}-O^-$	sodium carbonate
anhydride	$O=C=O$	carbon dioxide
ester	$CH_3CH_2-O-\overset{\overset{\displaystyle O}{\|}}{C}-O-CH_2CH_3$	diethyl carbonate
amide	$H_2N-\overset{\overset{\displaystyle O}{\|}}{C}-NH_2$	urea
acid chloride	$Cl-\overset{\overset{\displaystyle O}{\|}}{C}-Cl$	phosgene

The equilibrium between carbon dioxide and carbonic acid favors carbon dioxide, but this equilibrium is shifted away from carbon dioxide in basic solution because of the ionization of carbonic acid

$$\underset{\text{carbonic acid}}{\text{H}-\text{O}-\overset{\overset{\textstyle O}{\|}}{\text{C}}-\text{O}-\text{H}} \rightleftharpoons \text{H}^+ + \underset{\text{bicarbonate}}{{}^-\text{O}-\overset{\overset{\textstyle O}{\|}}{\text{C}}-\text{O}-\text{H}} \rightleftharpoons 2\text{H}^+ + \underset{\text{carbonate}}{{}^-\text{O}-\overset{\overset{\textstyle O}{\|}}{\text{C}}-\text{O}^-}$$

The hydration of carbon dioxide is a very important reaction in the body because of its role in transporting carbon dioxide and in maintaining pH in various body fluids. Thus, it is not surprising that even though the spontaneous hydration of carbon dioxide is rapid, all animals have an enzyme which is capable of catalyzing this reaction. This enzyme, called *carbonic anhydrase,* is an enigma because the reaction it catalyzes is so simple that it is difficult to picture a role for the enzyme in the reaction.

PROBLEM 10-20   Show the resonance forms of carbonate ion and bicarbonate ion.

REACTIONS OF CARBONIC ACID DERIVATIVES   The chemistry of carbonic acid and its derivatives is much like that of carboxylic acids but for the additional complexity arising from the presence of two hydroxyl groups. Some typical reactions of carbonic acid derivatives are shown below.

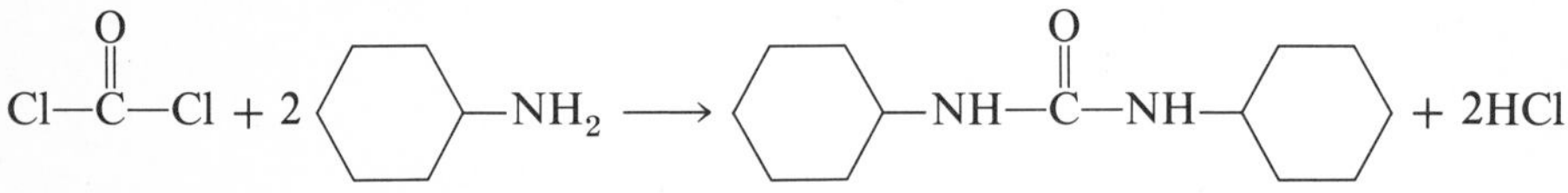

$$\underset{\text{phosgene}}{\text{Cl}-\overset{\overset{\textstyle O}{\|}}{\text{C}}-\text{Cl}} + 2\;\underset{\substack{\text{cyclohexyl}\\\text{amine}}}{\bigcirc-\text{NH}_2} \longrightarrow \underset{\text{dicyclohexyl urea}}{\bigcirc-\text{NH}-\overset{\overset{\textstyle O}{\|}}{\text{C}}-\text{NH}-\bigcirc} + 2\text{HCl}$$

The mechanism of this reaction is just like that of the reaction of ordinary acid chlorides with amines except that it involves *two* cycles of nucleophilic attack and carbon-chlorine bond breaking

$$\underset{\text{diethyl carbonate}}{\text{CH}_3\text{CH}_2-\text{O}-\overset{\overset{\textstyle O}{\|}}{\text{C}}-\text{O}-\text{CH}_2\text{CH}_3} + 2\text{NaOH} \longrightarrow 2\text{CH}_3\text{CH}_2-\text{OH} + \text{Na}_2\text{CO}_3$$

The mechanism of this reaction is like mechanism 10-1 taken twice.

PROBLEM 10-21   Give the mechanism of the above reaction.

Plants and many microorganisms contain an enzyme which is capable *UREASE* of hydrolyzing urea to ammonia and carbonic acid, a reaction which can also occur by means of hydroxide

$$\underset{\text{urea}}{H_2N-\overset{\displaystyle O}{\overset{\|}{C}}-NH_2} + 2H_2O \xrightarrow[\text{or enzyme}]{OH^-} 2NH_3 + H_2CO_3$$

This enzyme, called *urease,* is of particular interest because it was the first enzyme to be purified and crystallized (figure 1-1). It was the purification and study of this enzyme which finally permitted the true composition of enzymes to be deciphered (chapter 12).

When urea is heated with malonic acid, two molecules of water are *BARBITURATES* eliminated and the product is a cyclic amide called a *barbiturate*

| urea | malonic acid | barbituric acid (a barbiturate) |

Other barbiturates are made by the use of malonic acids which have substituent groups attached to the center carbon atom.

With the exception of barbituric acid itself, all of these compounds are hypnotics; that is, they depress the central nervous system. The first barbiturate to be used in medicine was diethylbarbituric acid, also called *Veronal,* which was introduced in 1903. This compound is still considered an excellent hypnotic and is still used occasionally, but it is too long-acting for most purposes.

Following the introduction of phenobarbital (also called *Luminal*) in 1912, an extensive search was made for barbiturates with desirable properties. Over 2500 barbiturates were synthesized and tested. Of the fifty or so that have been marketed at various times, only about twelve are in current use. The structures of some of these are shown in figure 10-9. Although barbituric acid is actually an amide, rather than an acid, the NH of barbituric acid is quite acidic (pK_a about 8) because the conjugate base of barbituric acid is resonance-stabilized. For solubility reasons, barbiturates are frequently administered in the form of their sodium salts.

Give a synthesis for each of the barbiturates in figure 10-9 from urea *PROBLEM 10-22* and an appropriately substituted malonic acid.

PROBLEM 10-23 Give the resonance forms of the conjugate base of barbital.

10-11 PHOSPHORIC ACID AND ITS DERIVATIVES

Phosphoric acid is a trihydroxy acid which has three acidic hydrogens

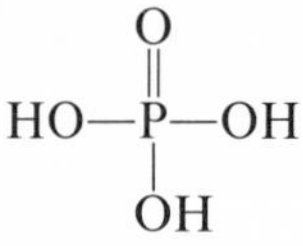

$$\begin{array}{c} O \\ \| \\ HO-P-OH \\ | \\ OH \end{array}$$

phosphoric acid

Phosphoric acid is capable of undergoing three successive dissociations:

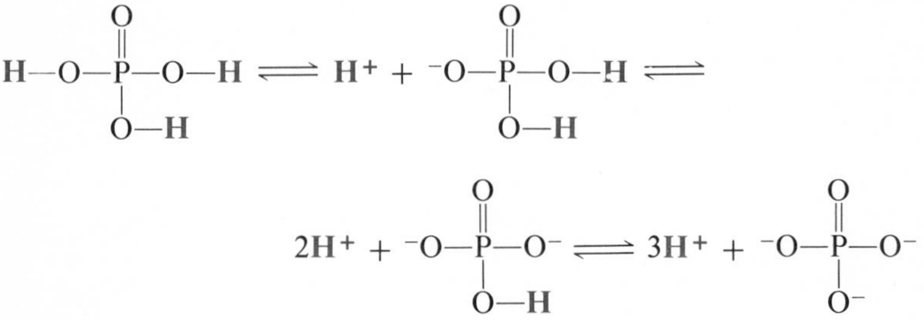

$$\begin{array}{c} O \\ \| \\ H-O-P-O-H \\ | \\ O-H \end{array} \rightleftharpoons H^+ + \begin{array}{c} O \\ \| \\ {}^-O-P-O-H \\ | \\ O-H \end{array} \rightleftharpoons$$

$$2H^+ + \begin{array}{c} O \\ \| \\ {}^-O-P-O^- \\ | \\ O-H \end{array} \rightleftharpoons 3H^+ + \begin{array}{c} O \\ \| \\ {}^-O-P-O^- \\ | \\ O^- \end{array}$$

The pK_a values for the three acid dissociations of H_3PO_4 are 2.1, 7.2, and 12.3.

PROBLEM 10-24 All the phosphate anions are resonance-stabilized. Draw the resonance forms of $H_2PO_4^-$, HPO_4^{2-}, and PO_4^{3-}.

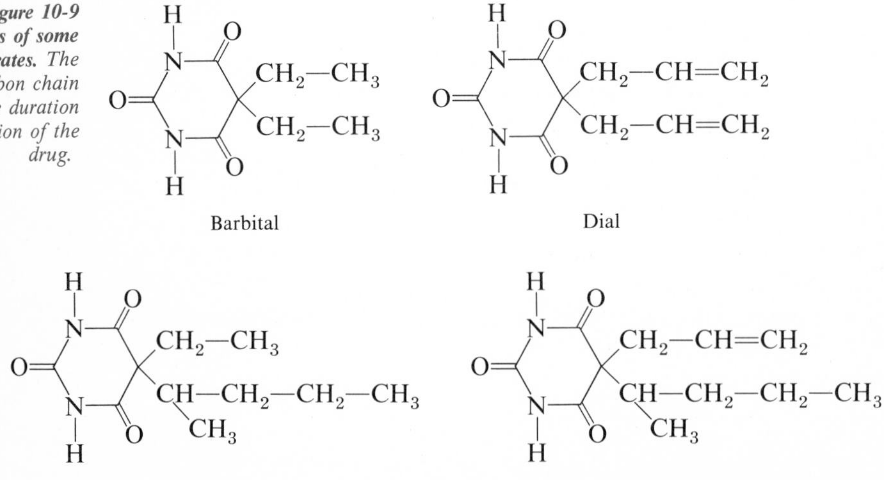

Figure 10-9
Structures of some
barbiturates. The
hydrocarbon chain
affects the duration
of action of the
drug.

Barbital

Dial

Nembutal

Seconal

Like carboxylic acids, phosphoric acid is capable of forming esters, amides, and anhydrides by combination with alcohols, amines, and itself, respectively. However, because of the existence of three acidic groups, phosphoric acid can form a greater variety of compounds than carboxylic acids can. Examples of some important types of phosphoric acid derivatives are given in table 10-2.

PHOSPHORIC ACID DERIVATIVES

A great variety of derivatives of phosphoric acid exist in nature. All types of compounds shown in table 10-2 occur except triesters and free triphosphoric acid. Most of these phosphates are quite soluble in water because at physiological pH one or more of the acidic hydrogens are dissociated and the resulting phosphate *ions,* like most ions, are soluble in water.

Derivatives of phosphoric acid undergo reactions which are analogous to those of carboxylic acids but more complex. A triester of phosphoric acid, for example, is capable of undergoing not one ester hydrolysis but three sequential ones

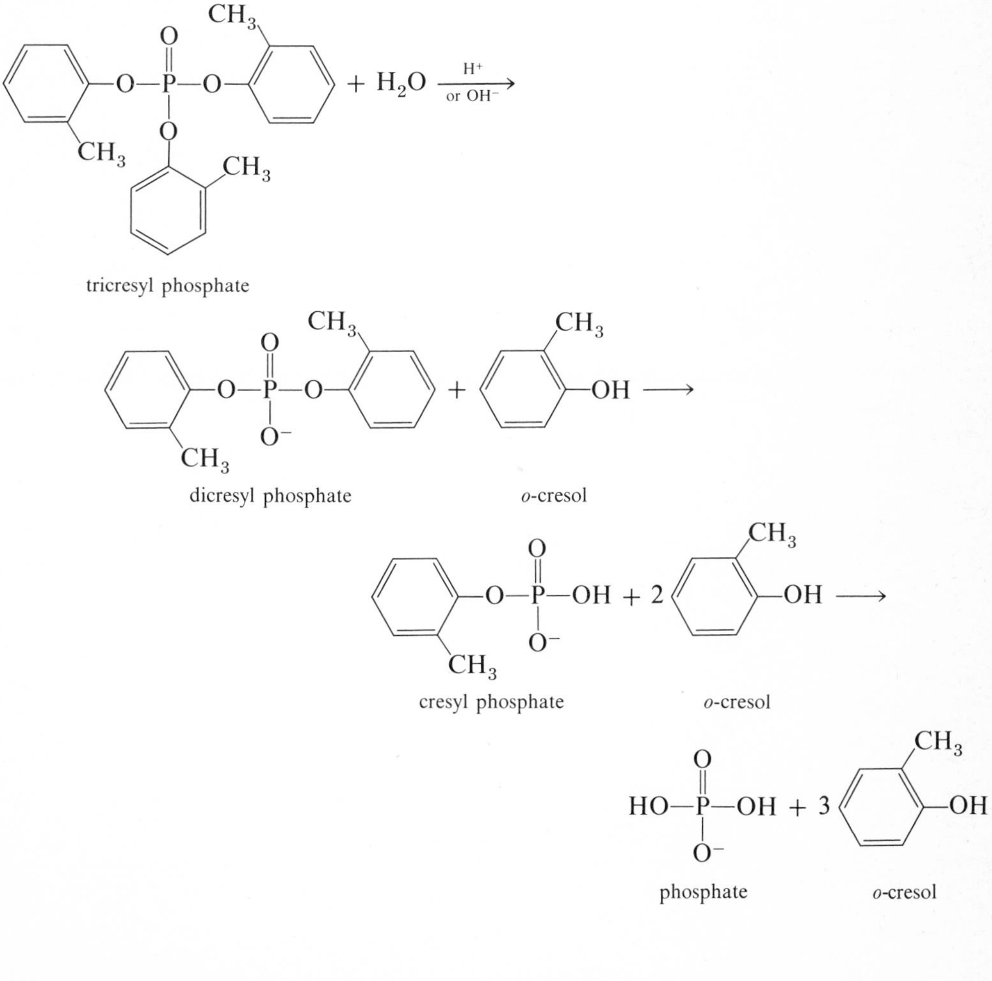

ESTER HYDROLYSIS

$$\text{tricresyl phosphate} + H_2O \xrightarrow[\text{or OH}^-]{H^+}$$

tricresyl phosphate

$$\text{dicresyl phosphate} + \text{o-cresol} \longrightarrow$$

dicresyl phosphate   *o*-cresol

$$\text{cresyl phosphate} + 2\ \text{o-cresol} \longrightarrow$$

cresyl phosphate   *o*-cresol

$$HO-P(=O)(O^-)-OH + 3\ \text{o-cresol}$$

phosphate   *o*-cresol

Table 10-2
Derivatives of Phosphoric Acid

Type	*Structure*	*Name*
esters		
monoesters	$CH_3—O—\overset{\overset{\textstyle O}{\|\|}}{\underset{\underset{\textstyle OH}{\|}}{P}}—OH$	methyl phosphate
diesters	$CH_3—O—\overset{\overset{\textstyle O}{\|\|}}{\underset{\underset{\textstyle OH}{\|}}{P}}—O—CH_3$	dimethyl phosphate
triesters	$CH_3—O—\overset{\overset{\textstyle O}{\|\|}}{\underset{\underset{\textstyle O—CH_3}{\|}}{P}}—O—CH_3$	trimethyl phosphate
amides	$HO—\overset{\overset{\textstyle O}{\|\|}}{\underset{\underset{\textstyle OH}{\|}}{P}}—NH_2$	phosphoramidate
anhydrides	$HO—\overset{\overset{\textstyle O}{\|\|}}{\underset{\underset{\textstyle OH}{\|}}{P}}—O—\overset{\overset{\textstyle O}{\|\|}}{\underset{\underset{\textstyle OH}{\|}}{P}}—OH$	pyrophosphoric acid
	$HO—\overset{\overset{\textstyle O}{\|\|}}{\underset{\underset{\textstyle OH}{\|}}{P}}—O—\overset{\overset{\textstyle O}{\|\|}}{\underset{\underset{\textstyle OH}{\|}}{P}}—O—\overset{\overset{\textstyle O}{\|\|}}{\underset{\underset{\textstyle OH}{\|}}{P}}—OH$	triphosphoric acid
ester anhydrides	$CH_3—O—\overset{\overset{\textstyle O}{\|\|}}{\underset{\underset{\textstyle OH}{\|}}{P}}—O—\overset{\overset{\textstyle O}{\|\|}}{\underset{\underset{\textstyle OH}{\|}}{P}}—OH$	methyl pyrophosphate
mixed anhydrides	$CH_3—\overset{\overset{\textstyle O}{\|\|}}{C}—O—\overset{\overset{\textstyle O}{\|\|}}{\underset{\underset{\textstyle OH}{\|}}{P}}—OH$	acetyl phosphate
	$H_2N—\overset{\overset{\textstyle O}{\|\|}}{C}—O—\overset{\overset{\textstyle O}{\|\|}}{\underset{\underset{\textstyle OH}{\|}}{P}}—OH$	carbamyl phosphate

SOURCE: E. M. Kosower, *Molecular Biochemistry*, McGraw-Hill Book Company, New York, 1962.

A variety of anhydrides serve central roles in metabolism. One such *ANHYDRIDES* compound is acetyl phosphate, which can hydrolyze to acetate and phosphate

$$CH_3-\overset{\overset{\displaystyle O}{\|}}{C}-O-\overset{\overset{\displaystyle O}{\|}}{\underset{\underset{\displaystyle O^-}{|}}{P}}-OH + H_2O \longrightarrow CH_3-\overset{\overset{\displaystyle O}{\|}}{C}-OH + HO-\overset{\overset{\displaystyle O}{\|}}{\underset{\underset{\displaystyle O^-}{|}}{P}}-OH$$

    acetyl phosphate   acetic acid   phosphate

Acetyl phosphate can also serve as an acyl transfer agent for the formation of acetyl coenzyme A

$$CH_3-\overset{\overset{\displaystyle O}{\|}}{C}-O-\overset{\overset{\displaystyle O}{\|}}{\underset{\underset{\displaystyle OH}{|}}{P}}-O^- + HS-CoA \longrightarrow CH_3-\overset{\overset{\displaystyle O}{\|}}{C}-S-CoA + HO-\overset{\overset{\displaystyle O}{\|}}{\underset{\underset{\displaystyle OH}{|}}{P}}-O^-$$

    acetyl phosphate   acetyl CoA

The hydrolysis of the simplest phosphate anhydride, pyrophosphate, is catalyzed by an enzyme called *inorganic pyrophosphatase*

$$HO-\overset{\overset{\displaystyle O}{\|}}{\underset{\underset{\displaystyle O^-}{|}}{P}}-O-\overset{\overset{\displaystyle O}{\|}}{\underset{\underset{\displaystyle O^-}{|}}{P}}-OH + H_2O \xrightarrow[\text{pyrophosphatase}]{\text{inorganic}} 2HO-\overset{\overset{\displaystyle O}{\|}}{\underset{\underset{\displaystyle O^-}{|}}{P}}-OH$$

    pyrophosphate

Adenosine triphosphate (ATP) is a reactive phosphate anhydride. Like *ATP* carboxylic anhydrides, it reacts with a variety of alcohols to form esters—in this case, phosphate esters. ATP is the most important donor of phosphate groups for ester formation in living systems. A wide variety of enzymes are available to catalyze various ester formations. A few examples are shown in figure 10-10.

When the triphosphate of ATP is broken down during phosphate transfer reactions, a large amount of energy is liberated. Many biochemical processes depend on this energy, including protein synthesis, photosynthesis, and muscle contraction.

Many compounds which contain hydroxyl groups occur in nature both *PHOSPHATE* as the free compound and as a phosphate ester. The phosphate ester *ESTERS IN* group acts as a biological handle for a number of purposes: it increases *NATURE* solubility, serves as a specific site for binding of the compound to enzymes, and often plays a role in chemical reactions as well.

The formation of phosphate esters from alcohols is catalyzed by a

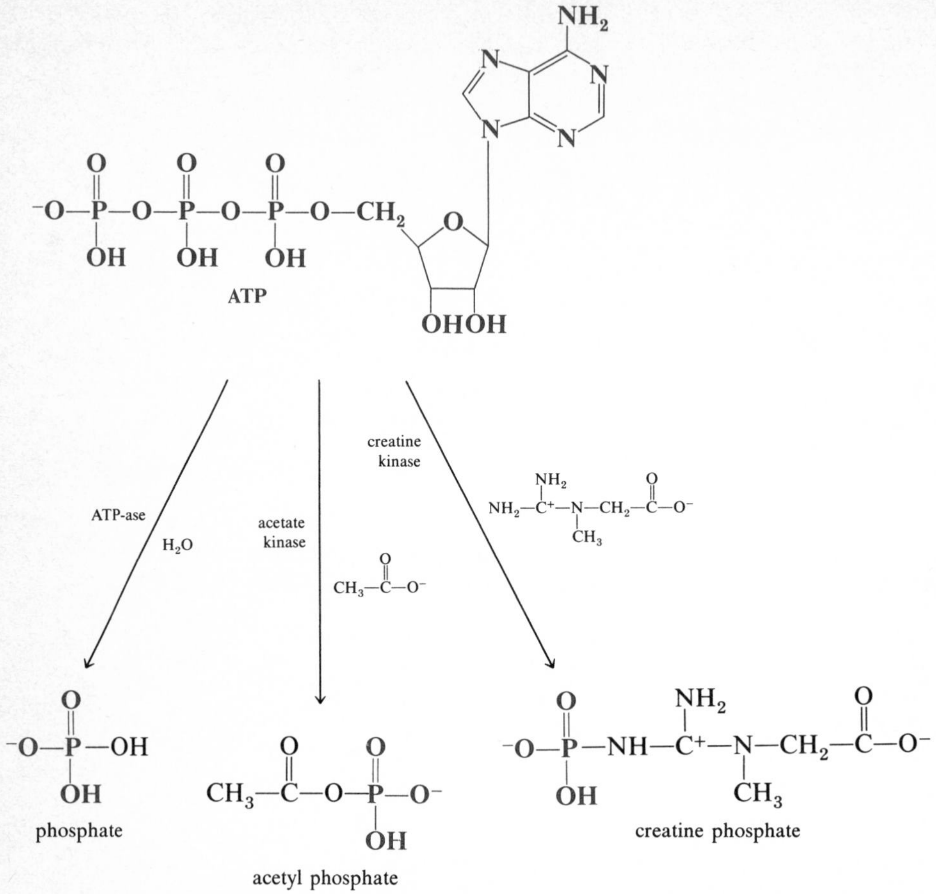

ATP

ATP-ase

H₂O

acetate
kinase

creatine
kinase

phosphate

acetyl phosphate

creatine phosphate

Figure 10-10
Reactions of ATP.

*A variety of enzymes called **kinases** catalyze the transfer of a phosphate group from ATP to a variety of alcohols,*

amines, acids, and other compounds. The other product is always adenosine diphosphate (ADP).

family of enzymes called *kinases*. The source of the phosphate for these reactions is ATP (figure 10-10)

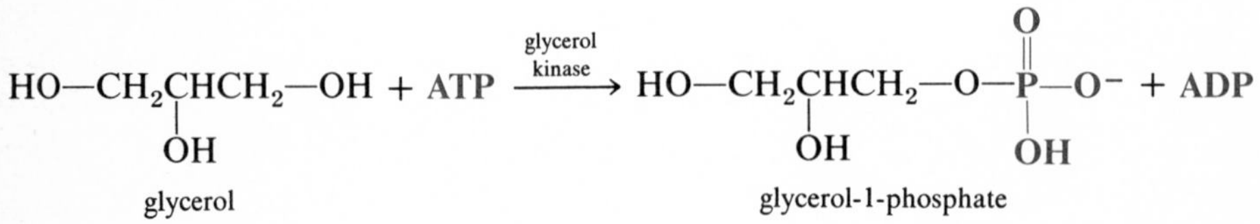

$$HO-CH_2CHCH_2-OH + ATP \xrightarrow[\text{kinase}]{\text{glycerol}} HO-CH_2CHCH_2-O-P-O^- + ADP$$

glycerol

glycerol-1-phosphate

Hydrolysis of these phosphate esters is catalyzed by a family of enzymes called *phosphatases*

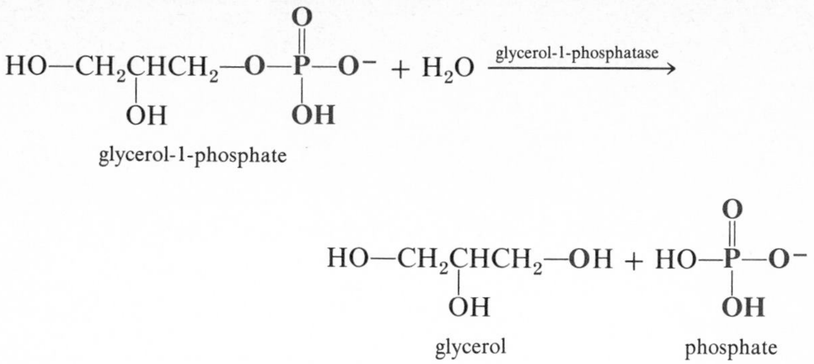

$$HO-CH_2CHCH_2-O-\overset{\overset{\displaystyle O}{\|}}{P}-O^- + H_2O \xrightarrow{\text{glycerol-1-phosphatase}}$$
$$\underset{\displaystyle OH}{|} \qquad\quad \underset{\displaystyle OH}{|}$$

glycerol-1-phosphate

$$HO-CH_2CHCH_2-OH + HO-\overset{\overset{\displaystyle O}{\|}}{P}-O^-$$
$$\underset{\displaystyle OH}{|} \qquad\qquad \underset{\displaystyle OH}{|}$$

glycerol   phosphate

Phosphatases and kinases play an especially important role in the metabolism of carbohydrates (chapter 13).

PROBLEMS

10-25 Give the structure of the acid chloride, the amide, the methyl ester, and the anhydride of propionic acid. Arrange these compounds in order of increasing reactivity toward NaOH.

10-26 Give the mechanism and product of each of the following reactions: (*a*) ethyl butyrate + NaOH; (*b*) ethyl butyrate + NH$_3$; (*c*) butyronitrile + NaOH; (*d*) ethyl butyrate +phenyl magnesium bromide, then + H$_2$O; (*e*) ethyl butyrate + dilute H$_2$SO$_4$; (*f*) ethyl butyrate + sodium ethoxide.

10-27 Choline is converted into choline phosphate by the enzyme choline kinase. Give the structure of choline phosphate.

10-28 Give the structure of each of the following compounds: (*a*) phenyl acetate; (*b*) *N-n*-propylacetamide; (*c*) lauronitrile; (*d*) isobutyl formate.

10-29 Acetyl phosphate is a mixed anhydride and can give two different sets of products on reaction with a nucleophile. Give the products of reaction of acetyl phosphate with sodium ethoxide.

10-30 Name the following compounds:

(*a*) $CH_3-\overset{\overset{\displaystyle O}{\|}}{C}-NH_2$

(*b*) $CH_3-\overset{\overset{\displaystyle CH_3}{|}}{\underset{\underset{\displaystyle CH_3}{|}}{C}}-\overset{\overset{\displaystyle O}{\|}}{C}-OCH_3$

(*c*) $CH_3CH_2\overset{\overset{\displaystyle O}{\|}}{C}-O\underset{\underset{\displaystyle CH_3}{|}}{C}HCH_3$

(*d*) $CH_3CH_2CH_2\underset{\underset{\displaystyle CH_3}{|}}{C}H\overset{\overset{\displaystyle O}{\|}}{C}-NH_2$

(*e*) $CH_3CH_2CH_2\overset{\overset{\displaystyle O}{\|}}{C}-O-\overset{\overset{\displaystyle O}{\|}}{C}CH_2CH_2CH_3$

(*f*) $H\overset{\overset{\displaystyle O}{\|}}{C}-NHCH_3$

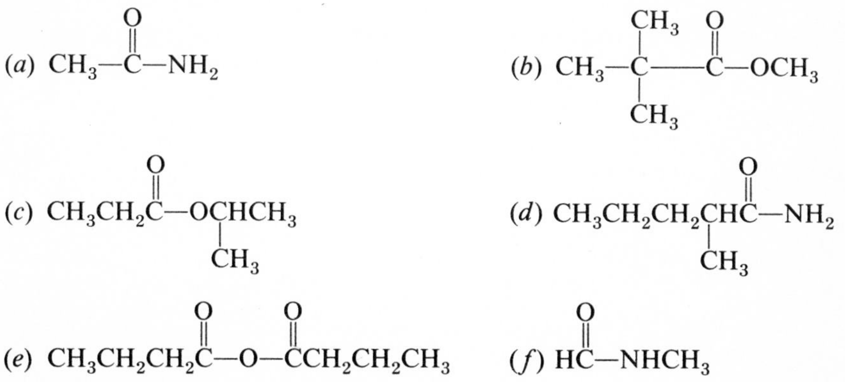

(g) $CH_3CH_2CH_2CH_2CH_2CH_2CH_2\overset{\displaystyle O}{\overset{\|}{C}}-Cl$

(h) $CH_3\overset{\displaystyle }{\underset{\displaystyle Cl}{CH}}CH_2CH_2CH_2\overset{\displaystyle O}{\overset{\|}{C}}-OCH_3$

10-31 The hydrogen atoms bound to nitrogen in amides are slightly acidic. Use resonance to explain why benzamide can lose a proton. Use resonance to explain why phthalimide is more acidic than benzamide.

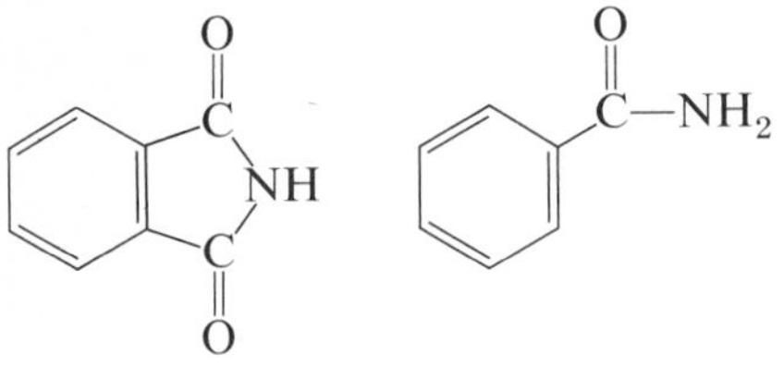

phthalimide   benzamide

10-32 Give a mechanism for the following reaction, which occurs during the hydrolysis of ethyl thionbenzoate:

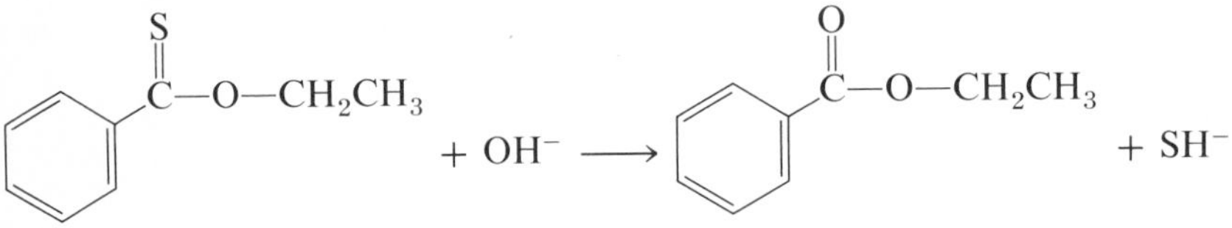

ethyl thionbenzoate   ethyl benzoate

10-33 Heroin, also called diacetyl morphine, is made by the reaction of morphine (figure 6-7) with acetic anhydride. Give the structure of heroin.

10-34 Sarin, diisopropyl fluorophosphate, is a phosphate diester acid fluoride. It was developed for use as a nerve gas in World War II. Sarin works by reacting with the hydroxyl group at the active site of acetylcholinesterase (figure 10-5). The product of the reaction is a triester. Give its structure.

10-35 Triacetin is an ester that is used as a fungistat. When it is applied to the skin, the enzymes present slowly catalyze the hydrolysis of triacetin to acetic acid and glycerol. Give the structure of triacetin.

10-36 Give the structure of the product of the reaction of each of the compounds on the left with each of the compounds on the right. If no reaction, so state.

$CH_3CH_2\overset{\displaystyle O}{\overset{\|}{C}}Cl$ H_2O

$CH_3CH_2\overset{\displaystyle O}{\overset{\|}{C}}OCH_3$ NH_3

$$CH_3CH_2\overset{\overset{\displaystyle O}{\|}}{C}NH_2 \qquad\qquad NaOH$$

$$CH_3CH_2\overset{\overset{\displaystyle O}{\|}}{C}O\overset{\overset{\displaystyle O}{\|}}{C}CH_2CH_3 \qquad CH_3OH$$
$$\qquad\qquad\qquad\qquad\qquad dil.\ H_2SO_4$$

10-37 Give the structure of the product of the reaction of methyl butyrate with (*a*) phenylmagnesium bromide, then water; (*b*) LiAlH$_4$, then water; (*c*) sodium methoxide; (*d*) sodium hydroxide.

10-38 Give the structure of the product of each of the following reactions.

(*a*) [benzoyl chloride structure] + HOCH$_2$CH$_2$OH $\longrightarrow$

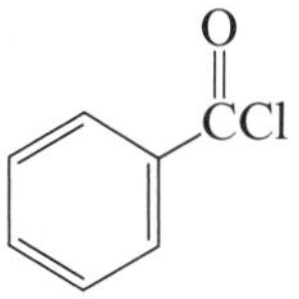

(excess)

(*b*) CH$_3$CH$_2$CH$_2$$\overset{\overset{\displaystyle O}{\|}}{C}NH_2$ + CH$_3$OH $\xrightarrow{\ H^+\ }$

(*c*) [cyclic phosphate structure] + NaOH $\longrightarrow$

(*d*) [phthalic anhydride structure] + 1-decanol $\longrightarrow$

(*e*) [isochromanone structure] + NaOH $\longrightarrow$

(*f*) Cl$-\overset{\overset{\displaystyle O}{\|}}{\underset{\underset{\displaystyle Cl}{|}}{P}}-$Cl + [aniline structure]$-NH_2$ $\longrightarrow$

(excess)

(*g*) CH$_3$CH$_2$CH$_2$CH$_2$$\overset{\overset{\displaystyle O}{\|}}{C}$Cl + NaCN $\longrightarrow$

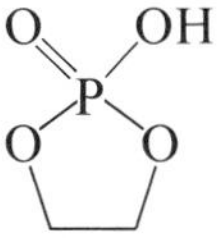

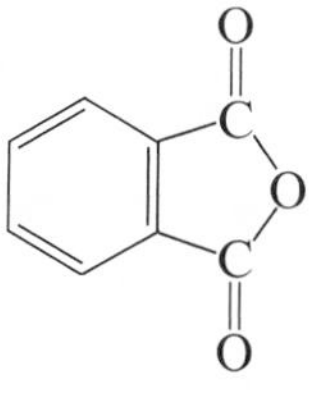

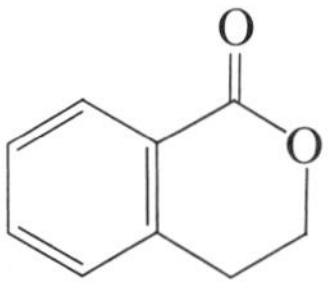

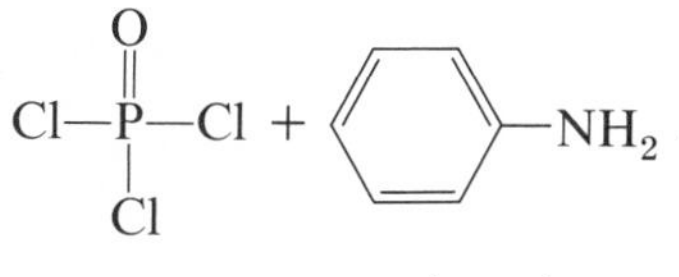

(h)

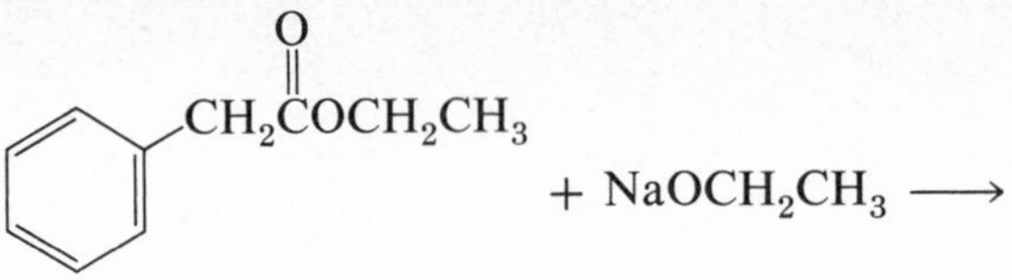

$$CH_2\overset{\overset{\displaystyle O}{\|}}{C}OCH_2CH_3$$

$+ \; NaOCH_2CH_3 \longrightarrow$

(i)

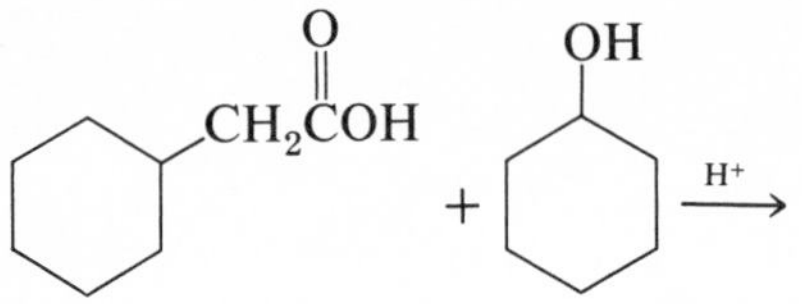

$$CH_2\overset{\overset{\displaystyle O}{\|}}{C}OH$$

$+$

$$\overset{\displaystyle OH}{\bigcirc}$$

$\overset{H^+}{\longrightarrow}$

(j) $CH_3CH_2O\overset{\overset{\displaystyle O}{\|}}{C}OCH_2CH_3 + CH_3NH_2 \longrightarrow$
(excess)

(k)

$$CH_2\overset{\overset{\displaystyle O}{\|}}{C}OCH_2CH_3$$

$+ \; NaOH \longrightarrow$

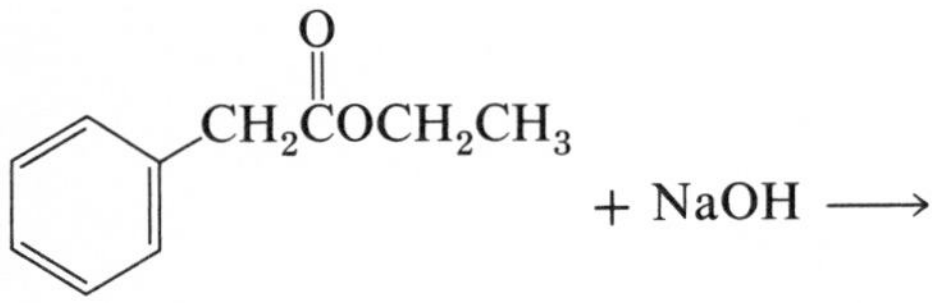

(l) $H\overset{\overset{\displaystyle O}{\|}}{C}OCH_2CH_3 + CH_3CH_2CH_2CH_2MgBr \longrightarrow \overset{then}{\underset{H_2O}{\longrightarrow}}$

(m)

$$\overset{\overset{\displaystyle O}{\|}}{C}OCH_2CH_2CH_3$$

$+ \; LiAlH_4 \longrightarrow \overset{then}{\underset{H_2O}{\longrightarrow}}$

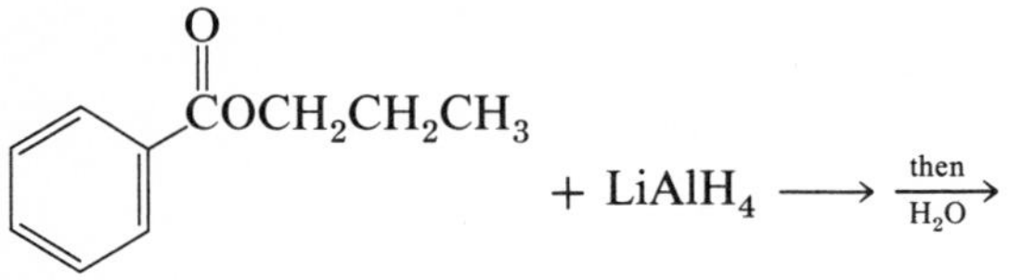

(n) $CH_3\underset{\underset{\displaystyle CH_3}{|}}{CH}-\overset{\overset{\displaystyle O}{\|}}{C}-OCH_3 + NaOCH_2CH_2CH_3 \longrightarrow$

10-39 Give the reagents needed to carry out each of the following transformations. You can use any inorganic reagents and any organic reagents with four or fewer carbon atoms. More than one step may be required.

(a)

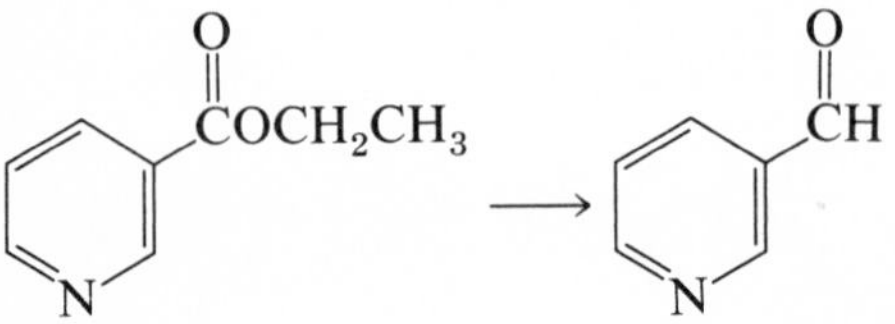

$$\overset{\overset{\displaystyle O}{\|}}{C}OCH_2CH_3$$ on pyridine ring $\longrightarrow$ $$\overset{\overset{\displaystyle O}{\|}}{C}H$$ on pyridine ring

(b)

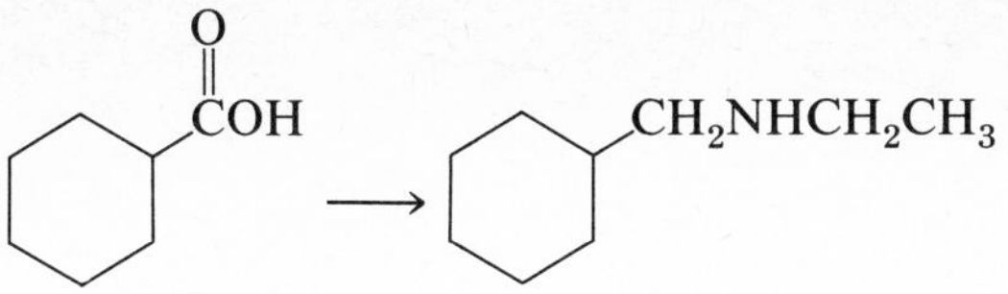

$$\text{(cyclohexane)}-\overset{\overset{\displaystyle O}{\|}}{C}OH \longrightarrow \text{(cyclohexane)}-CH_2NHCH_2CH_3$$

(c)

$$\text{(cyclopentane)}-\overset{\overset{\displaystyle O}{\|}}{C}OH \longrightarrow \text{(cyclopentane)}-CH_2O\overset{\overset{\displaystyle O}{\|}}{C}CH_2CH_3$$

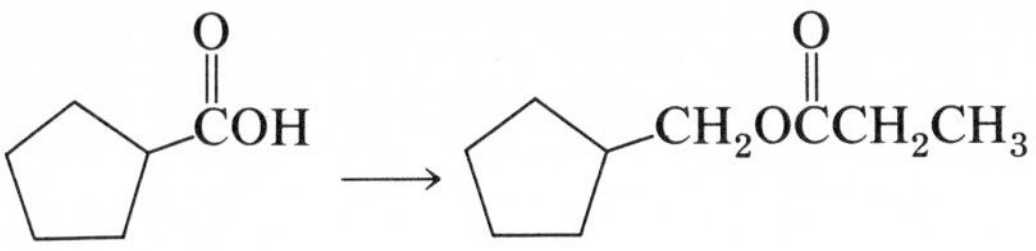

(d)

$$\text{(benzene)}-\overset{\overset{\displaystyle O}{\|}}{C}OCH_3 \longrightarrow \text{(benzene)}-\overset{\overset{\displaystyle CH_3}{|}}{C}=CH_2$$

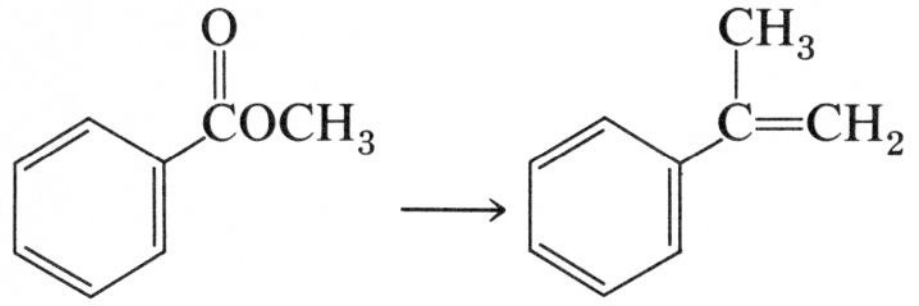

(e) $CH_3CH_2CH_2CH_2CH_2\overset{\overset{\displaystyle O}{\|}}{C}OCH_3 \longrightarrow CH_3CH_2CH_2CH_2CH_2CH_3$

SUGGESTED READINGS

Adams, E.: Barbiturates, p. 81 in *Organic Chemistry of Life: Readings from Scientific American,* Freeman, San Francisco, 1973.

Green, D. E.: The Metabolism of Fats, p. 300 in *Organic Chemistry of Life: Readings from Scientific American,* Freeman, San Francisco, 1973. The synthesis and degradation of fatty acids in cells.

Lehninger, A. L.: *Bioenergetics,* 2d ed., Benjamin, Menlo Park, Calif., 1971. A good discussion of ATP and its relationship to energy production in the cell.

Rose, A. H.: New Penicillins, *Sci. Am.,* March 1961, p. 66.

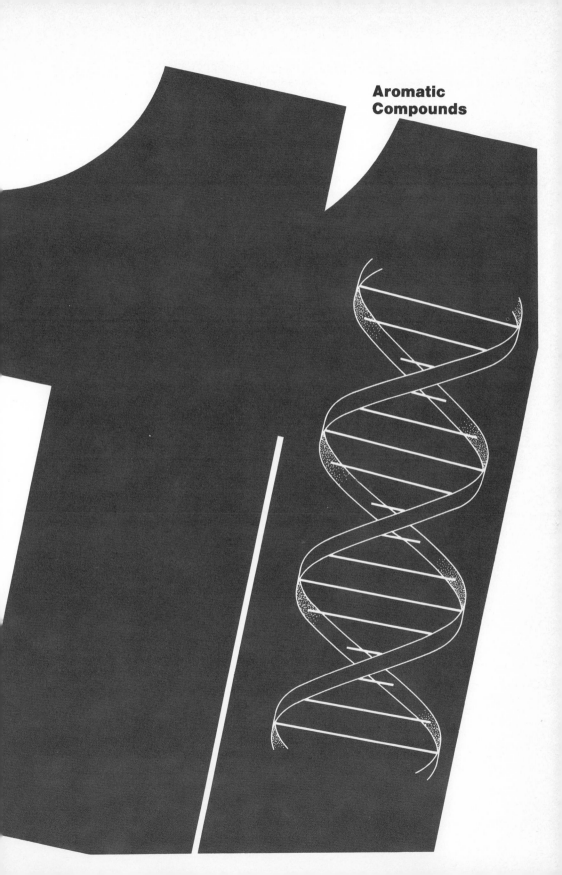

**Aromatic
Compounds**

This chapter might be called "The Importance of Having Resonance" because this property is so important for the structures and chemistry of aromatic compounds. In this chapter we will consider how resonance affects the structures and reactivities of a variety of aromatic compounds.

Compounds containing aromatic rings are found everywhere in nature—in DNA, in RNA, in proteins, in many pigments, in vitamins, and in many other places. The fact that a compound has an aromatic ring may be overlooked because of the attention paid to its other functional groups. However, these aromatic rings have reactions and properties all their own; it is the purpose of this chapter to explore these.

11-1 STRUCTURES AND NOMENCLATURE OF SUBSTITUTED BENZENES

Benzene is a highly symmetrical molecule with six equivalent carbon-carbon bonds. Because the six carbon atoms of benzene are equivalent, there is only one isomer of any monosubstituted benzene. Methylbenzene, for example, is called *toluene*

toluene

Three different isomers are possible for a disubstituted benzene. They are called *ortho, meta,* and *para* isomers, abbreviated *o-, m-,* and *p-*. Thus, there are three structural isomers of xylene (dimethylbenzene):

o-xylene   *m*-xylene   *p*-xylene

Systematic names of substituted benzenes are used less frequently than a large group of trivial names, some of which are given in figure 11-1.

PROBLEM 11-1 Draw the structures of all isomers of (*a*) methylnaphthalene; (*b*) dichloronaphthalene; (*c*) trichlorobenzene. The structure of naphthalene is

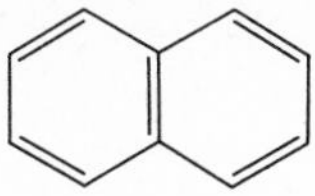

naphthalene

If a hydroxyl group is substituted on the ring of toluene, the resulting compound is called a *cresol*. Draw and name the three isomers of cresol.

PROBLEM 11-2

11-2 PHENOL AND ANILINE

Aromatic compounds which have a hydroxyl group attached to a benzene ring are called *phenols,* after the parent compound, phenol. Compounds with an amino group attached to a benzene ring are called *anilines*.

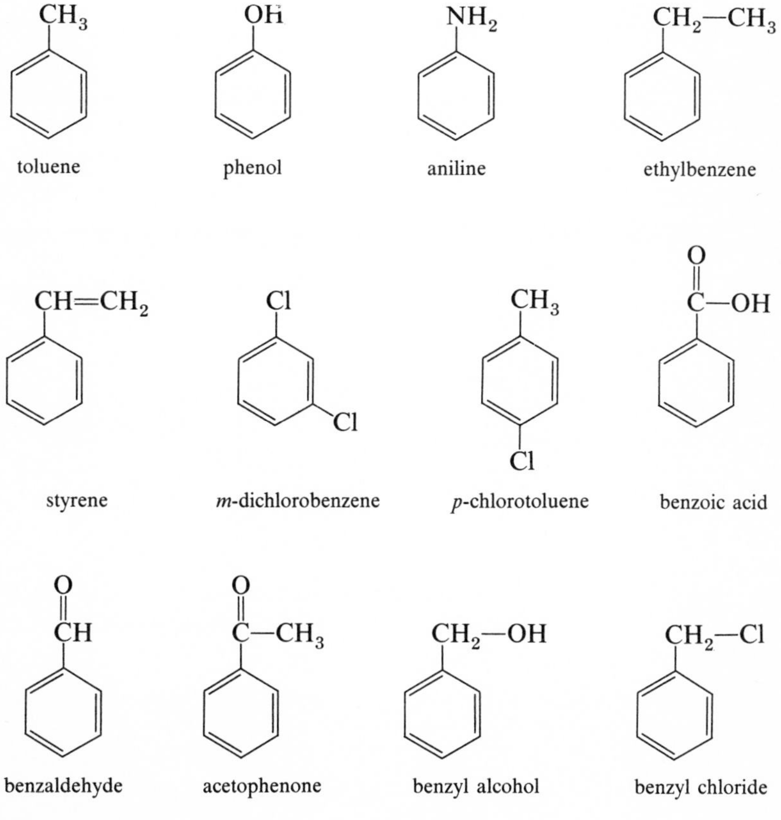

toluene   phenol   aniline   ethylbenzene

Figure 11-1
Names and
structures of some
substituted benzenes.

styrene   *m*-dichlorobenzene   *p*-chlorotoluene   benzoic acid

benzaldehyde   acetophenone   benzyl alcohol   benzyl chloride

PHENOL Because the hydroxyl group of phenol is attached to a benzene ring, resonance interactions play an important role in determining the acidity of phenol. The resonance forms of phenol are analogous to those of benzene (figure 11-2), and analogous resonance forms can be written for phenoxide ion. However, additional resonance forms can be written for phenoxide ion which place the negative charge in the ortho and para positions of the benzene ring. As a result of this extra resonance in the phenoxide anion, phenol is about a million times more acidic than a simple alcohol like ethanol.

Because phenol is more acidic than water or simple alcohols, it can be converted quantitatively into its conjugate base by reaction with sodium hydroxide or a sodium alkoxide

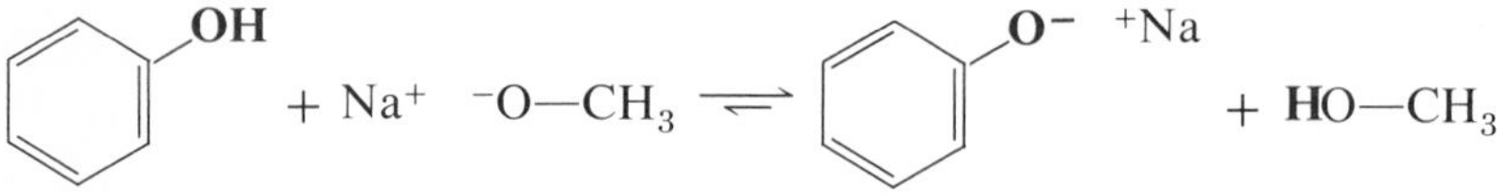

phenol   sodium phenoxide

PHENOL IN HISTORY Phenol is an acid, albeit a weaker acid than mineral acids or carboxylic acids, and it is often called *carbolic acid*. Carbolic acid was the first common antiseptic, having been put to this use by Lister in 1867. The use of phenol as an antiseptic in surgery reduced fatalities due to infection by a large margin, and phenol soon became the standard by which the potency of other antiseptics was measured. Since phenol is slightly toxic, it is seldom used today.

A number of substituted phenols are currently used as antiseptics.

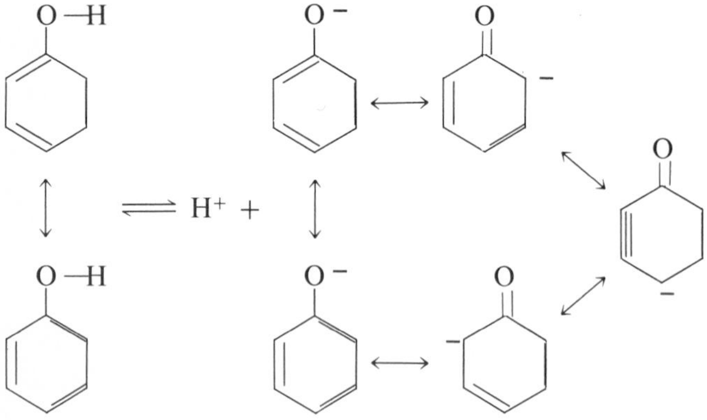

phenol   phenoxide ion

Figure 11-2 *Both phenol and phenoxide ion are* *para positions of the ring. This resonance*
Resonance in phenol *stabilized by resonance. The negative* *makes phenol more acidic than it would*
and phenoxide ion. *charge in phenoxide ion is* *otherwise be.*
resonance-stabilized over the ortho and

One such compound is *hexachlorophene,* which until recently was the principal active ingredient of a number of soaps and germicides

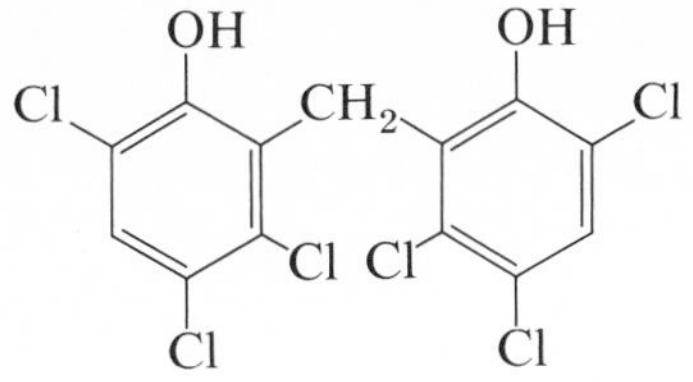

hexachlorophene

However, like most potent chemicals, this one has occasional side effects, and it has been withdrawn from nonprescription use. It is still used in certain circumstances by the medical profession.

Aniline, like other amines, is basic, but it is considerably less basic than the simple aliphatic amines considered in chapter 6. Although this change in basicity is the result of resonance (figure 11-3), the workings of resonance are more complex here than in the case of phenol.

ANILINE

Both aniline and its conjugate acid, anilinium ion, have a pair of resonance structures analogous to those of benzene, phenol, and other simple benzene derivatives. No further resonance is possible for the anilinium ion. However, aniline has three doubly charged resonance structures in which the pair of electrons on nitrogen is involved in bonding to the benzene ring. Although these charged resonance struc-

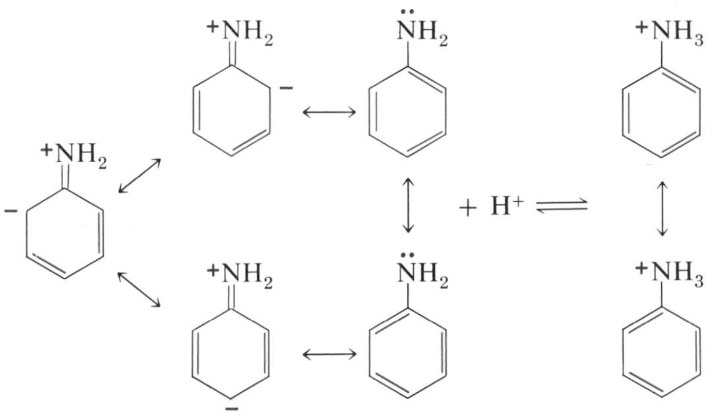

aniline

anilinium ion

Both aniline and anilinium ion are stabilized by resonance, but aniline is additionally stabilized by charged resonance forms involving the unshared

electron pair on nitrogen. Because of these charged forms, aniline is less basic than simple amines.

Figure 11-3
Resonance in aniline and anilinium ion.

tures are less important than the two uncharged resonance structures, the small amount of additional resonance is sufficient to make aniline about a millionfold *less* basic than simple amines like ethylamine.

11-3   THE BENZYLIC EFFECT

Both aniline and the phenoxide ion are resonance-stabilized to a greater extent than their conjugate acids. This kind of extra resonance occurs any time an unshared pair of electrons or an empty *p* orbital is adjacent to an aromatic ring.

This resonance is important because it invariably stabilizes molecules in which it occurs. In phenol, the phenoxide anion is stabilized by resonance, and phenol is more acidic than it would otherwise be. In aniline, extra resonance stabilization of aniline (but not of anilinium ion) occurs, and aniline is less basic than it would otherwise be. In other cases, reactions may be accelerated, or their mechanisms may be changed, or the reaction products may be changed. A number of cases where this effect operates are shown in figure 11-4. Additional examples are given in the following discussion.

Figure 11-4
Examples of
resonance
stabilization.
Unshared electron
pairs and empty p
orbitals adjacent to
an aromatic ring can
undergo resonance
stabilization.

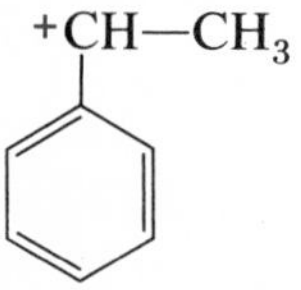

resonance-stabilized
carbonium ion formed during
electrophilic addition to
olefins and during carbonium-
ion-forming reactions of alkyl
halides and alcohols

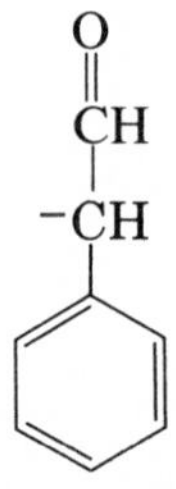

resonance-stabilized
enolate ion formed by
loss of a proton from an
aldehyde or ketone

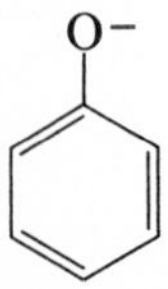

resonance-stabilized
phenoxide anion; as a result
of this resonance, phenol is
more acidic than ordinary
alcohols (figure 11-2)

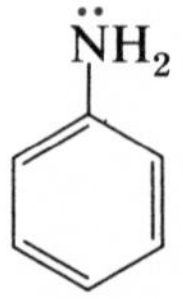

aniline is resonance-
stabilized; as a result
of this resonance, aniline
is less basic than
aliphatic amines (figure 11-3)

If a carbonium-ion intermediate can be stabilized by resonance, that *EXAMPLES* carbonium ion will be formed particularly easily. For example,

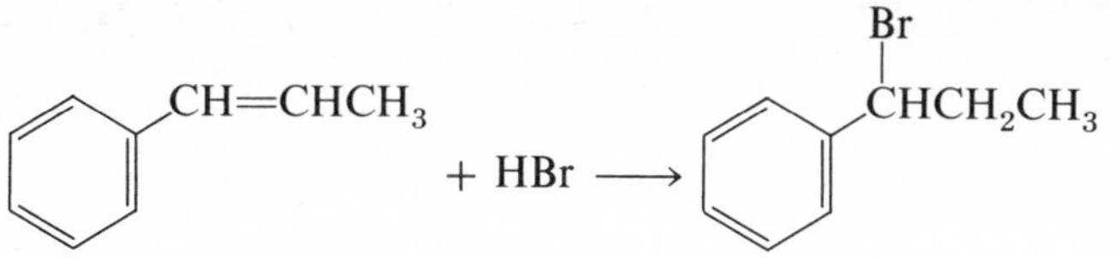

The addition proceeds exclusively in the indicated direction because of resonance stabilization of the carbonium-ion intermediate. The reaction is much faster than the addition of HBr to 2-butene, also because of resonance stabilization.

Enolate anions can be stabilized by resonance. For example, phenyl-acetone enolizes exclusively in the —CH_2— group because the enolate thus formed can participate in resonance with the aromatic ring. Enolate reactions involving this carbon are particularly easy

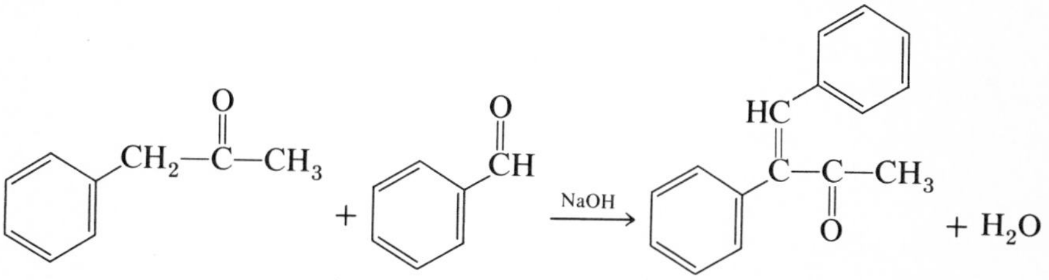

phenylacetone

Reactivity can either increase or decrease because of resonance, depending on the reaction being considered. The reactivity of phenox-ide ion in nucleophilic displacements, for example, is lower than that of a simple alkoxide (such as ethoxide) because the negative charge is not localized on the oxygen atom of the phenoxide ion but delocal-ized over the ring. Since this reactivity depends on the availability of that pair of electrons, the reactivity of phenoxide is reduced by resonance. The reactivity of aniline is lower than that of ethylamine for the same reason.

This effect is called the *benzylic effect* because it occurs only with orbitals that are benzylic, that is, on an atom attached to an aromatic ring.

Give the mechanism of the following reaction, and show why this *PROBLEM 11-3* reaction happens more easily than elimination reactions not involving aromatic rings:

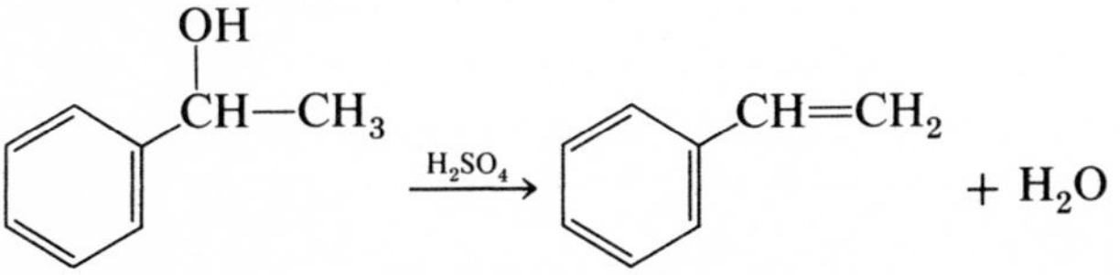

PROBLEM 11-4 Show that the benzylic effect does not operate in the reaction

$$\text{C}_6\text{H}_5\text{CH}_2-\text{CH}_2-\text{OH} \xrightarrow{\text{H}_2\text{SO}_4} \text{C}_6\text{H}_5\text{CH}=\text{CH}_2 + \text{H}_2\text{O}$$

11-4 ELECTROPHILIC AROMATIC SUBSTITUTION

Because of their resonance, aromatic compounds are very stable. They react only with very reactive reagents, and the course of such reactions is governed by the tendency of the molecule to maintain its resonance. Reactions of aromatic compounds usually do not result in destruction of the aromatic ring; such destruction would at the same time destroy resonance, and this is a very undesirable situation. Where such reactions occur, extremely reactive reagents are required.

The most important reaction of benzene is *electrophilic substitution,* attack on the ring by a very reactive electron-poor reagent (the electrophile), leading to displacement of a proton from the ring. The intermediate formed in this reaction is resonance-stabilized, and the product of the reaction has resonance like that of the starting compound. In the following sections we will consider examples of this reaction.

NITRATION Benzene reacts with a mixture of concentrated nitric and sulfuric acids to form nitrobenzene

$$\text{benzene} + \text{HNO}_3 \xrightarrow{\text{H}_2\text{SO}_4} \text{nitrobenzene (NO}_2\text{)} + \text{H}_2\text{O}$$

The sulfuric acid functions as a catalyst and is not consumed in the reaction.

The three-step mechanism of this reaction is typical of electrophilic aromatic substitution reaction mechanisms (mechanism 11-1). In the first step, a highly reactive electrophile is formed (in this case, NO_2^+); in the second step, this electrophile attacks the aromatic ring and a resonance-stabilized intermediate is formed; finally, loss of a proton from the intermediate completes the reaction.

Mechanism 11-1
Nitration of Benzene

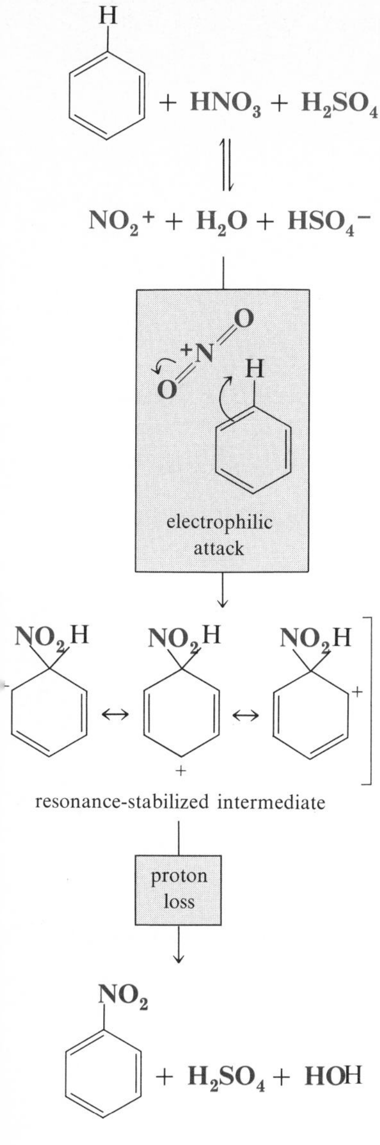
H
+ HNO3 + H2SO4
NO2+ + H2O + HSO4−
O
+N
O
H
electrophilic
attack
NO2H
NO2H
NO2H
resonance-stabilized intermediate
proton
loss
NO2
+ H2SO4 + HOH

*A mixture of concentrated sulfuric and nitric acids forms a small amount of the very reactive **nitronium ion,** NO_2^+. Although the concentration of this species is always low, it reacts readily with the benzene present.*

The nitrogen atom of the nitronium ion is electron-deficient and attacks the benzene ring. In this step, a σ bond is formed between the positively charged nitrogen and a pair of π electrons from the aromatic ring.

The intermediate which is formed by this attack has lost part of the resonance energy of the benzene ring, but a large amount of resonance still exists. The positive charge in the intermediate is resonance-stabilized over three carbon atoms.

The last step in the reaction is loss of a proton from the ring and regeneration of the aromatic ring.

The products of the reaction are nitrobenzene and a proton. The proton immediately combines with the HSO_4^- formed in the first step.

Nitration of *p*-xylene proceeds similarly:

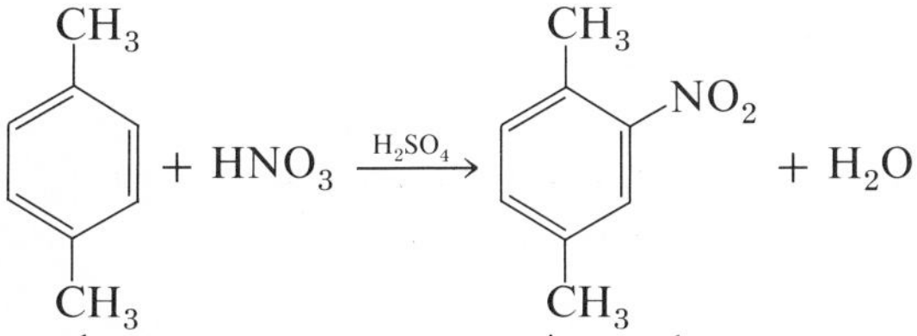
CH3
+ HNO3
H2SO4
CH3
NO2
+ H2O
CH3
CH3

p-xylene   nitro-*p*-xylene

One reason for the tremendous importance of nitration in chemistry is that nitro groups can easily be reduced to amino groups by zinc and HCl or by hydrogen with a platinum catalyst

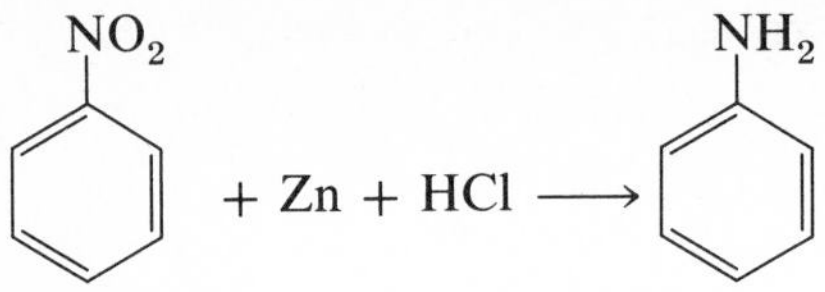

NO$_2$ NH$_2$

+ Zn + HCl $\longrightarrow$

nitrobenzene   aniline

PROBLEM 11-5   Show the resonance structures of the intermediate formed during the nitration of *p*-xylene.

SULFONATION   Concentrated sulfuric acid can react with benzene to form a sulfonic acid. Such a reaction is called *sulfonation:*

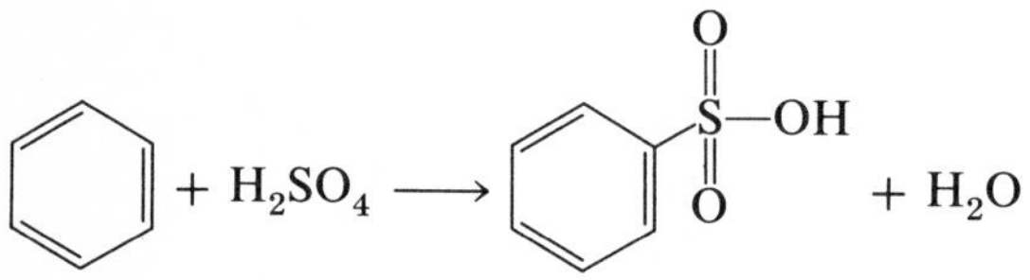

+ H$_2$SO$_4$ $\longrightarrow$ + H$_2$O

benzene   benzenesulfonic
acid

The reactive reagent in sulfonation is the highly reactive HSO$_3^+$ ion, which is formed to a small extent in concentrated sulfuric acid

$$2H_2SO_4 \rightleftharpoons HSO_3^+ + H_2O + HSO_4^-$$

The sulfur atom in the HSO$_3^+$ ion bears the positive charge, and attack of this ion on the benzene ring results in the formation of a new carbon-sulfur bond. The intermediate formed by this attack is resonance-stabilized

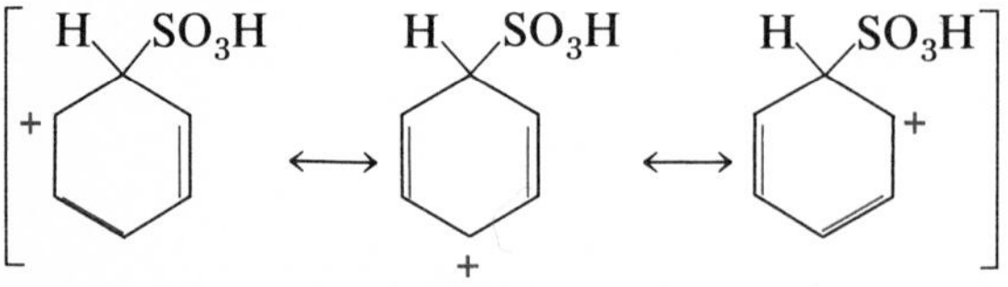

resonance-stabilized intermediate

The product formed by loss of a proton is a sulfonic acid. Sulfonic acids are moderately strong organic acids. Benzenesulfonic acid has a pK$_a$ of 0.8 and thus is stronger than benzoic acid (pK$_a$ 4.2) or phosphoric acid (pK$_a$ 2.1).

Give the mechanism for the formation of *p*-toluenesulfonic acid from toluene and sulfuric acid.   *PROBLEM 11-6*

Soaps are produced by the hydrolysis of animal fat (chapter 10). In the early part of the twentieth century the supply of animal fat became inadequate to meet the rising demand for soap, and industrial processes for producing artificial soaps, called *synthetic detergents,* were developed. The production of these synthetic materials reached the level of 5 billion pounds per year in 1966.   *SYNTHETIC DETERGENTS*

In the 1940s and 1950s the most important synthetic detergent was the mixture of sulfonic acid salts (mostly ortho and para, with a small amount of meta) produced by the sulfonation of a branched-chain alkyl benzene which had been produced by the reaction of benzene with propene

SO_3^- Na^+

CHCH$_2$CHCH$_2$CHCH$_2$CHCH$_3$
| | | |
CH$_3$ CH$_3$ CH$_3$ CH$_3$

a tetrapropylene benzenesulfonate
(not biodegradable because of the
branched carbon chain)

These compounds had an adverse effect on the environment because they were not degraded during sewage treatment and soon caused widespread pollution of rivers and lakes.

The degradation of these compounds in the environment (called *biodegradation*) proceeds along a saturated carbon chain by oxidizing two carbons at a time until the carbon chain has been converted completely into carbon dioxide. However, this degradation ceases when a branch point is reached in the chain.

In the mid-1960s the branched-chain alkyl benzenesulfonates were replaced by linear alkyl benzenesulfonates

SO_3^- Na^+

CH$_3$CHCH$_2$CH$_2$CH$_2$CH$_2$CH$_2$CH$_2$CH$_2$CH$_2$CH$_3$

a linear alkyl benzenesulfonate (biodegradable)

The unbranched chain makes these compounds readily degradable in the environment.

HALOGENATION   Most aromatic compounds do not react with bromine or chlorine unless a catalyst is present. Bromine and chlorine are relatively unreactive electrophiles which react with isolated double bonds (section 3-9), but they are not sufficiently reactive to disturb the benzene ring. In the presence of a catalyst, reaction occurs, but the result is substitution analogous to that observed with nitration and sulfonation

$$\text{benzene} + Br_2 \xrightarrow{FeBr_3} \text{bromobenzene} + HBr$$

This reaction can be understood if we assume that Br_2 reacts with $FeBr_3$ to form the reactive electrophile Br^+. Whether this idea is actually correct is still a matter of speculation. In some reactions, it appears that free Br^+ might be formed, but in other cases this ion is certainly not formed.

Chlorination can be carried out with an $FeCl_3$ catalyst:

$$\text{benzene} + Cl_2 \xrightarrow{FeCl_3} \text{chlorobenzene} + HCl$$

PROBLEM 11-7   Write out the mechanism for the chlorination of benzene.

CHLORO-   It was pointed out in chapter 5 that the only halogenated compounds
PEROXIDASE   which occur commonly in living things are the iodinated compounds related to thyroxine. A few microorganisms produce other halogenated compounds, and the mechanism by which they do so is very interesting.

The enzyme *chloroperoxidase* catalyzes chlorination, bromination, and iodination of a variety of organic substrates. The starting materials for this reaction, however, are not the organic substrate and free halogen but the organic substrate, halide ion, and hydrogen peroxide

$$\text{anisole} + \text{Cl}^- + \text{H}_2\text{O}_2 + \text{H}^+ \xrightarrow{\text{chloroperoxidase}} p\text{-chloroanisole} + o\text{-chloroanisole} + 2\text{H}_2\text{O}$$

anisole   *p*-chloroanisole   *o*-chloroanisole

Free chlorine is not involved in such reactions. Instead, it is believed that hydrogen peroxide oxidizes the chloride ion to enzyme-bound Cl^+, the same chlorinating agent that occurs in chlorinations in organic chemistry. Formation of Cl^+ in an enzyme-bound state provides a means of natural control over this reaction. However, even in the enzyme-bound state, this intermediate is highly reactive. If a substrate is not available for chlorination, the enzyme will eventually chlorinate itself and thus destroy the potential of the cell to perform further chlorinations* (until the cell synthesizes more of the enzyme).

Many aromatic compounds react with acid chlorides or anhydrides in the presence of an aluminum chloride catalyst to form ketones

FRIEDEL-CRAFTS ACYLATION

$$\text{benzene} + \text{CH}_3-\overset{\text{O}}{\underset{}{\text{C}}}-\text{Cl} \xrightarrow{\text{AlCl}_3} \text{acetophenone} + \text{HCl}$$

Like other electrophilic aromatic substitutions, this one proceeds in three steps: (1) reaction of the acid chloride with aluminum chloride to produce the highly reactive ion $\text{CH}_3-\text{C}^+\!=\!\text{O}$; (2) attack of this ion on the aromatic ring; and (3) loss of a proton from the ring.

This reaction is frequently used to make cyclic compounds

$$\text{C}_6\text{H}_5\text{CH}_2\text{CH}_2\text{CH}_2-\overset{\text{O}}{\underset{}{\text{C}}}-\text{Cl} \xrightarrow{\text{AlCl}_3} \text{tetralone} + \text{HCl}$$

*It is not known whether this constitutes murder or suicide, that is, whether an enzyme molecule chlorinates another enzyme molecule or itself.

PROBLEM 11-8   Show two resonance structures for $CH_3—C^+=O$.

PROBLEM 11-9   Give the mechanism of the last reaction above.

SUMMARY OF
ELECTROPHILIC
AROMATIC
SUBSTITUTION

All the electrophilic aromatic substitution reactions discussed in this section proceed by the same three-step mechanism:

1 Formation of a highly reactive positively charged reagent
2 Attack of this reagent on the aromatic ring, forming a resonance-stabilized intermediate
3 Loss of a proton to form the product

The difference between the various reactions is in the nature of the highly reactive reagent. The reactions and reagents are summarized in table 11-1.

11-5 DIRECTIVE EFFECTS IN ELECTROPHILIC AROMATIC SUBSTITUTION

When a substituted benzene is the substrate in an electrophilic aromatic substitution reaction, the substituent group in the benzene ring

Table 11-1
Summary of Electrophilic Aromatic Substitution

Reaction	Reagents	Active Species	Product of Reaction with Benzene
bromination	$Br_2 + FeBr_3$	Br^+	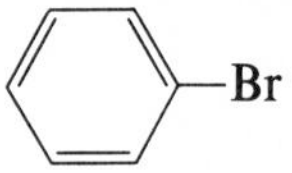 —Br
nitration	conc. $HNO_3 + H_2SO_4$	NO_2^+	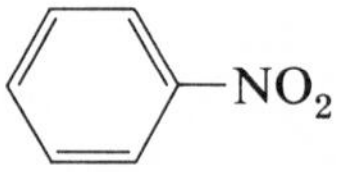 —NO_2
sulfonation	H_2SO_4	HSO_3^+	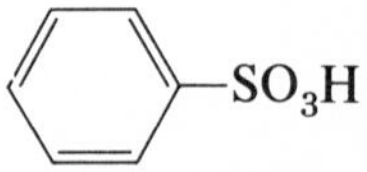 —SO_3H
acylation	$CH_3—\overset{O}{\overset{\|}{C}}—Cl^* + AlCl_3$	$CH_3—C^+=O$	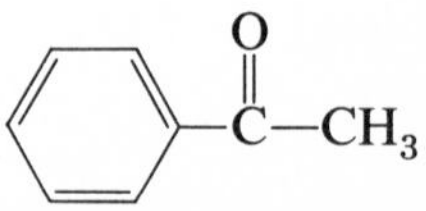 —$\overset{O}{\overset{\|}{C}}—CH_3$

*Or other acid chloride.

controls the position of attack by the electrophile. For example, because of the influence of the methyl group, the chlorination of toluene produces only the ortho and para isomers

toluene   *o*-chlorotoluene,   *p*-chlorotoluene,
                         60%   40%

Because of the influence of the nitro group, the chlorination of nitrobenzene produces only the meta isomer

nitrobenzene   *m*-chloronitrobenzene

The explanation for these effects lies in the properties of the resonance-stabilized intermediate formed during the substitution.

The intermediate in electrophilic aromatic substitution is a resonance-stabilized carbonium ion. Factors which affect the stability of ordinary carbonium ions (section 5-6) also affect the stability of these resonance-stabilized carbonium ions. For example, when an electrophile attacks toluene in the position ortho to the methyl group, one of the resonance forms of the intermediate is a *tertiary* carbonium ion, rather than the secondary carbonium ion which occurs when the carbon is unsubstituted. Note the first resonance form below:

ORTHO-PARA-DIRECTING GROUPS

Because tertiary carbonium ions are more stable than secondary carbonium ions, this mode of attack is favorable. Similar enhancement occurs following attack on the position para to the methyl group. When attack occurs meta to the methyl group, no direct resonance interactions with the methyl group are possible. Thus, electrophilic substitution reactions of toluene result in ortho and para substitution; a methyl group is said to be an *ortho-para-directing group*.

Amino groups, the hydroxyl group, and alkoxy groups are also ortho-para-directing. Additional resonance interactions are possible in these cases because of direct resonance interactions with an unshared electron pair on oxygen or nitrogen. The nitration of anisole produces ortho and para isomers

because of the resonance interactions involving the $—OCH_3$ group

Reactions of anilines are similar, provided that the reaction is not conducted in acidic solution. In acidic solution, the $—NH_2$ group is transformed into an $—NH_3^+$ group, which is a meta-directing group.

META-DIRECTING GROUPS The nitro group, $—NO_2$, can also have important effects on carbonium-ion intermediates, but the nitro group destabilizes the carbonium ion. The nitro group in nitrobenzene has two resonance forms,

both of which place a positive charge on the nitrogen atom. Attack of an electrophile on an ortho or para position of nitrobenzene would produce unfavorable interactions between the positive charge of the carbonium ion and the positive charge of the nitro-group nitrogen. Because of this, attack is meta, rather than ortho or para. Thus, a nitro group is a *meta-directing group*. However, even in the case of meta attack, the nitro group still exerts a deactivating effect on the reaction.

Other meta-directing groups include the ammonium group, the carbonyl group, and the cyano group. The directing effects of a variety of groups are summarized in table 11-2.

Ortho-Para-Directing	Meta-Directing	Table 11-2 Directive Effects in Electrophilic Aromatic Substitution

Ortho-Para-Directing

—CH$_3$
—Cl, —Br, —I
—OH, —OCH$_3$
—NH$_2$, —NHCH$_3$

Meta-Directing

$\overset{\text{O}}{\overset{\|}{—\text{C}}}$—CH$_3$
—NO$_2$
—SO$_3$H
—NH$_3^+$

Additional examples of the directing effects of substituents are the following: **EXAMPLES**

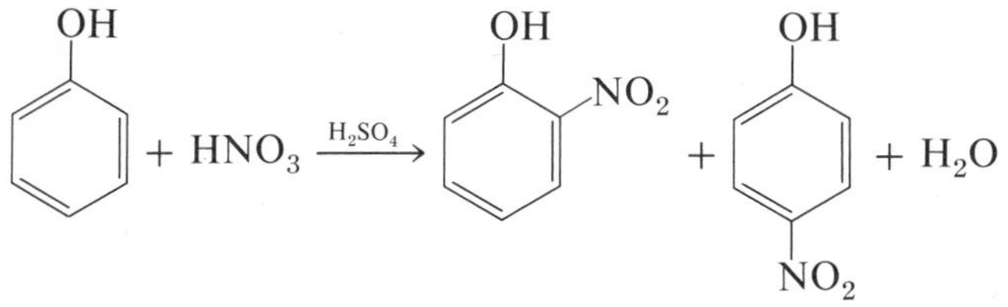

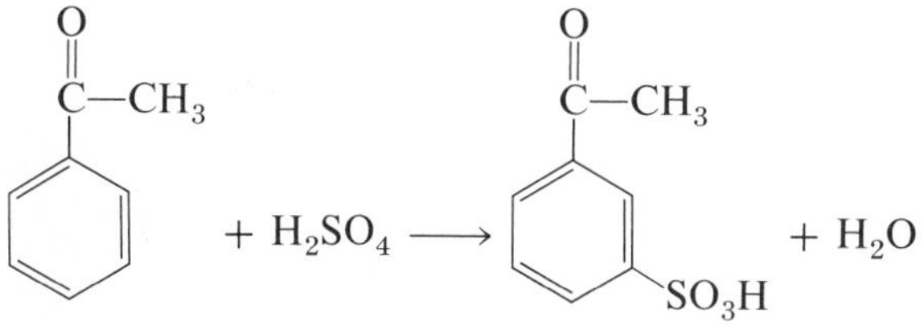

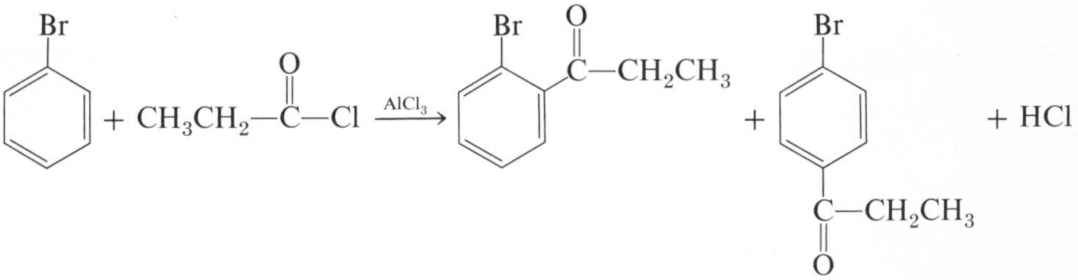

Give the products of the reaction of anisole with (a) H$_2$SO$_4$; **PROBLEM 11-10** (b) HNO$_3$ + H$_2$SO$_4$; (c) Cl$_2$ + FeCl$_3$. (d) Give the products of the reaction of nitrobenzene with the reagents in parts (a) to (c).

The nitro group has a strong effect on the acidity of phenol. Draw **PROBLEM 11-11** resonance structures of p-nitrophenol and its conjugate base, and predict whether this compound is more acidic or less acidic than phenol.

11-6 QUINONES

Aromatic compounds which have two hydroxyl groups oriented either ortho or para can easily be oxidized to ketonelike compounds called *quinones*. Mild oxidizing agents such as Ag^+, O_2, or an electric current can be used:

catechol   *o*-quinone

hydroquinone   *p*-quinone

1,4-naphthalenediol   1,4-naphthoquinone

The reverse reaction, reduction of the quinone, also occurs readily. 1,3-Dihydroxybenzene does not undergo oxidation to a quinone.

Because of the facility with which quinones undergo oxidation-reduction reactions, they are involved in many oxidation-reduction reactions in biochemistry. A great variety of quinones exist; they are found chiefly in higher plants, fungi, and bacteria. The ubiquinones (also called coenzyme Q) are important electron carriers in the oxidation-reduction reactions in living systems. A number of naturally occurring quinones are shown in figure 11-5.

BIOLOGICAL TRANSFORMA-TIONS LEADING TO QUINONES   The degradation of aromatic compounds in living systems usually proceeds by way of a series of oxidations of the aromatic ring. These oxidations lead to quinones and then to other degradation products.

One such route of particular importance is the route from the amino acid phenylalanine to *melanin,* the color pigment of skin (figure 11-6). Phenylalanine is first converted to tyrosine; both these amino acids are necessary for life and are constituents of proteins (chapter 12). Tyrosine is then further oxidized to dihydroxyphenylalanine (usually

*Figure 11-5
Some naturally
occurring quinones.*

lawsone, yellow
isolated from the tropical
shrub henna; Mohammed is
said to have dyed his beard
with henna

embelin, orange-yellow
found in the berries of an
Indian shrub; note the
long carbon chain

coenzyme Q
important in the oxidation-
reduction reactions of metabolism in
living things; the long chain is made out
of isoprene units; different organisms
have different numbers of isoprene units
in the chain

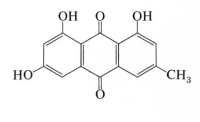

perezone, orange
isolated from the root
of the Mexican plant
Perezonia; it shares
with coenzyme Q the
unusual derivation from
isoprene units

emodin
widely distributed
in molds, higher fungi,
lichens, flowering plants,
and insects; this is the
most common of the quinones
having the carbon skeleton of
anthracene

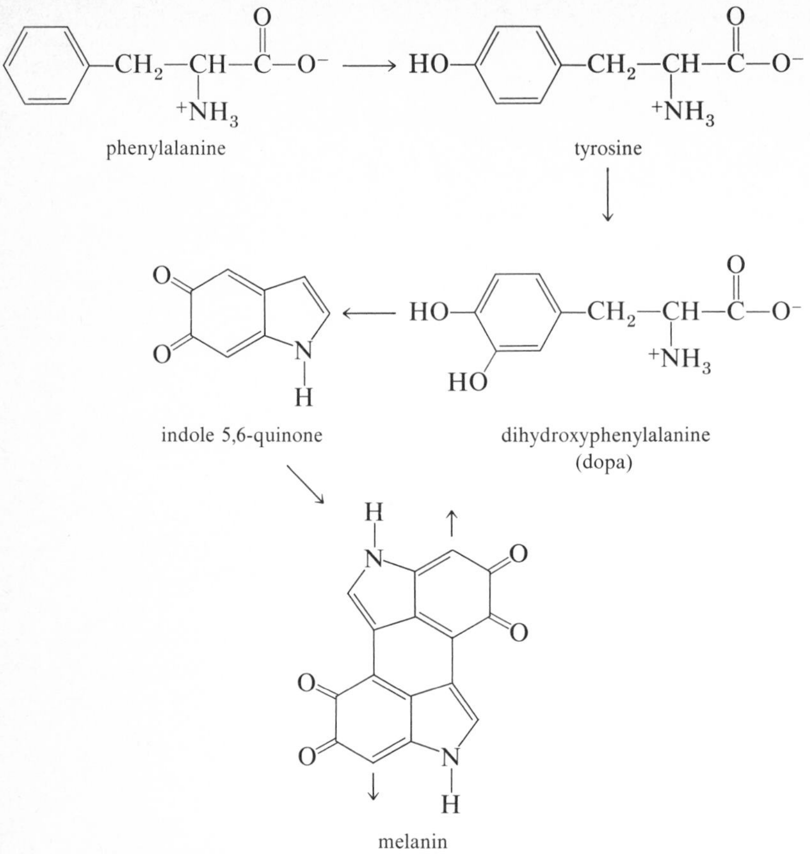

phenylalanine   tyrosine

indole 5,6-quinone   dihydroxyphenylalanine
(dopa)

melanin

Figure 11-6
Formation of
melanin from
phenylalanine.
In the first step phenylalanine is oxidized to tyrosine by the enzyme phenylalanine hydroxylase. Tyrosine is then transformed into dopa by the enzyme tyrosinase. The conversion of dopa into melanin is not enzyme-catalyzed. Melanin is a very large molecule formed by the condensation of a large number of the aromatic units shown. Adjacent units are connected at the points shown by the arrows.

called *dopa*), which is not a constituent of proteins but which is an important amino acid in the central nervous system (section 6-7). From dopa, a multistep reaction occurs which finally results in the formation of melanin. This sequence is of particular interest because it is not enzyme-catalyzed; the mechanism is unknown.

The enzyme which catalyzes the formation of tyrosine from phenylalanine is called *phenylalanine hydroxylase*. About one in every eighty people of European origin carries a recessive gene which prevents the formation of this enzyme* and leads to phenylketonuria (PKU), a

---

*Note that since this trait is recessive, a child may have PKU only if *both* of his parents have this recessive gene; thus, only about 1 in 6000 individuals might show signs of PKU.

condition in which large amounts of aromatic ketones are secreted in the urine. These compounds are toxic, and their presence may cause severe mental retardation. PKU can be controlled by restricting the dietary intake of phenylalanine.

The transformation of tyrosine into dopa is catalyzed by the enzyme *tyrosinase*. Absence of this enzyme leads to the hereditary condition known as *albinism,* in which the skin, hair, and eyes lack pigmentation.

11-7 HETEROCYCLIC COMPOUNDS

Cyclic compounds containing atoms other than carbon in the ring are called *heterocyclic compounds*. If the ring is not aromatic, the properties of such a compound are not appreciably different from those of noncyclic analogs. For example, pyrrolidine and tetrahydrofuran are

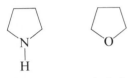

pyrrolidine   tetrahydrofuran

similar to other amines and ethers, respectively. On the other hand, if the ring is aromatic, there may be great differences in properties between these compounds and others we have considered.

Heterocyclic compounds are marked by diversity of both structures and reactions. Some important aromatic heterocyclic compounds are shown in figure 11-7.

Like benzene, heterocyclic aromatic compounds are characterized by *resonance*. For pyridine, two resonance forms can be written which are similar to those of benzene (figure 11-8), but for pyrrole, only one

PYRIDINE AND PYRROLE

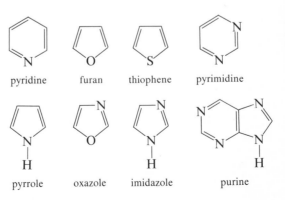

pyridine   furan   thiophene   pyrimidine

pyrrole   oxazole   imidazole   purine

Figure 11-7
Some heterocyclic
aromatic compounds.

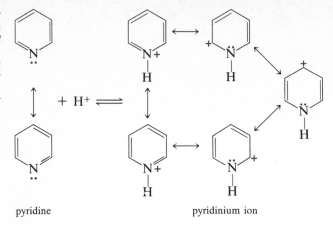

Figure 11-8
Resonance in
pyridine and
pyridinium ion.
Despite the extra
resonance of the
pyridinium ion,
pyridine is less basic
than simple amines.

pyridine   pyridinium ion

uncharged resonance form can be written (figure 11-9); the remaining resonance forms are charged. Nevertheless, these resonance forms are very important for the structure of pyrrole; pyrrole has a large amount of resonance stabilization.

REACTIONS
OF PYRIDINE

Pyridine is an amine, albeit a somewhat unusual one, and it can be converted by protonation into a conjugate acid, the pyridinium ion. A variety of chemical and physical studies demonstrate that the positive charge in the pyridinium ion resonates around the ring, as shown in figure 11-8. However, the nitrogen atoms in pyridine and in the pyridinium ion are sp^2-hybridized, and such nitrogen atoms are less basic than sp^3-hybridized nitrogens. Thus, in spite of the extensive resonance shown in figure 11-8, pyridine is about a millionfold *less* basic than simple tertiary amines such as triethylamine.

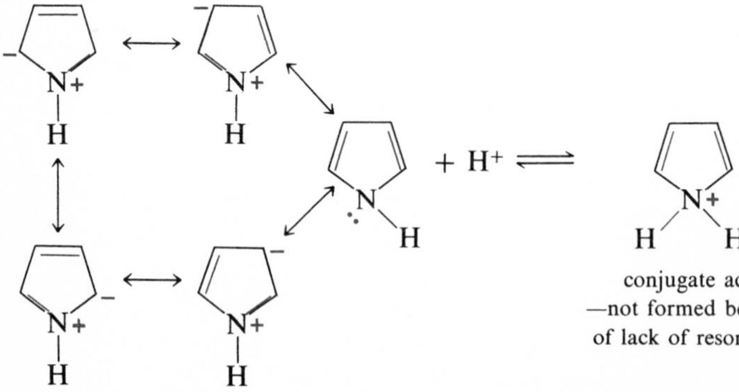

Figure 11-9
Resonance in
pyrrole. *The*
conjugate acid of
pyrrole is not formed
because it can have
no resonance.

+ H$^+$ ⇌

conjugate acid
—not formed because
of lack of resonance

pyrrole

The nitrogen atom of pyridine is also nucleophilic, and it can participate in a variety of displacement reactions. The products of such reactions are *N*-substituted pyridinium ions

pyridine $\qquad$ *N*-ethylpyridinium iodide

A biological example of this reaction has already been cited (section 5-12).

Unlike pyridine, pyrrole is hardly basic at all. If pyrrole were to form a conjugate acid, the nitrogen atom would have four σ bonds to different atoms and it would be unable to participate in resonance with the adjacent π electrons; all resonance would thus be destroyed. As a result, pyrrole can be protonated only in concentrated acid, and protonation occurs on carbon, rather than nitrogen. The product of this protonation still has limited possibilities for resonance.

    For the same reason that pyrrole is not very basic, the nitrogen atom in pyrrole is not nucleophilic, and pyrrole does not react with alkyl halides.

PYRROLE

Write resonance structures of (*a*) pyrimidine; (*b*) the conjugate acid of pyrimidine; (*c*) imidazole; (*d*) the conjugate acid of imidazole. Would you expect imidazole to be more basic or less basic than pyrrole? Why?

PROBLEM 11-12

Draw the resonance structures of the conjugate acid of pyrrole formed by protonation on the carbon adjacent to nitrogen. Draw the conjugate acid of pyrrole formed by protonation on nitrogen and show that no resonance is possible. Use this to explain why protonation of pyrrole occurs on carbon rather than on nitrogen.

PROBLEM 11-13

A great variety of heterocyclic compounds occur in nature. Among the important ones are a number of the vitamins, organic compounds that are needed in small amounts by living things in order to survive. Vitamins are not synthesized by the organism itself or are not synthesized in sufficient quantities; as a result, they must be present in the diet. The isolation and identification of these substances and the explanation of their roles in various deficiency diseases have been

VITAMINS

accomplished since the turn of the century and constitute some of the most significant examples of the interaction between chemistry and medicine.

Vitamins function in metabolism as *coenzymes,* small molecules that help enzymes catalyze reactions. The functions of many of these coenzymes have been elucidated in detail in recent years. Isolation and identification of these compounds were difficult both because of the complexity of many of the structures and because of the very small amounts present in living things (the daily requirement of vitamin B_{12} for adults, for example, is only 5 micrograms). In addition, many of these compounds undergo decomposition readily during isolation. This has posed a tremendous problem for the food industry because both long-term storage and extensive heating destroy at least some of the vitamins present in fresh foods. The balance between preserving vitamins and preventing spoilage is difficult.

A number of vitamins are heterocyclic compounds (figure 11-10). The functions of many of these compounds depend on the presence of heterocyclic rings.

11-8 REDUCTION OF AROMATIC COMPOUNDS

Reduction in organic chemistry is the addition of a hydrogen atom, H·, or a hydride ion, H^-. The simplest form of reduction is hydrogenation, a reaction which olefins undergo in the presence of molecular hydrogen and a metal catalyst (section 3-9).

The hydrogenation of benzene is possible, but it is considerably more difficult than the hydrogenation of olefins. At high temperature and high hydrogen pressure, benzene can be converted into cyclohexane

benzene   (high   cyclohexane
            pressure)

REDUCTIONS IN
BIOCHEMISTRY

Although the reaction above is important to the organic chemist, it is much harsher than is allowed in biochemical systems. The reduction of certain heterocyclic compounds in living systems occurs readily and has an important role in many biochemical oxidation-reduction reactions. The best-known case is nicotinamide adenine dinucleotide (NAD^+) (figure 7-7), which is derived from the vitamin niacin (figure 11-10). Reduction of NAD^+ is accompanied by the oxidation of some other substrate. Such reactions are reversible and are mediated by the action of a large number of enzymes. The reduced form of NAD^+, called NADH, has a considerable amount of resonance energy.

The reduction of NAD^+ is like the reductions involving lithium

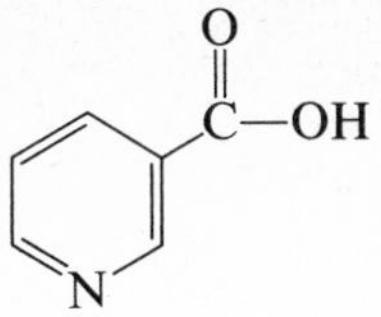

niacin

its absence causes pellagra—a
disease which has been known
for several hundred years;
niacin functions (as NAD⁺) in
many oxidation-reduction
reactions (figure 7-7)

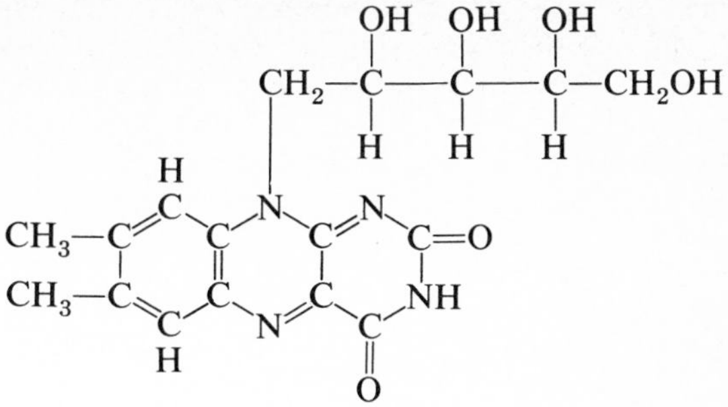

riboflavin (vitamin B₂)

occurs commonly in many plants and microorganisms,
but is not synthesized by animals; because of its
widespread occurrence, riboflavin deficiency is not
a serious nutritional problem; riboflavin is involved
in many oxidation-reduction reactions

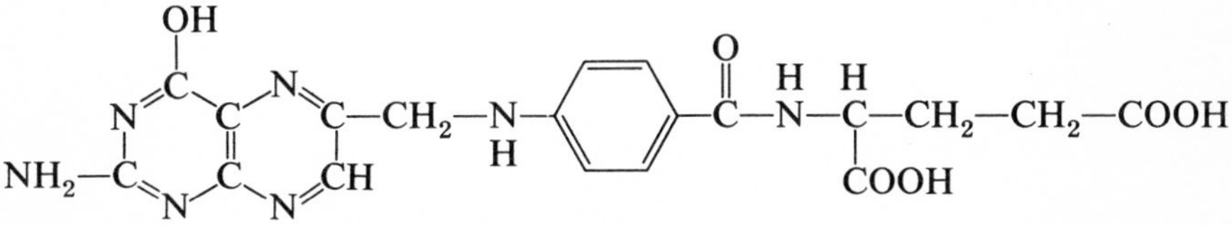

folic acid

important in DNA and RNA synthesis; absence
of folic acid causes a type of anemia; in
living systems folic acid serves as a carrier of
one-carbon fragments

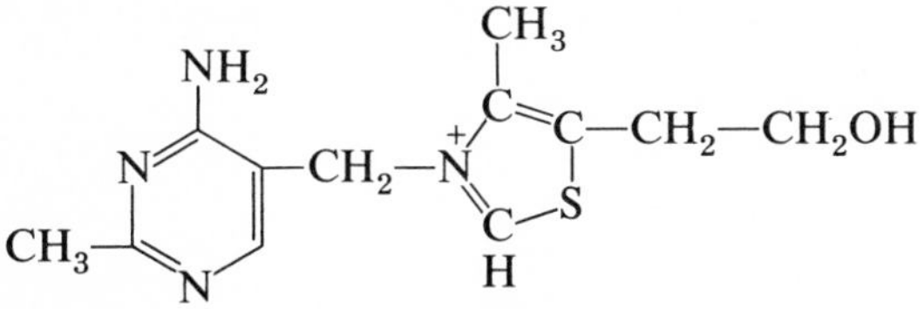

thiamine (vitamin B₁)

absence of thiamine in the diet causes beriberi;
the shells of wheat, rice, and other grains are a
rich source of this vitamin; in living systems
thiamine functions (in the form of a pyrophosphate
ester) as a coenzyme for a number of carbon-carbon
bond-making and bond-breaking reactions

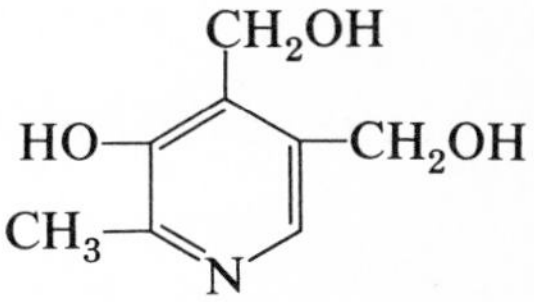

pyridoxine (vitamin B₆)

occurs in microorganisms
and in many green plants;
pyridoxine deficiency can
cause convulsions; an
oxidized form of pyridoxine,
pyridoxal 5′-phosphate, is
important in the metabolism
of amino acids

Figure 11-10
Some vitamins
containing
heterocyclic rings.

aluminum hydride and sodium borohydride considered in chapter 7, in that the reduction occurs by the transfer of a hydride ion (a hydrogen nucleus and a pair of electrons) from one compound to another.

PROBLEM 11-14 Do you think that the analog of NAD^+ in which the pyridine ring is replaced by a benzene ring would be capable of serving as an oxidizing agent in biochemical systems? Why?

PROBLEMS **11-15** Like other highly reactive, positively charged ions, carbonium ions can attack a benzene ring. Give the structure of the product of the following reaction, and then give the mechanism.

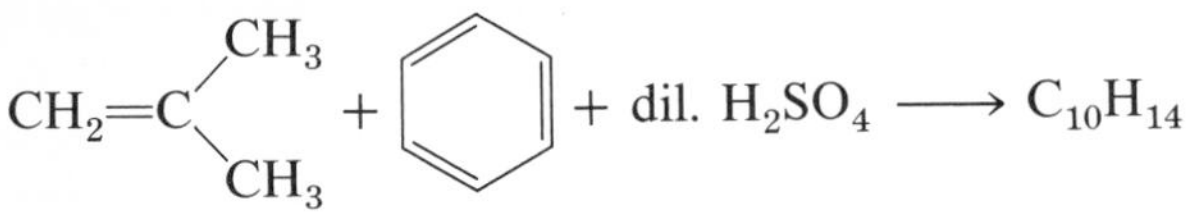

$$CH_2{=}C\overset{CH_3}{\underset{CH_3}{\diagup}} + \bigcirc + \text{dil. } H_2SO_4 \longrightarrow C_{10}H_{14}$$

11-16 Give the resonance forms of purine.

11-17 Do you think that pyrimidine would react readily with CH_3I? Why?

11-18 Pyrrole undergoes electrophilic aromatic substitution very much more readily than benzene. Give the mechanism of the following reaction, and show the resonance structures of the intermediate.

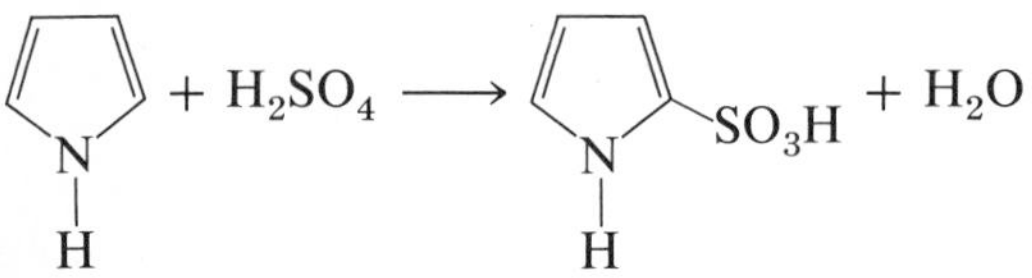

$$\underset{\underset{H}{|}}{\overset{\diagdown}{N}} + H_2SO_4 \longrightarrow \underset{\underset{H}{|}}{\overset{\diagdown}{N}}{-}SO_3H + H_2O$$

11-19 Give the mechanism of the reaction

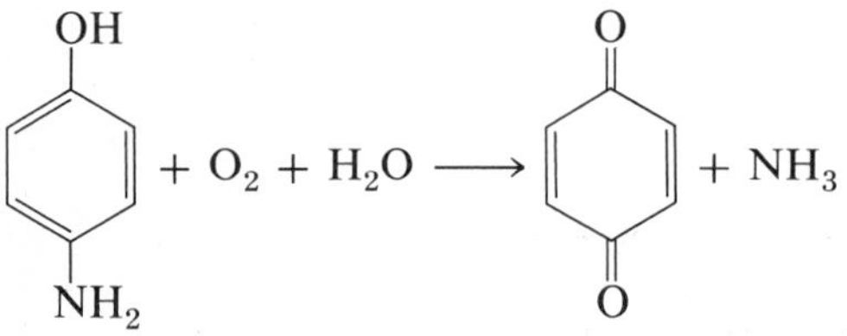

$$\underset{NH_2}{\overset{OH}{\bigcirc}} + O_2 + H_2O \longrightarrow \underset{O}{\overset{O}{\bigcirc}} + NH_3$$

11-20 Arrange the following compounds in order of increasing acidity:

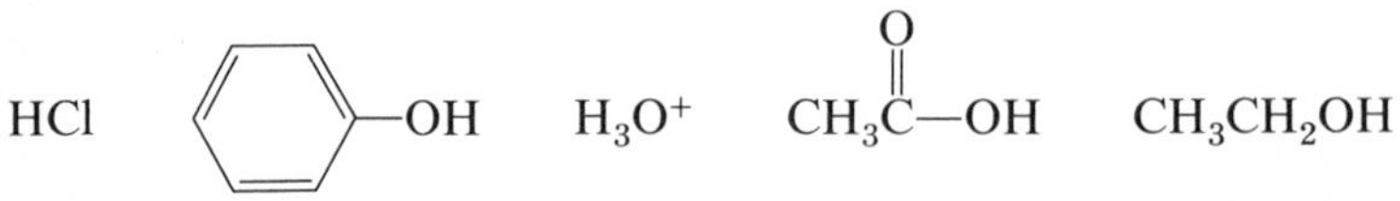

$$HCl \quad \bigcirc{-}OH \quad H_3O^+ \quad CH_3\overset{O}{\overset{\|}{C}}{-}OH \quad CH_3CH_2OH$$

11-21 Arrange the following compounds in order of increasing basicity:

NaOH ⬡(pyridine) (CH₃CH₂)₃N ⬠(pyrrole with N-H)

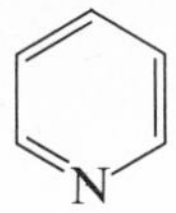

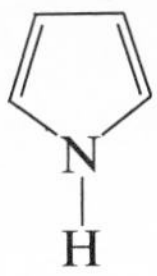

11-22 Triphenylmethyl chloride, $(C_6H_5)_3CCl$, forms a carbonium ion readily when it is dissolved in a polar solvent. Show the resonance structures of the carbonium ion.

11-23 Do you think that phenyl mercaptan, C_6H_5SH, is more acidic than ethyl mercaptan, C_2H_5SH? Why?

11-24 Give the structure of the product or products expected from the reaction of toluene with (*a*) H_2SO_4; (*b*) $H_2SO_4 + HNO_3$; (*c*) $C_6H_5COCl + AlCl_3$; (*d*) HCl; (*e*) $Cl_2 + FeCl_3$; (*f*) $H_2 + Pd$.

11-25 Certain aromatic compounds readily undergo acid-catalyzed hydrogen exchange. Give the mechanism of the following reaction, and explain why hydrogen exchange occurs much more rapidly in the para position than in the meta position:

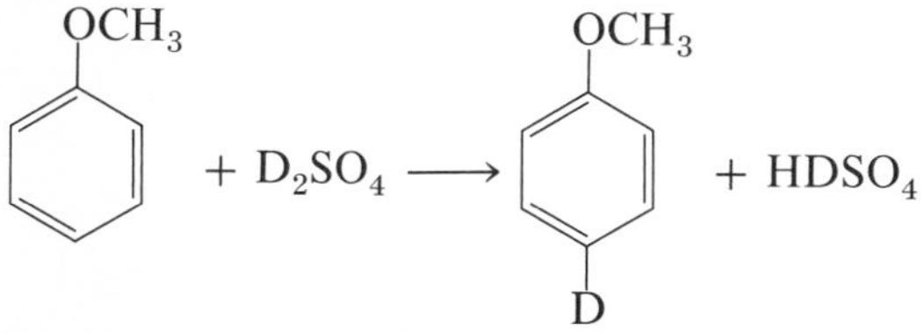

OCH₃ (on benzene) $+ D_2SO_4 \longrightarrow$ OCH₃ (on benzene with D at para) $+ HDSO_4$

11-26 The hydrogen exchange shown in problem 11-25 does not occur with trimethylanilinium chloride. Why not?

11-27 Give the structure of the product or products of each of the following reactions.

(*a*) (pyridine) $+ CH_3CH_2CH_2I \longrightarrow$

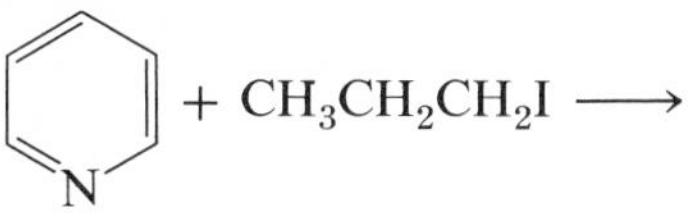

(*b*) (naphthalene) $+ H_2SO_4 \longrightarrow$ two products

(*c*) (1,2-dihydroxynaphthalene, OH, OH) $+ O_2 \longrightarrow$

(*d*) (aniline, NH₂) $+ CH_3CH_2I \longrightarrow$
(excess)

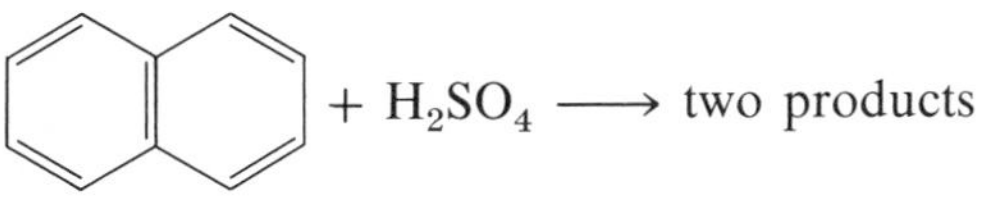

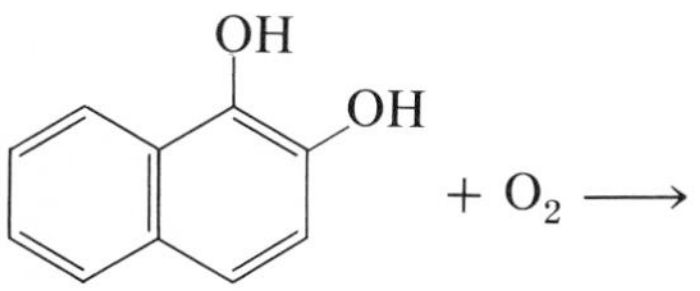

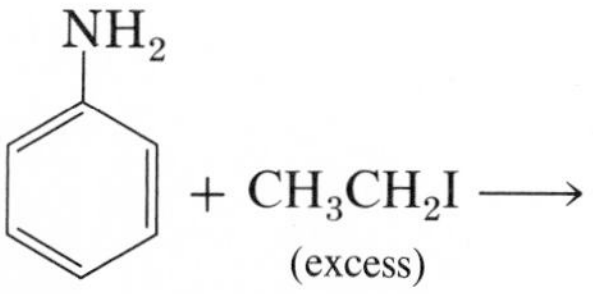

(e)

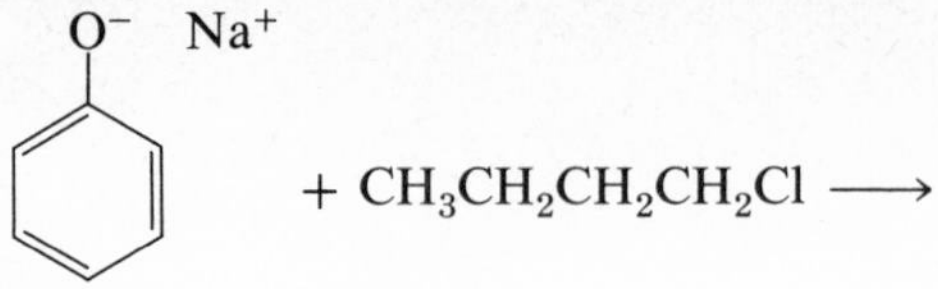

$+ CH_3CH_2CH_2CH_2Cl \longrightarrow$

(f)

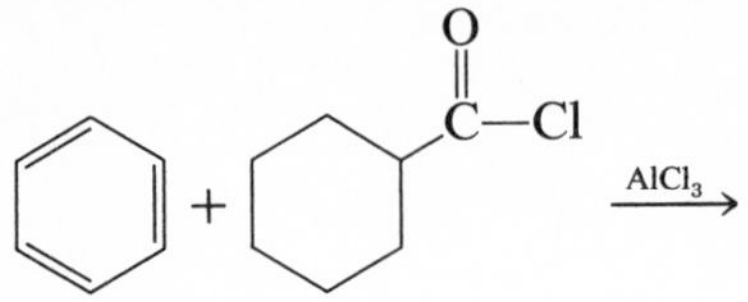

$\xrightarrow{AlCl_3}$

(g)

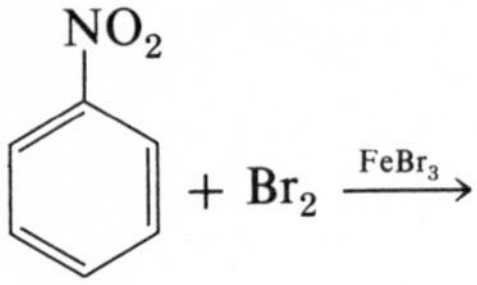

$+ Br_2 \xrightarrow{FeBr_3}$

(h)

$+ H_2SO_4 \longrightarrow$

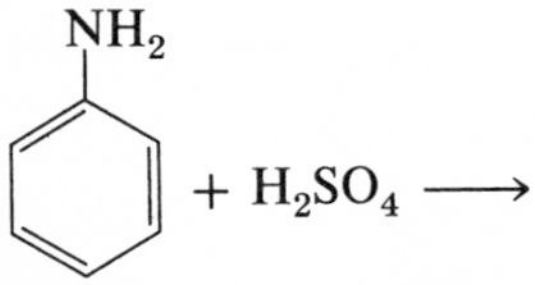

11-28   When it is necessary to attach several substituents to a benzene ring, the order in which the substituents are attached is important because of the directive effects of the substituents. Give the sequence of steps needed to synthesize the following compounds from benzene:

(a)

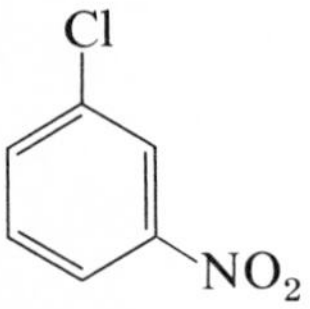

(b)

(c)

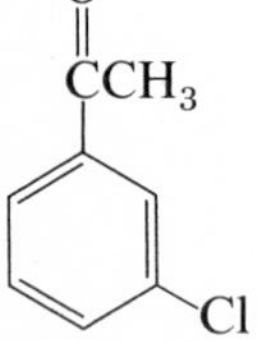

(d)

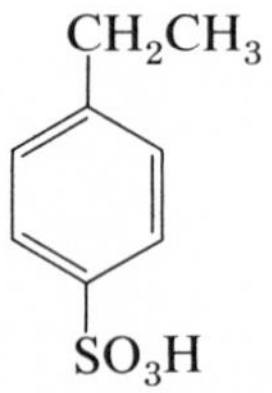

SUGGESTED READINGS   Breslow, R.: Aromatic Compounds, *Sci. Am.,* August 1972, p. 32.

Shea, K. P.: PCB, *Environment,* **15:** 25 (November 1973). The fate of halogenated aromatic compounds in the environment.

White-Stevens, R.: *Pesticides in the Environment,* Marcel Dekker, New York, 1971. Discusses the chemistry and biochemistry of DDT and other pesticides.

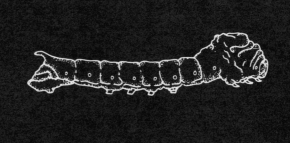

Amino Acids,
Peptides,
and Proteins

In spite of the efforts of generations of early chemists, the alchemist's dream of turning base metal into gold has not become a reality. The lowly silkworm comes closer to this goal than people have been able to: it turns leaves into silk. Silk is not gold, of course, but its beauty and quality are clear even to the casual observer, and it is a substance of great value. Silk is a protein, a large molecule made up of amino acids, and its virtue lies in the precision with which this protein is constructed.

Proteins are important components of every cell. All cells contain a wide variety of different proteins, which serve an equally wide variety of functions for the cell. Enzymes are proteins; hair and silk are proteins; cell walls are partly protein; the list could go on and on.

The purpose of this chapter is to consider the amino acids and then to examine how they are put together to form peptides and proteins. In spite of the great size and complexity of proteins, the rules which govern their structures are simple and can be understood in terms of concepts with which we are already familiar.

12-1 THE BUILDING BLOCKS: AMINO ACIDS

As the name implies, amino acids are both amines and acids. Those occurring in proteins have the amino group on the carbon atom adjacent to the carboxyl group; thus, they are called *α-amino acids*. Except for proline, all amino acids which occur in proteins have the structure

$$R-\overset{\displaystyle |}{\underset{\displaystyle NH_2}{CH}}-\overset{\displaystyle O}{\overset{\displaystyle \|}{C}}-OH$$

The difference in the various amino acids is in the group R, frequently called the *side chain,* which in some cases has an additional functional group.

A central group of about twenty amino acids occurs as components of the proteins in every living organism. Their structures are shown in figure 12-1, together with the commonly used abbreviations for their names.

With the exception of glycine, all the amino acids in figure 12-1 have at least one asymmetric carbon. In proteins these amino acids occur exclusively in the L configuration. D-amino acids occur occasionally in nature (section 4-7) but never as components of proteins.

THE SIDE CHAINS A variety of functional groups occur in amino acid side chains. Amino acids are classified according to the properties of these side chains:

1 Those which are *hydrophobic* (hydrocarbon side chains and a few others)

2 Those which are *polar* but *uncharged* (side chains containing hydroxyl groups, amide groups, and a few others)

3 Those which are *polar* and *charged* (side chains containing carboxylate anions, ammonium cations, and a few others)

All three types of side chains have important roles in proteins. By their shapes, sizes, and polarities they help determine the structures of proteins. By their chemistry they help determine the chemistry of proteins.

Several of the amino acid side chains shown in figure 12-1 are capable of resonance. Show resonance structures of the side chains of (*a*) tyrosine; (*b*) tryptophan; (*c*) glutamic acid; (*d*) histidine; (*e*) arginine.

PROBLEM 12-1

Proteins are an essential part of the diet of all animals. Many common foods—notably meat, eggs, milk, and certain grains—contain large amounts of protein. When these protein-containing foods are eaten, enzymes in the alimentary tract break down the proteins into their component amino acids. These amino acids are then absorbed through the intestinal wall and transmitted throughout the body to be used for the synthesis of proteins and for other functions.

PROTEINS AND AMINO ACIDS IN NUTRITION

Although twenty different amino acids are necessary for the formation of proteins, not all twenty need be included in the diet; some can be synthesized by the animal. On this basis, biochemists have divided the twenty amino acids into two groups. The *essential* amino acids must be supplied as part of the diet, and the *nonessential* amino acids can be synthesized by the animal. Different animals have different amino acid requirements (table 12-1).

The nutritional requirements of plants are different from those of animals. Plants are able to synthesize all twenty amino acids from CO_2, NH_3, H_2O, and light.

Although there are many possible sources of dietary protein, not all sources are equally good. It has long been known that animal protein is superior to plant protein as food for human beings. Plant proteins are frequently deficient in the essential amino acids lysine, methionine, and tryptophan. In underdeveloped areas of the world, children frequently suffer from *kwashiorkor,* a protein-deficiency disease characterized by anemia, growth retardation, and other symptoms. This condition usually appears in children who are fed mostly corn and other cereal grains. Such a diet provides inadequate protein and is also deficient in other essential dietary factors. The low lysine level

Figure 12-1
The common amino
acids. The ionic
forms shown are
those which
predominate at pH 7.

Hydrophobic Amino Acids

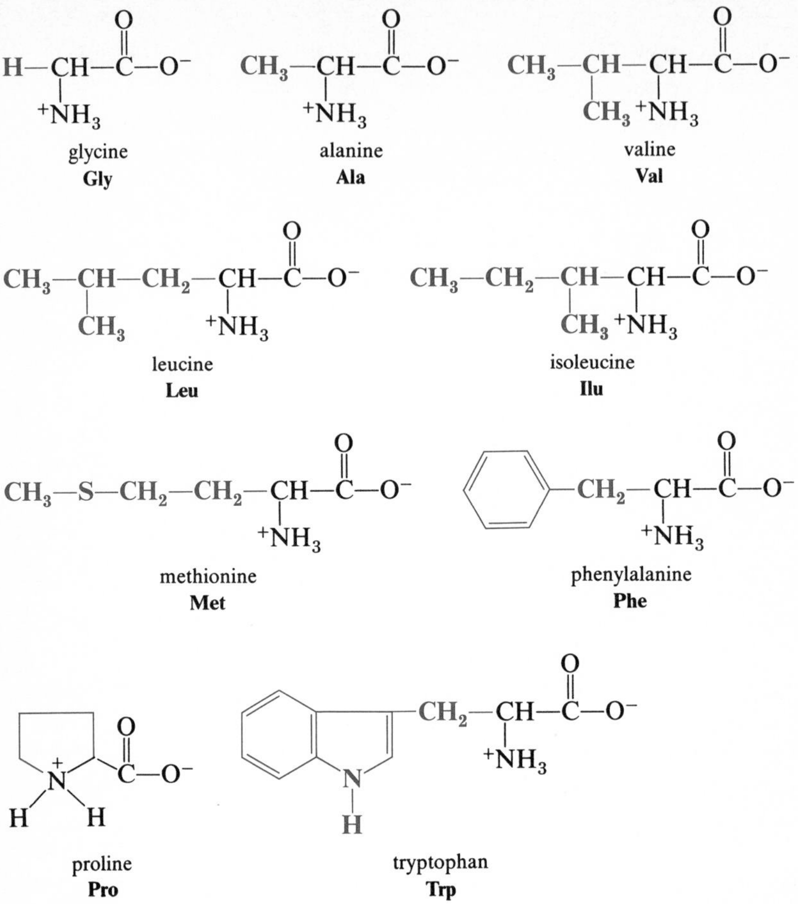

$$H-\underset{\underset{+NH_3}{|}}{CH}-\overset{\overset{O}{\|}}{C}-O^-$$

glycine
Gly

$$CH_3-\underset{\underset{+NH_3}{|}}{CH}-\overset{\overset{O}{\|}}{C}-O^-$$

alanine
Ala

$$CH_3-\underset{\underset{CH_3}{|}}{CH}-\underset{\underset{+NH_3}{|}}{CH}-\overset{\overset{O}{\|}}{C}-O^-$$

valine
Val

$$CH_3-\underset{\underset{CH_3}{|}}{CH}-CH_2-\underset{\underset{+NH_3}{|}}{CH}-\overset{\overset{O}{\|}}{C}-O^-$$

leucine
Leu

$$CH_3-CH_2-\underset{\underset{CH_3}{|}}{CH}-\underset{\underset{+NH_3}{|}}{CH}-\overset{\overset{O}{\|}}{C}-O^-$$

isoleucine
Ilu

$$CH_3-S-CH_2-CH_2-\underset{\underset{+NH_3}{|}}{CH}-\overset{\overset{O}{\|}}{C}-O^-$$

methionine
Met

phenylalanine
Phe

proline
Pro

tryptophan
Trp

of these grains is perhaps the most serious problem. Recent progress in breeding strains of corn which have a higher lysine content holds promise of alleviating this deficiency.

PROBLEM 12-2 The dietary requirement for tyrosine depends on the amount of phenylalanine in the diet. Why?

AMINO ACID NUTRITION AND CANCER Injection of the serum of the guinea pig or certain other animals into the bloodstream of some (though not all) patients with leukemia causes a temporary remission of the disease. The active ingredient of the

Polar Uncharged Amino Acids

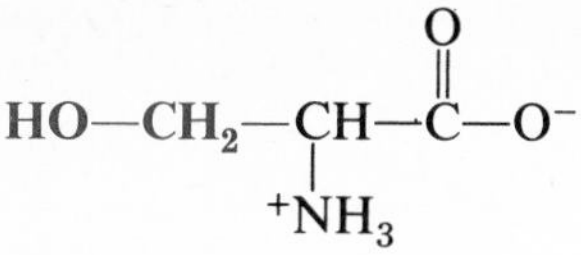

$$HO-CH_2-\underset{\underset{+NH_3}{|}}{CH}-\overset{\overset{O}{\|}}{C}-O^-$$

serine
Ser

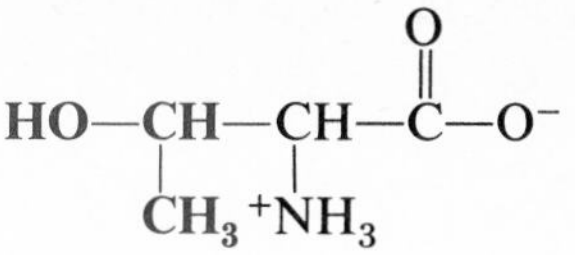

$$HO-\underset{\underset{CH_3}{|}}{CH}-\underset{\underset{+NH_3}{|}}{CH}-\overset{\overset{O}{\|}}{C}-O^-$$

threonine
Thr

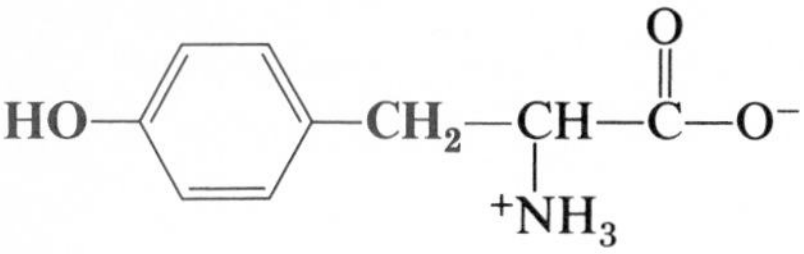

$$HO-\!\!\bigcirc\!\!-CH_2-\underset{\underset{+NH_3}{|}}{CH}-\overset{\overset{O}{\|}}{C}-O^-$$

tyrosine
Tyr

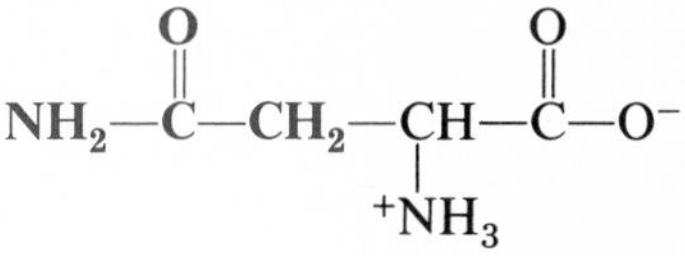

$$NH_2-\overset{\overset{O}{\|}}{C}-CH_2-\underset{\underset{+NH_3}{|}}{CH}-\overset{\overset{O}{\|}}{C}-O^-$$

asparagine
Asn

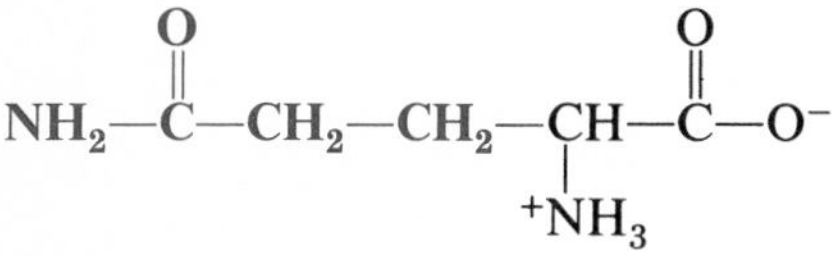

$$NH_2-\overset{\overset{O}{\|}}{C}-CH_2-CH_2-\underset{\underset{+NH_3}{|}}{CH}-\overset{\overset{O}{\|}}{C}-O^-$$

glutamine
Gln

$$HS-CH_2-\underset{\underset{+NH_3}{|}}{CH}-\overset{\overset{O}{\|}}{C}-O^-$$

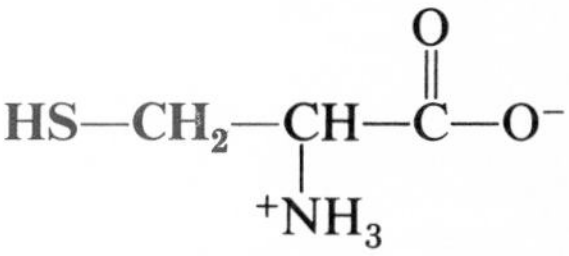

cysteine
Cys

Polar Charged Amino Acids

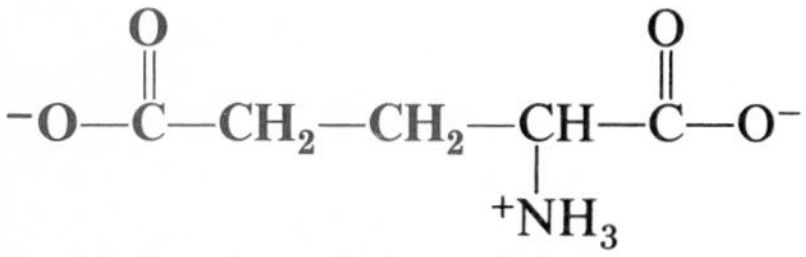

$$^-O-\overset{\overset{O}{\|}}{C}-CH_2-CH_2-\underset{\underset{+NH_3}{|}}{CH}-\overset{\overset{O}{\|}}{C}-O^-$$

glutamic acid
Glu

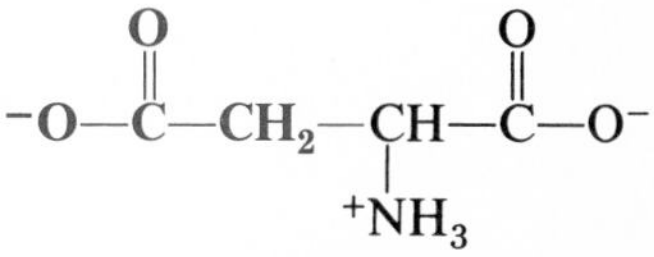

$$^-O-\overset{\overset{O}{\|}}{C}-CH_2-\underset{\underset{+NH_3}{|}}{CH}-\overset{\overset{O}{\|}}{C}-O^-$$

aspartic acid
Asp

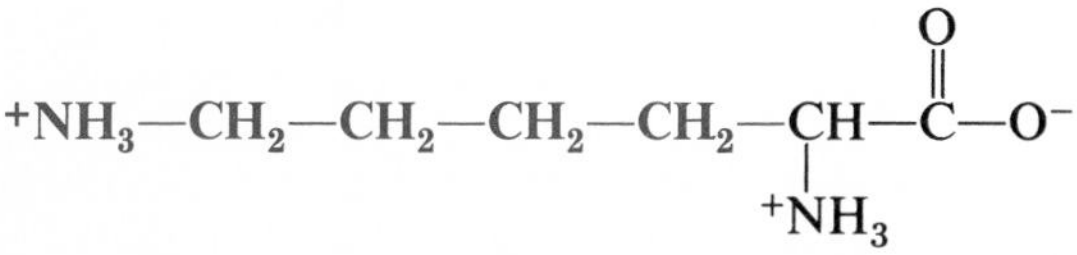

$$^+NH_3-CH_2-CH_2-CH_2-CH_2-\underset{\underset{+NH_3}{|}}{CH}-\overset{\overset{O}{\|}}{C}-O^-$$

lysine
Lys

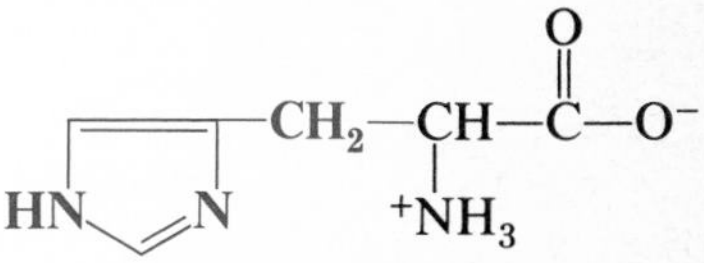

$$HN\!\!\diagdown\!\!N-CH_2-\underset{\underset{+NH_3}{|}}{CH}-\overset{\overset{O}{\|}}{C}-O^-$$

histidine
His

$$NH_2-\underset{}{\overset{\overset{NH_2}{|}}{C}}{+}-NH-CH_2-CH_2-CH_2-\underset{\underset{+NH_3}{|}}{CH}-\overset{\overset{O}{\|}}{C}-O^-$$

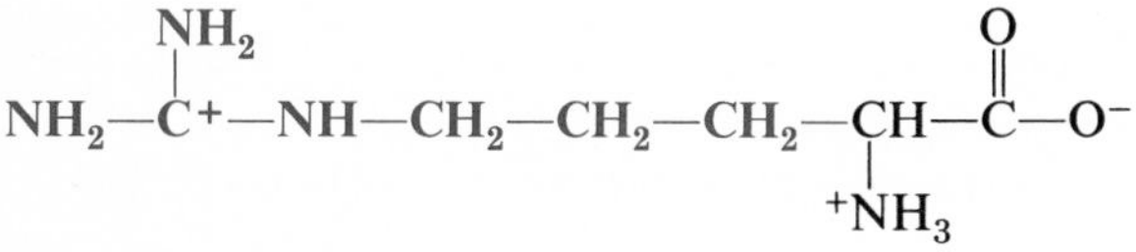

arginine
Arg

Table 12-1 Amino Acid Requirements	Amino Acid	Man	Dog	Honeybee	Plants
	alanine	0	0	0	0
	arginine	0	0	+	0
	aspartic acid	0	0	0	0
	cysteine	0	0	0	0
	glutamic acid	0	0	0	0
	glycine	0	0	0	0
	histidine	0	+	+	0
	isoleucine	+	+	+	0
	leucine	+	+	+	0
	lysine	+	+	+	0
	methionine	+	+	+	0
	phenylalanine	+	+	+	0
	proline	0	0	0	0
	serine	0	0	0	0
	threonine	+	+	+	0
	tryptophan	+	+	+	0
	tyrosine	0	0	0	0
	valine	+	+	+	0

SOURCE: A. Meister, *Biochemistry of the Amino Acids,* 2d ed., Academic Press, Inc., New York, 1965.
KEY: + = essential; 0 = nonessential.

guinea pig serum is the enzyme *asparaginase,* which catalyzes the hydrolysis of the amide group of asparagine

$$NH_2-\overset{\overset{O}{\|}}{C}-CH_2\underset{\underset{NH_2}{|}}{CH}-\overset{\overset{O}{\|}}{C}-OH + H_2O \xrightarrow{\text{asparaginase}} HO-\overset{\overset{O}{\|}}{C}-CH_2\underset{\underset{NH_2}{|}}{CH}-\overset{\overset{O}{\|}}{C}-OH + NH_3$$

asparagine   aspartic acid

Asparagine is an essential amino acid for these leukemic cells; since they cannot synthesize asparagine, they are unable to synthesize proteins and are unable to grow when asparaginase is present.

Asparaginase therapy has frequently been useful in the treatment of leukemia, but it is not successful against all types of leukemia and it is usually successful for only a limited time. This is a good example of a medical treatment which is made possible by differences between the metabolism of the host and that of the invader. Many other such examples exist.

12-2 ACID-BASE REACTIONS OF AMINO ACIDS

Although amino acids are frequently written as if they were neutral, uncharged compounds, for example,

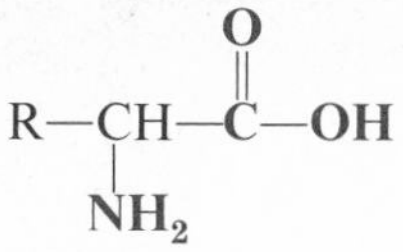

$$R-CH-\overset{\overset{\displaystyle O}{\|}}{C}-OH$$
$$|$$
$$NH_2$$

this representation is incorrect because it neglects the acidity of the carboxyl group and the basicity of the amino group. Because of this acidity and basicity, amino acids exist near neutral pH as internal salts, or *zwitterions,*

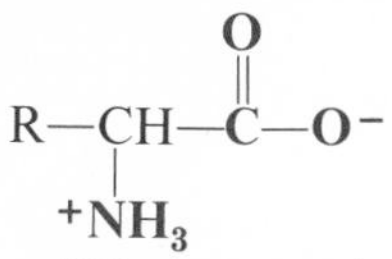

$$R-CH-\overset{\overset{\displaystyle O}{\|}}{C}-O^-$$
$$|$$
$$^+NH_3$$

When an amino acid is dissolved in an acidic or basic solution, the ionization properties of the amino and carboxyl groups are like those of simple amines and carboxylic acids (figure 12-2). At low pH, the amino group and the carboxyl group are both protonated, and the amino acid has a charge of $+1$. At neutral pH, the amino acid is a zwitterion and has an overall charge of zero. At high pH, neither the carboxyl group nor the amino group is protonated, and the amino acid has a net charge of -1.

Some amino acids have ionizable groups in the side chain. In those cases, the charge on the amino acid may sometimes be $+2$ at low pH, or -2 at high pH. The pK_a values for amino acids are similar to those of simple amines and carboxylic acids (table 12-2). The structures shown in figure 12-1 are those which predominate at pH 7.

The pH at which the amino acid has no net charge is called the isoelectric pH or the *isoelectric point*. Because of the differences in pK_a values for various amino acids and because of the presence of additional ionizable groups in some cases, the isoelectric points are different for different amino acids (table 12-2). Below the isoelectric point, an amino acid has a positive charge; above the isoelectric point, it has a negative charge. As a result, at a particular pH, different amino acids are charged to different extents. This difference is the basis of the most important method for analysis of amino acids (section 12-6).

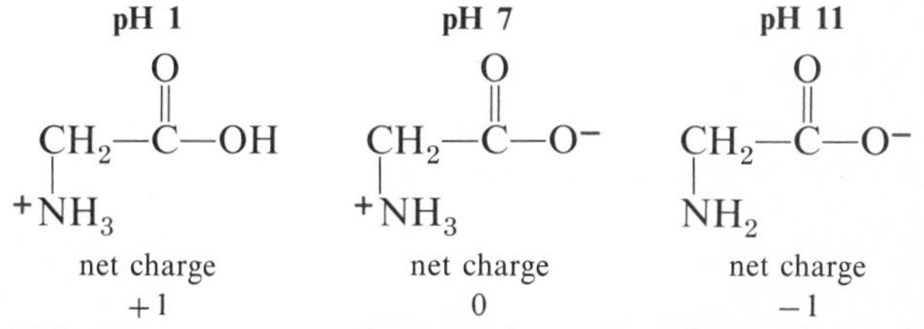

pH 1	pH 7	pH 11
$$CH_2-\overset{\overset{\displaystyle O}{\|}}{C}-OH$$	$$CH_2-\overset{\overset{\displaystyle O}{\|}}{C}-O^-$$	$$CH_2-\overset{\overset{\displaystyle O}{\|}}{C}-O^-$$
$^+NH_3$	$^+NH_3$	NH_2
net charge	net charge	net charge
$+1$	0	-1

For amino acids which have no ionizable groups in the side chain, the *charge on the amino acid varies from $+1$ at low pH to -1 at high pH.*

Figure 12-2
The pH dependence of amino acid structure.

	pK$_a$			
Amino Acid	Carboxyl	Amino	Side Chain	Isoelectric Point
glycine	2.3	9.6	—	6.0
alanine	2.3	9.7	—	6.0
phenylalanine	1.8	9.1	—	5.5
lysine	2.2	8.9	10.3	9.6
glutamic acid	2.2	9.7	4.3	3.2
histidine	1.8	9.0	6.0	7.5
serine	2.2	9.2	—	5.7
tyrosine	2.2	9.1	10.1	5.7

*Table 12-2 Ionization Constants and Isoelectric Points of Some Amino Acids**

*Data from J. P. Greenstein and M. Winitz, *Chemistry of the Amino Acids,* John Wiley & Sons, Inc., New York, 1961.
†The pK$_a$ values given for amino groups are actually the pK$_a$ values for the conjugate acid, the ammonium ion. This is the same convention as that used in chapter 6.

PROBLEM 12-3 Write the structure of the predominant form of each of the following amino acids at pH 2, pH 7, and pH 12: (*a*) valine; (*b*) glutamic acid; (*c*) lysine; (*d*) tyrosine.

12-3 THE CHEMICAL SYNTHESIS OF AMINO ACIDS

Ever since the occurrence of amino acids in nature was recognized in the early part of the nineteenth century, chemists have been interested in the synthesis of these materials, in order to have authentic compounds for comparison with the materials of unknown structure isolated from nature and to test the effects of these synthetic materials on living organisms.

When amino acids are needed in large quantities, they are sometimes chemically synthesized and sometimes isolated from natural sources. Certain bacteria produce useful quantities of amino acids. The advantage of this procedure is that the materials so obtained are optically active and of the proper L configuration. Chemical synthesis invariably produces racemic amino acids. Two important synthetic routes are given in the following discussion.

THE STRECKER SYNTHESIS Reaction of an aldehyde with a mixture of ammonia and hydrogen cyanide produces the nitrile of an amino acid

$$CH_3CH-\underset{O}{\overset{CH_3}{C}}H + NH_3 + HCN \longrightarrow CH_3CH-\underset{\underset{C\equiv N}{|}}{\overset{CH_3\ NH_2}{C}}H + H_2O$$

valine nitrile

This reaction proceeds in two stages: (1) the aldehyde reacts with ammonia to form an imine (mechanism 7-2); (2) HCN reacts with the imine to form the nitrile (mechanism 7-1).

If the nitrile is hydrolyzed with aqueous acid, an amino acid is formed

$$\underset{\underset{C\equiv N}{|}}{\overset{\overset{CH_3\ NH_2}{|\ \ \ |}}{CH_3CH-CH}} + H_2O + HCl \longrightarrow \underset{\underset{NH_2}{|}}{\overset{CH_3}{CH_3CH-CH}}\overset{\overset{O}{\|}}{-C}-OH + NH_4Cl$$

valine

Another example of this same synthesis is

phenylalanine

This synthesis, called the *Strecker synthesis,* is a versatile and widely used method for the chemical synthesis of amino acids. Its primary disadvantage is the necessity of using HCN, a highly toxic gas, as a reagent.

The adaptability of lower forms of life to various environments and nutritional states is amazing. In spite of the extreme toxicity of HCN to all living things, a number of fungi, known as *cyanogenic fungi,* actually make alanine by the reaction of acetaldehyde, ammonia, and HCN without killing themselves*

$$\overset{\overset{O}{\|}}{CH_3-CH} + NH_3 + HCN + H_2O \xrightarrow[\text{fungi}]{\text{cyanogenic}} \underset{\underset{NH_2}{|}}{CH_3-CH}\overset{\overset{O}{\|}}{-C}-OH + NH_3$$

alanine

---

*The concentration of free cyanide ion in the growing fungus is low, presumably because the cyanide is converted into alanine as fast as it is formed. Most of the cyanide is probably present as sugar cyanohydrins. Cyanide also appears to play a defensive role similar to that in certain millipedes (page 178). When the fungus is injured, say by an insect starting to eat it, cyanide production increases rapidly.

Write out the mechanism for the Strecker synthesis of valine.

Show how you would use the Strecker synthesis to make (*a*) leucine; (*b*) glutamic acid.

AMINATION OF BROMOACIDS The bromination of saturated hydrocarbon chains is usually of little synthetic use because it is completely random (section 3-8). However, carboxylic acids can be specifically brominated on the α carbon by using a trace of phosphorus as a catalyst

$$CH_3CH_2-\overset{\displaystyle O}{\overset{\|}{C}}-OH + Br_2 \xrightarrow{P} CH_3\underset{\underset{Br}{|}}{CH}-\overset{\displaystyle O}{\overset{\|}{C}}-OH + HBr$$

Reaction of this bromoacid with ammonia results in nucleophilic displacement (mechanism 6-1) and formation of an amino acid

$$CH_3\underset{\underset{Br}{|}}{CH}-\overset{\displaystyle O}{\overset{\|}{C}}-OH + NH_3 \longrightarrow CH_3\underset{\underset{NH_2}{|}}{CH}-\overset{\displaystyle O}{\overset{\|}{C}}-OH + HBr$$

(excess)　　　　　alanine

Another example of this method of synthesis is

$$CH_3\underset{\underset{CH_3}{|}}{CH}CH_2CH_2-\overset{\displaystyle O}{\overset{\|}{C}}-OH + Br_2 \xrightarrow{P} CH_3\underset{\underset{CH_3}{|}}{CH}CH_2\underset{\underset{Br}{|}}{CH}-\overset{\displaystyle O}{\overset{\|}{C}}-OH + HBr$$

$$CH_3\underset{\underset{CH_3}{|}}{CH}CH_2\underset{\underset{Br}{|}}{CH}-\overset{\displaystyle O}{\overset{\|}{C}}-OH + NH_3 \longrightarrow CH_3\underset{\underset{CH_3}{|}}{CH}CH_2\underset{\underset{NH_2}{|}}{CH}-\overset{\displaystyle O}{\overset{\|}{C}}-OH + HBr$$

(excess)　　　　　leucine

One difficulty with this reaction is that the amino group of the amino acid is nucleophilic and may react with another molecule of the bromo acid. This difficulty can be prevented by the use of a large excess of ammonia.

Why can't you make lysine by this route?

Show how you would use the bromoacid amination method to make (*a*) phenylalanine; (*b*) serine.

Allylglycine,

$$CH_2{=}CH{-}\underset{\underset{NH_2}{|}}{CH}{-}\overset{\overset{O}{\|}}{C}OH$$

is a synthetic amino acid that has been used in studies of the central nervous system. It can be made by one of the above syntheses but not by the other. Explain why and show the reactions.

The amino acids produced in the Strecker synthesis and in the bromo acid amination are invariably racemic. Since most enzymes are specific for the L isomers of amino acids, the question naturally arises as to the effect of having a racemic mixture. *USE OF SYNTHETIC AMINO ACIDS IN NUTRITION*

It is possible to resolve these racemic amino acids, thereby eliminating the unwanted enantiomer (section 4-6), but since this procedure is time-consuming and expensive, for commercial purposes it is to be avoided whenever possible.

Extensive nutrition studies have been conducted using racemic amino acids and using D-amino acids in order to determine whether the presence of D isomers is detrimental. D-Amino acids are not in general toxic to living things. If ingested by animals or people, D-amino acids suffer one of two fates: they may be excreted unchanged, or they may be oxidized by enzymes present in the liver (figure 4-12), thereby producing keto acids which are not chiral and can be used by the organism. However, the amounts of these oxidizing enzymes in the liver are usually not high, and D-amino acids have very limited nutritional value.

12-4   THE CHEMISTRY OF AMINO ACIDS

To a first approximation, we can consider that amino acids are amines and acids and their reactivity derives from these two functional groups. However, in some cases we will see that the functional groups interact and cannot be considered separately.

Because amino acids are both acids and bases, they can form salts by reaction with both bases and acids. Metal salts, for example, are formed on reaction with hydroxides

$$CH_3\underset{\underset{{}^+NH_3}{|}}{CH}{-}\overset{\overset{O}{\|}}{C}{-}O^- + NaOH \longrightarrow CH_3\underset{\underset{NH_2}{|}}{CH}{-}\overset{\overset{O}{\|}}{C}{-}O^-\ \ Na^+ + HOH$$

alanine   sodium alaninate

Chloride salts can be formed by reaction with hydrochloric acid

$$\underset{\underset{+NH_3}{|}}{CH_2}-\overset{\overset{O}{\|}}{C}-O^- + HCl \longrightarrow \underset{\underset{+NH_3}{|}}{CH_2}-\overset{\overset{O}{\|}}{C}-OH \quad Cl^-$$

glycine   glycine hydrochloride

Amino acids undergo reactions typical of carboxylic acids; for example, esterification

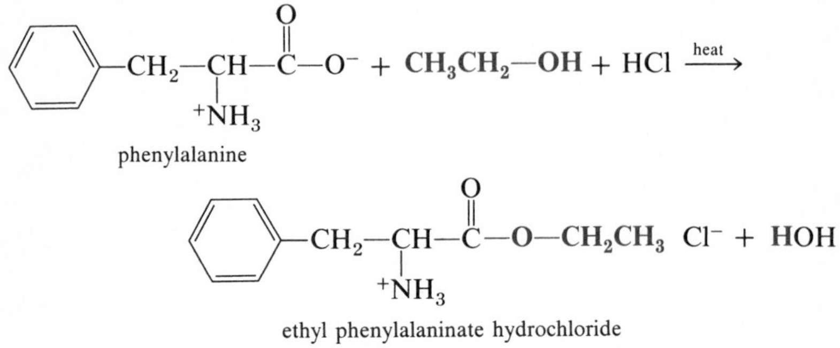

$$\bigcirc\!\!-CH_2-\underset{\underset{+NH_3}{|}}{CH}-\overset{\overset{O}{\|}}{C}-O^- + CH_3CH_2-OH + HCl \xrightarrow{\text{heat}}$$

phenylalanine

$$\bigcirc\!\!-CH_2-\underset{\underset{+NH_3}{|}}{CH}-\overset{\overset{O}{\|}}{C}-O-CH_2CH_3 \quad Cl^- + HOH$$

ethyl phenylalaninate hydrochloride

The amino group in amino acids is nucleophilic. Reactions with alkyl halides occur readily in alkaline solution

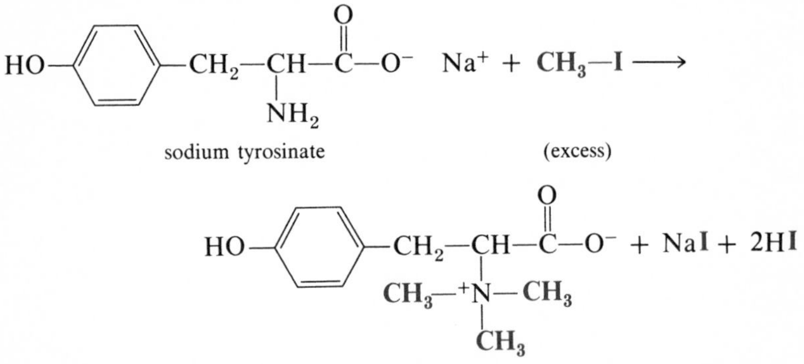

$$HO-\bigcirc\!\!-CH_2-\underset{\underset{NH_2}{|}}{CH}-\overset{\overset{O}{\|}}{C}-O^- \quad Na^+ + CH_3-I \longrightarrow$$

sodium tyrosinate   (excess)

$$HO-\bigcirc\!\!-CH_2-\underset{\underset{\underset{CH_3}{|}}{CH_3-{}^+N-CH_3}}{CH}-\overset{\overset{O}{\|}}{C}-O^- + NaI + 2HI$$

Amides can be formed by reaction with acid chlorides or anhydrides

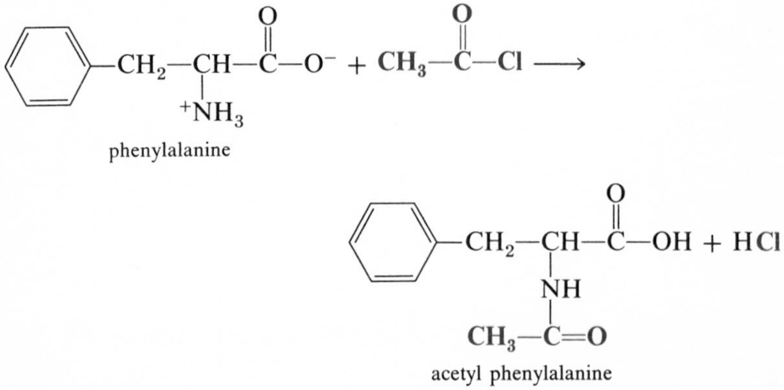

$$\bigcirc\!\!-CH_2-\underset{\underset{+NH_3}{|}}{CH}-\overset{\overset{O}{\|}}{C}-O^- + CH_3-\overset{\overset{O}{\|}}{C}-Cl \longrightarrow$$

phenylalanine

$$\bigcirc\!\!-CH_2-\underset{\underset{\underset{CH_3-C=O}{|}}{NH}}{CH}-\overset{\overset{O}{\|}}{C}-OH + HCl$$

acetyl phenylalanine

However, there are some differences between amino acids and simple compounds. For example, the hydrochloride salt of an amino acid ester is a stable compound, but if the hydrochloride is neutralized by addition of a weak base, such as $NaHCO_3$, the ammonium group becomes a primary amino group, which is nucleophilic, and over a period of a few hours this amino group will react with the ester function of another molecule (mechanism 10-2). The final product results from the combination of two molecules of amino acid into a cyclic *diketopiperazine*

$$2CH_3-\underset{\underset{+NH_3}{|}}{CH}-\overset{\overset{O}{\|}}{C}-O-CH_2CH_3 \quad Cl^- + 2NaHCO_3 \longrightarrow$$

$$+ 2CO_2 + 2H_2O + 2NaCl + 2HO-CH_2CH_3$$

a diketopiperazine

By the same token, certain derivatives of amino acids cannot be made because they would immediately react with themselves. The acid chloride of alanine, for example,

$$CH_3-\underset{\underset{NH_2}{|}}{CH}-\overset{\overset{O}{\|}}{C}-Cl$$

alaninyl chloride
(not a stable compound)

cannot be made because the amino group of the compound would react almost instantly with the acid chloride. However, the acid chloride of acetyl alanine

$$CH_3-\overset{\overset{O}{\|}}{C}-NH-\underset{\underset{CH_3}{|}}{CH}-\overset{\overset{O}{\|}}{C}-Cl$$

acetyl alaninyl chloride

can be made because the nitrogen of the amide is not sufficiently nucleophilic to react with the acid chloride.

PROBLEM 12-9

If optically active phenylalanine ethyl ester is used to form a diketopiperazine, only one stereoisomer is formed, but if racemic phenylalanine ethyl ester is used, two stereoisomers are formed. Explain, and give the structures of the compounds.

PROBLEM 12-10   Amino acids with functional groups in the side chain may undergo slightly more complex reactions than those given above. Give the product of each of the following reactions: (a) lysine + excess CH_3I; (b) glutamic acid + excess ethanol + H^+ (reflux); (c) serine + acetic anhydride; (d) ethyl lysinate hydrochloride + $NaHCO_3$; (e) diethyl glutamate hydrochloride + $NaHCO_3$.

12-5   STRUCTURES OF PEPTIDES AND PROTEINS

Amino acids serve two types of function in living systems: they are key intermediates in a number of metabolic pathways, and they are the building blocks from which peptides and proteins are made. Peptides and proteins are linear chains of amino acids ranging in size from a few amino acids to over a thousand amino acids in a single chain. The smaller members of the class, containing up to about fifty amino acids, are called *peptides;* larger compounds are called *proteins.*

Every cell contains many different peptides and proteins, each having a particular function for the cell. Even such simple organisms as the intestinal bacterium *Escherichia coli* contain several thousand different peptides and proteins. In the following sections we will explore the structures and properties of these important materials.

THE PEPTIDE   Adjacent amino acids along a peptide chain are held together by amide
LINK   linkages which connect the amino group of one amino acid to the carboxyl group of its neighbor (figure 12-3). Such an amide linkage is called a *peptide link.* Only the amino acids shown in figure 12-1 occur in proteins, and these amino acids are inevitably of the L configuration.

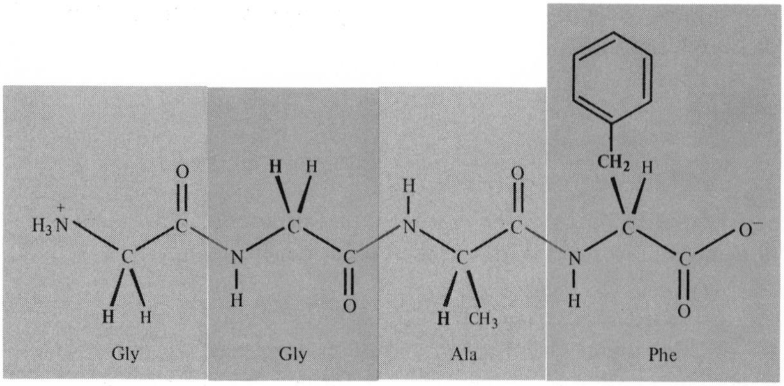

Figure 12-3
The tetrapeptide
Gly-Gly-Ala-Phe.
Peptide bonds are
shown in red.

The amino acid *sequence* of a protein or peptide is a listing of the order in which the amino acids occur along the peptide chain. For brevity, the abbreviations in figure 12-1 are usually used to specify the amino acids. The amino nitrogen of each amino acid is assumed to be on the left of the amino acid, and the carboxyl group is on the right. The farthest amino acid on the left (the *amino-terminal* amino acid) has a free amino group, and the farthest amino acid on the right (the *carboxy-terminal* amino acid) has a free carboxyl group.

Draw the structure of the peptide Phe-Glu-Ala-Met, and show that it is different from the peptide Met-Ala-Glu-Phe.

PROBLEM 12-11

Large molecules with many single bonds are frequently quite flexible and can adopt a variety of conformations. A peptide containing only five amino acids has a chain of sixteen atoms connected by single bonds and might adopt a seemingly endless variety of conformations. Larger peptides have correspondingly more possibilities for conformational isomerism. However, the geometry of peptides and proteins is very restricted, and the possibilities for conformational isomerism are limited.

GEOMETRY OF THE PEPTIDE LINK

The amide carbon, nitrogen, and oxygen of the peptide link are all involved in resonance (figure 10-2). This means that the carbon-nitrogen bond is to some extent a double bond, and its rotation is restricted. As a result, the α carbon (adjacent to the carbonyl carbon), the carbonyl carbon, the carbonyl oxygen, the amide nitrogen, its hydrogen, and the carbon of the next amino acid to which it is attached are all constrained to lie in the same plane (figure 12-4). Only two bonds in each amino acid are able to rotate, and only a small number of rotational configurations are stable. The results of this restriction will become clear when we consider the structures of proteins.

The amide carbon-nitrogen bond is a partial double bond and might

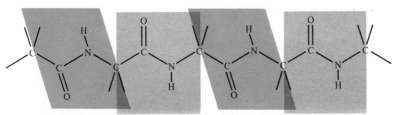

Because of the resonance of the peptide link, successive groups of four atoms along the peptide backbone are constrained to lie in the same plane. The carbonyl oxygen and the amide hydrogen also lie in this plane.

*Figure 12-4
Geometry of the peptide link.*

have two configurations, a cis configuration and a trans configuration:

cis configuration   trans configuration

The trans configuration is almost invariably observed in natural materials, although a few cases of cis configurations are known.

PROBLEM 12-12 Build a model of the peptide Gly-Gly-Gly-Gly-Gly, arrange the peptide links in the cis configuration, and draw a picture of it. Do the same with the peptide links in the trans configuration. Does this help to explain why peptides are usually trans?

PROBLEM 12-13 Build a model of the peptide Gly-Gly-Pro-Gly-Gly with all peptide links trans in order to show the effect of proline on conformation. Draw a picture of the structure.

12-6 ANALYSIS OF PEPTIDES AND PROTEINS

Proteins are long chains of amino acids held together by peptide links. Because proteins contain a hundred or more amino acids in a single chain, determining the complete structure of a protein is difficult. The process of protein-structure determination can be divided into three distinct parts: amino acid analysis (determining the amino acids present, with no regard for their sequence along the protein chain), amino acid sequence (specifying the order of amino acids along the chain), and tertiary-structure determination (determining the conformation of the chain).

ACID HYDROLYSIS OF PROTEINS Proteins are made up of amino acids, and one of the important reactions of proteins is hydrolysis of the protein to form a mixture of amino acids. Hot hydrochloric acid is generally used

$$\text{Protein} + H_2O \xrightarrow[\text{heat}]{\text{HCl}} \text{amino acid hydrochlorides}$$

The advantage of this procedure is that structure analysis of the amino acids is simpler than structure analysis of the intact protein. Although the structure of the protein cannot be reconstructed from the analysis of the hydrolyzed protein, the simplicity of this method of structure analysis has made it an important starting point for complete structure determination.

Until the mid-1940s, the separation and identification of individual amino acids was a slow and laborious procedure requiring large amounts of time and material. Today, the analysis of a mixture of amino acids can be accomplished by machine in a matter of a few hours and requires less than a milligram of material. The progress made in protein chemistry is the result of the development of *chromatography,* a rapid method for the separation and analysis of mixtures.

Chromatography is conducted by adsorbing the mixture to be separated to one end of a solid surface (frequently a finely powdered solid in a long tube called a *column*). A flow of liquid through the column gradually washes the material down the column. Different materials traverse the column at different rates and therefore reach the exit at different times. A sensitive detector placed at the exit measures the separated compounds as they emerge and records this information on a graph.

The quantitative analysis of mixtures of amino acids is rapid and routine today because of the amino acid analyzer (figure 12-5), a machine which carries out a chromatographic separation of amino acids according to their isoelectric points and measures the amounts of the separated amino acids.

AMINO ACID ANALYSIS

The amino acid sequence of a peptide or protein, called its *primary structure,* determines its biological function. Sequences for proteins having the same function in different animals are usually similar, though not identical.*

Determination of sequences is accomplished by cleaving the protein into a set of peptides containing from two to perhaps twenty amino acids each. The sequence of amino acids in such peptides can be determined by chemical methods. More than 300 amino acid sequences of proteins have been reported, and new ones are reported at the rate of several dozen each year.

AMINO ACID SEQUENCE

To illustrate the complexity of peptide sequencing, consider the following problem: a peptide was found to contain eight amino acids, one each of glycine, valine, arginine, aspartic acid, proline, phenylalanine, leucine, and isoleucine. How many possible amino acid sequences are there for this peptide?

PROBLEM 12-14

---

*The similarity of amino acid sequences for a particular protein from different species has been used as a criterion of evolutionary relationship between the two species. The differences in sequence indicate how far back in evolutionary time their common ancestor is; the more unlike the sequences, the farther back the common ancestor.

Figure 12-5
(a) *A modern amino
acid analyzer.* (b)
*The record of amino
acids produced by
this instrument.*
(*Beckman
Instruments.*)

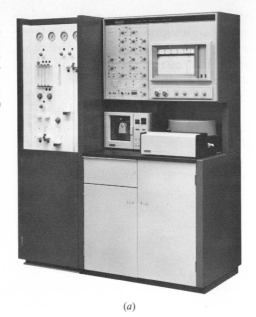

(a)

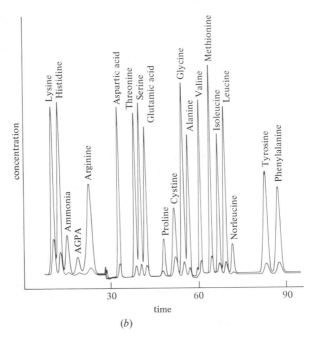

(b)

The chemical reactions of metabolism are capable of proceeding at very high rates because of the catalytic power of enzymes. However, these rates are carefully controlled in order to meet the needs of the organism for energy and for chemical reactants. The substances which control the rates of metabolic reactions are called *hormones*.

Hormones are produced by the endocrine glands, which include the thyroid, the adrenals, the pancreas, the testes, the ovaries, and others. These glands thus control the rates of a wide variety of metabolic processes; for example, oxygen consumption (by the thyroid), utilization of sugars (by the pancreas), pulse rate and blood pressure (by the adrenals), and sexual functions (by the testes and ovaries). Hormones are of two types, steroids (section 5-4) and peptides.

Insulin is typical of the peptide hormones secreted by these glands. Its amino acid sequence is shown in figure 12-6. Insulin has fifty-one amino acids in two peptide chains held together by disulfide bridges (section 12-7).

The sizes of the polypeptide hormones vary considerably. Insulin is one of the larger ones, with fifty-one amino acids. A number of hormones have ten or fewer amino acids, and at least one hormone is known which contains only three amino acids.

HORMONES

Each protein in a cell has a particular sequence of amino acids, and the function of the protein is closely related to that amino acid sequence. But how close is this connection? Does the organism have some ability to withstand sequence changes, or is the sequence-function connection so close that no sequence variation is possible? The answer to this question is beginning to emerge as a result of studies of the variation in amino acid sequence for different proteins having the same function.

Limited tolerance for sequence changes exists. The amino acid sequence of human insulin differs from that of bovine insulin shown in figure 12-6 by three amino acid replacements, and the sequence of porcine insulin differs from human insulin by one amino acid replacement. However, since all three species have survived, it is clear

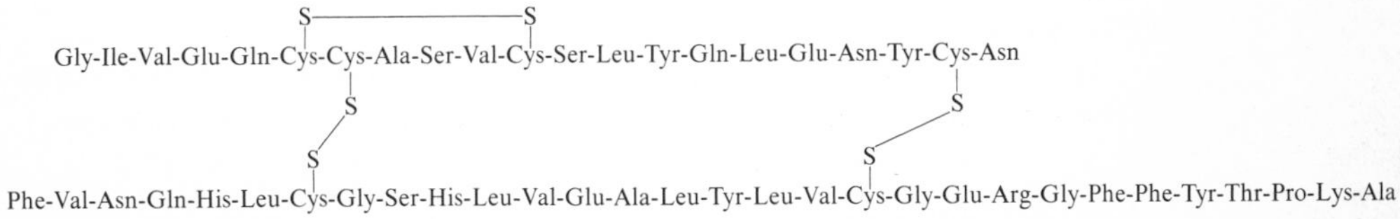

VARIATIONS IN AMINO ACID SEQUENCE

Gly-Ile-Val-Glu-Gln-Cys-Cys-Ala-Ser-Val-Cys-Ser-Leu-Tyr-Gln-Leu-Glu-Asn-Tyr-Cys-Asn

Phe-Val-Asn-Gln-His-Leu-Cys-Gly-Ser-His-Leu-Val-Glu-Ala-Leu-Tyr-Leu-Val-Cys-Gly-Glu-Arg-Gly-Phe-Phe-Tyr-Thr-Pro-Lys-Ala

The two peptide chains are connected by two disulfide bridges. F. Sanger won the 1958 Nobel prize for the determination of this sequence.

Figure 12-6
The amino acid sequence of bovine insulin.

that these amino acid changes are of no significance. The insulin used to treat human diabetes mellitus is usually a mixture of bovine and porcine insulin, and this material functions satisfactorily in spite of the amino acid replacements.

Much has been learned about the structure-function relationship as a result of studies of hemoglobin, the oxygen-carrying protein of blood. A large number of slightly altered forms of hemoglobin have been isolated. The best-known of these is sickle-cell hemoglobin, which is responsible for sickle-cell anemia. This disorder is characterized by erythrocytes (red blood cells) of an abnormal sickle shape, rather than the usual disk shape. People with this affliction usually suffer from anemia. The condition is hereditary and is observed most often in persons of African descent. Sickle-cell hemoglobin differs from normal hemoglobin in one single amino acid replacement: a glutamic acid of the normal material is replaced by valine. It is striking that one single change could have such profound effects on the properties of hemoglobin.

More than a hundred abnormal human hemoglobins have been reported. In most of them the sequence of amino acids differs by only a single amino acid replacement from the common sequence. It is clear that slight latitude is possible in the relationship between amino acid sequence and biological function, but it is likely that this latitude is not great.

PROBLEM 12-15 The sequence variations which can be tolerated best are those in which nonpolar amino acids replace nonpolar amino acids, etc. Which of the following amino acid replacements would probably be well tolerated and which would not: (*a*) glycine for valine; (*b*) glycine for glutamic acid; (*c*) glutamine for glutamic acid; (*d*) aspartic acid for glutamic acid; (*e*) lysine for glutamic acid; (*f*) methionine for lysine?

12-7 CONFORMATIONS OF PROTEINS

In living organisms, proteins adopt specific conformations. It is only after these conformations are assumed that proteins can function. The genetic code dictates the sequence of amino acids, but except for the sequence itself, there is no information in the genetic code which specifies conformation. That information is somehow inherent in the amino acids themselves and in their sequence along the protein chain.

Conformations of proteins are governed by several forces:

1 The amide linkage is planar because of resonance (section 12-5).

2 Carbonyl oxygens and amide hydrogens like to form hydrogen bonds. The $C{=}O \cdots\cdot H{-}N$ hydrogen bond occurs very commonly.

3 Hydrophobic amino acid side chains like to avoid contact with the aqueous environment; as a result, they usually aggregate in the interior of the structure in contact with other nonpolar side chains.

4 Polar side chains like to be on the exterior of the structure in contact with other polar side chains and with water.

5 Sulfhydryl groups of cysteine residues sometimes form disulfide linkages.

How these rules operate to produce a particular conformation is complex, and although we understand the rules, we still cannot predict three-dimensional structures from amino acid sequences. Nonetheless, nature does it exactly that way: sequence determines structure.

The next level of protein structure which we must consider is *secondary structure,* the arrangement of hydrogen bonds between different amino acids in the protein chain. There are two common types of secondary structure, the α helix and the β sheet.

A chain of L-amino acids (or a chain of glycines, but not a chain of prolines) can adopt a regular spiral (or helical) structure, as shown in figure 12-7. Each turn of the helix contains about 3.6 amino acids. The striking thing about this structure is its regularity; all amino acids (except those at the very ends of the helix) have identical environments. Every carbonyl oxygen atom is hydrogen-bonded to an amide hydrogen in the adjacent turn of the helix. The amino acid side chains stick out into the surrounding solution. The whole helix thus becomes a long, stiff rod and may contain hundreds of amino acids. Some proteins contain short regions of α helix made up of only a dozen or fewer amino acids.

THE α HELIX

The α helix occurs frequently in nature. Wool, for example, is a protein in the α-helical conformation. Because of the structure of the α helix, wool can stretch to twice its normal length without breaking. When wool is stretched, hydrogen bonds are broken but covalent bonds remain intact. The helix is restored and the hydrogen bonds are reformed when stretching is stopped.

If several polypeptide chains are stretched out (rather than coiled) and arranged with adjacent chains running in opposite directions, the result is a β sheet (figure 12-8). Because the α carbon atoms are tetrahedral, this sheet is puckered, or *pleated,* rather than planar. The β sheet structures are often formed by several polypeptide chains together. The collected chains form a relatively flat sheet, adjacent chains being held together by hydrogen bonds. If a single polypeptide chain is to form a pleated sheet, that chain must turn back upon itself once or more in order to form hydrogen bonds between adjacent sections of chain.

THE β SHEET

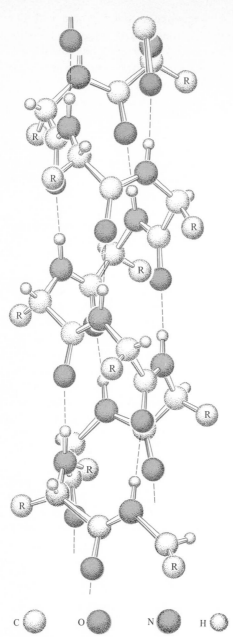

C ◯ O ⬤ N ⬤ H ◯

Figure 12-7
The α helix.

*Adjacent turns of the helix are connected
by hydrogen bonds (shown as red dotted
lines). All amide oxygens and nitrogens
are involved in hydrogen bonding. The
side chains (R) protrude into the
surrounding solution. (Reprinted from*

*Linus Pauling, The Nature of the
Chemical Bond, 3d ed. Copyright ©
1939 and 1940, 3d edition copyright ©
1960, by Cornell University. Used by
permission of Cornell University Press.)*

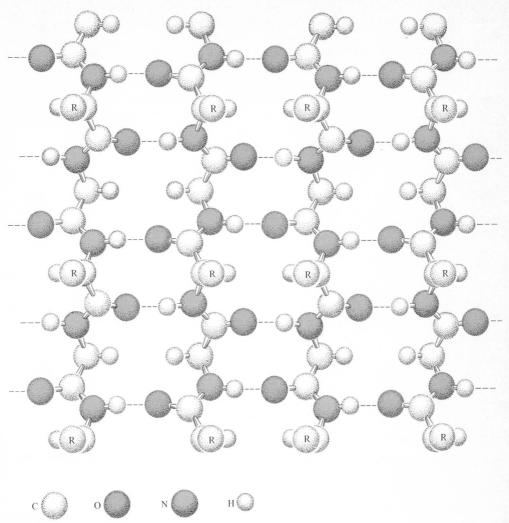

C ⚪ O ⚫ N ⚫ H ⚪

Polypeptide chains are stretched out in this structure, rather than coiled as in the α helix. Adjacent chains are connected by hydrogen bonds (shown as red dotted lines). (Reprinted from Linus Pauling,

The Nature of the Chemical Bond, 3d ed. Copyright © 1939 and 1940, 3d edition copyright © 1960, by Cornell University. Used by permission of Cornell University Press.)

Figure 12-8
The β sheet.

The β-sheet structure also occurs frequently in nature. Silk is a protein made up of many repetitions of the six–amino acid sequence

-Gly-Ser-Gly-Ala-Gly-Ala-

arranged in the β-sheet conformation. Because the protein chains in silk are stretched out, silk is not nearly as extensible as wool.

PROBLEM 12-16   Build molecular models of an α helix and a β sheet.

PROBLEM 12-17   What would be the effect of having a single D-amino acid in an α helix made up of L-amino acids?

PROBLEM 12-18   Would it be possible to make an α helix exclusively from D-amino acids? If so, what would it look like?

PROBLEM 12-19   Why can't proline be included in an α helix?

THE DISULFIDE BRIDGE

The conformations of proteins are determined primarily by hydrogen bonds and by interactions between polar and nonpolar side chains. However, conformations may also be determined in part by the disulfide bridge, a covalent bond between two cysteines.

The presence of cysteine in proteins is dictated by the genetic code. After a protein is formed and has adopted a conformation, two cysteines which are close together in the conformation can undergo oxidation to form a *disulfide bridge*, —S—S— (figure 12-9). Disulfide bridges are an effective means for stabilizing and stiffening a particular conformation. Structural proteins which are relatively rigid (such as hair, horns, wool, and hooves) have a high content of disulfide bridges.* Enzymes which exist in a hostile environment, like those in the mammalian digestive system, have disulfide bridges to hold them together and to protect them against loss of conformation.

*The permanent waving of hair involves the rearrangement of disulfide bridges. The bridges are temporarily broken, the hair is arranged into the desired configuration, and the bridges are reformed. They then hold the hair in this arrangement.

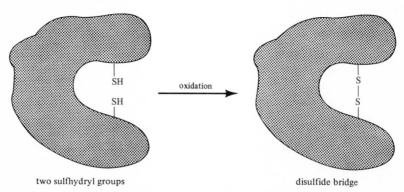

two sulfhydryl groups   disulfide bridge

Figure 12-9 Disulfide bridge formation.   *Oxidation of two nearby —SH groups in a protein results in formation of an —S—S— linkage, called a **disulfide***   ***bridge.** The double amino acid formed by this reaction is called **cystine.***

We cannot hope to understand the functioning of a protein until we know the amino acid sequence of the protein and the arrangement of the protein in space. This arrangement, called the *tertiary structure* of the protein, represents the final step in our definition of protein structure.

Two advances in protein chemistry which occurred around 1960 revolutionized our understanding of protein structures: perfection of the amino acid analyzer (figure 12-5) and the first determination of a protein structure by x-ray crystallography. The latter technique provides a complete picture of the arrangement of the protein in space and thus provides key information on how proteins function.

Myoglobin, the oxygen-storage protein of muscle, was the first protein whose structure was determined by x-ray crystallography. Myoglobin is a very small molecule by the standards of protein chemistry (it contains only 153 amino acids), but by the standards of organic chemistry it is very complex (it contains over 2400 atoms).

The three-dimensional structure of myoglobin is shown in figure 12-10. The most striking feature of this structure is the occurrence of large amounts of α helix: 121 of the 153 amino acids are in helical regions. The eight sections of α helix range in length from only 7 amino acids to 26 amino acids. The junctions between the helices are characterized by sharp bends, most of which contain proline.

In its center, myoglobin carries a heme, a complex heterocyclic compound consisting of an iron atom surrounded by four connected

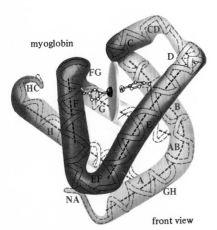

front view

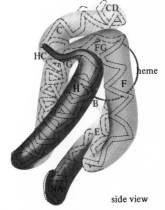

side view

The heme is a disk in the center of the molecule that is surrounded by eight stretches of α helix. The imidazole rings of two histidines interact with the iron atom of the heme on the left and right. Helices E and F form the walls of a box for the heme; B, G, and H are the floor; the C-D corner closes the open end. (Reproduction with permission from R. E. Dickerson and I. Geis, The Structure and Action of Proteins, W. A. Benjamin Company, Menlo Park, Calif. Copyright © 1969 by the authors.)

Figure 12-10
Schematic view of myoglobin.

pyrrole rings. The iron atom in the center of the heme is the oxygen-binding site.

12-8 ENZYMES

Throughout this book we have referred to the functioning of enzymes, the natural biological catalysts which accelerate nearly all reactions in metabolism. The catalytic power of enzymes is enormous; many chemical reactions proceed more than a billion times faster in the presence of an enzyme than in its absence. In this section we will consider the structures of enzymes and the forces which make them work.

STRUCTURE OF ENZYMES

Enzymes are proteins. Their sizes vary from slightly over a hundred amino acids in a single protein chain to several thousand amino acids in several chains. Amino acid sequences have been determined for a large number of enzymes and three-dimensional structures for a much smaller number. Although the three-dimensional structures of enzymes seem very complex, the same rules govern the structures of these proteins as govern the structures of other proteins.

In addition to the protein, some enzymes also contain small molecules called *coenzymes,* which help in the catalytic process. Some examples of coenzymes have already been given (figure 11-10). Coenzymes are synthesized from vitamins in the cell. Some enzymes also contain metal ions.

THE ACTIVE SITE AND ENZYME FUNCTION

Each enzyme has an *active site,* a particular confined space on the surface of the enzyme where the substrate (the reactive molecule) binds and reaction occurs. The shape and nature of this active site are crucial to the operation of the enzyme.

Enzyme function involves three factors:

1 *Catalysis,* the ability of the enzyme to increase the rate of the reaction

2 *Specificity,* the ability of the enzyme to choose its substrate and reject unwanted molecules

3 *Control,* the ability of the enzyme to increase or decrease its reaction rate according to the needs of the organism

These three factors will be considered in the following three sections.

CATALYSIS

Enzymes obey the laws of chemistry. In spite of the arguments of some past students of enzyme mechanisms, there is nothing magical about how enzymes catalyze reactions; their incredible catalytic power is due to the presence at the active site of amino acid functional groups that can act as catalysts. The arrangement of these groups within the active site has been carefully optimized through millions of years of evolution.

The majority of the chemical reactions we have considered involve one or more proton-transfer steps in their mechanisms. Enzymes can accelerate such reactions by providing acid or base catalysts to increase the rates of those reaction steps. The side chains of glutamic acid, aspartic acid, and histidine are particularly well suited for this role.

Acid-base catalysis is made all the more effective by the influence of *orientation*. The substrate binds firmly to a particular place on the enzyme surface, and the enzyme is so arranged that the acidic or basic groups needed for reaction are located exactly where they are needed.

Orientation plays many roles in enzymatic catalysis. In a number of cases, a functional group of the enzyme forms a covalent bond with the substrate during the course of the catalytic reaction (for example, acetoacetate decarboxylase, figure 7-6, and acetylcholinesterase, figure 10-5). The functional group involved is invariably located in precisely the right orientation to optimize the reaction.

An enzyme must be specific; it must be able to choose its proper *SPECIFICITY* substrate from among many similar molecules in the cell. Otherwise the cell would be unable to function.

Specificity is the result of shape complementarity. The active site of an enzyme is a hole on the surface of the enzyme whose shape is precisely complementary to that of the substrate, much as the shape of a lock is complementary to the shape of the key which fits it. Several examples of this complementarity have already been given (figures 3-7, 4-12, 5-13, and 10-5).

Specificity is not enough. In many cases, the cell must be able to *CONTROL* increase or decrease the rate of a chemical reaction according to its needs at that particular time. Several mechanisms are available by which the cell can do this.

The most important factor affecting the rate of an enzyme-catalyzed reaction is the concentration of the substrate. Within the range of concentrations usually found in the cell, increasing substrate concentration causes increasing reaction rate. However, at high substrate concentrations the reaction rate is independent of substrate concentration.

The activities of many enzymes are changed by the mediating effects of *hormones* (section 12-6). These mediators work by causing the cell to synthesize additional molecules of enzyme. The effects of these hormones are highly specific; only one or two enzymes are sensitive to the mediating effects of a particular hormone.

Table 12-3
Types of Enzymes

Type	Reactions Catalyzed
oxidoreductases	oxidations and reductions of carbonyl compounds and other compounds
transferases	transfer of an alkyl or acyl group from one molecule to another
hydrolases	hydrolysis of esters, amides, and other compounds
lyases	cleavage of bonds between carbon and some other atom
isomerases	isomerization

Other metabolites may also affect the rates of enzyme-catalyzed reactions. The final product of the operation of a sequence of metabolic reactions frequently affects the first enzyme in the metabolic sequence.

THE VARIETY OF ENZYMES Metabolism is a complex process involving a large number of chemical reactions. Each reaction is catalyzed by a particular enzyme; thus, it is not surprising that several thousand separate and distinct enzymes are known. These enzymes are often classified according to the type of reaction catalyzed. A summary of the types of enzymes, together with references to examples that have been given elsewhere in this book, is given in table 12-3.

PROBLEMS **12-20** Define and give an example of: (*a*) zwitterion; (*b*) enzyme; (*c*) α helix; (*d*) β pleated sheet; (*e*) essential amino acid; (*f*) polar amino acid; (*g*) hormone; (*h*) peptide.

12-21 Give the structure of the product of the reaction of leucine with (*a*) HI; (*b*) NaOH; (*c*) propionyl chloride; (*d*) excess isopropyl iodide; (*e*) excess methanol + HCl; (*f*) excess methanol + HCl, then acetyl chloride.

12-22 Discuss three essential properties of enzymes which enable them to catalyze reactions.

12-23 Asparagine and glutamine are not included in the amino acid requirements list (table 12-1). Do you think these amino acids are essential or nonessential? Why?

alcohol dehydrogenase and aldehyde dehydrogenase (figure 7-7)

methyltransferases (figures 5-16 and 5-17)
acetyl-CoA transferases (figure 10-6)
kinases (figure 10-10)
acetylcholinesterase (figure 10-5)
inorganic pyrophosphatase (page 273)
urease (page 269)
fumarase (figure 5-13)
acetoacetate decarboxylase (figure 7-6)
carbonic anhydrase (page 268)
amino acid decarboxylases (page 166)
amino acid racemases (page 99)

12-24   Lysine can react with itself under certain conditions to form a cyclic amide. Give the structure of the amide.

12-25   The amino acid analyzer (figure 12-5) usually has two columns for separating amino acids. The short column separates amino acids which have a positive charge at pH 6 but does not distinguish between other amino acids. The long column separates the remaining amino acids. What amino acids would be separated on the short column?

12-26   Design a synthesis of proline based on the Strecker synthesis.

12-27   What would be the effect of a single proline in the middle of a β-sheet structure?

12-28   The chemical synthesis of peptides has long been an important area of chemical investigation. Why is the following reaction not practical for the synthesis of Phe-Gly?

Phenylalanine ethyl ester + Gly $\longrightarrow$ Phe-Gly + ethanol

12-29   Many enzymes which catalyze reactions of glutamic acid and aspartic acid have very rigid and specific binding sites for the two carboxyl groups of these amino acids. Some enzymes are able to bind both amino acids, and some enzymes are able to bind only glutamic acid. Use molecular models to explain the difference between these two classes of enzymes.

12-30   Estimate the total charge of insulin (figure 12-6) (*a*) at pH 1; (*b*) at pH 12.

12-31   Life would probably not have been possible if nature had tried to make proteins by using ester linkages instead of amide linkages. Why?

12-32 Monosodium glutamate is frequently used as a seasoning because it enhances the taste of food without having much flavor of its own. Do you think this practice is safe? Why?

SUGGESTED READINGS

Dickerson, R. E.: The Structure and History of an Ancient Protein, p. 189 in *Organic Chemistry of Life: Readings from Scientific American,* Freeman, San Francisco, 1973. Discusses the relationship between amino acid sequence and genetic ancestry for cytochrome *c*. Interesting pictures of the tertiary structure of this protein.

——— and I. Geis: *The Structure and Action of Proteins,* Benjamin, Menlo Park, Calif., 1969. A well-illustrated, readable account of amino acid, peptide, and protein structure. Highly recommended.

Doty, P.: Proteins, p. 139 in *Organic Chemistry of Life: Readings from Scientific American,* Freeman, San Francisco, 1973.

Old, L. J., E. A. Boyse, and H. A. Campbell: L-Asparaginase and Leukemia, p. 102 in *Organic Chemistry of Life: Readings from Scientific American,* Freeman, San Francisco, 1973.

Phillips, D. C.: The Three-dimensional Structure of an Enzyme Molecule, p. 170 in *Organic Chemistry of Life: Readings from Scientific American,* Freeman, San Francisco, 1973. One of the early and very successful applications of x-ray crystallography to determination of protein structure. Excellent pictures.

Stein, W. H., and S. Moore: The Chemical Structure of Proteins, p. 147 in *Organic Chemistry of Life: Readings from Scientific American,* Freeman, San Francisco, 1973. Determination of the amino acid sequence of proteins.

Thompson, E. O. P.: The Insulin Molecule, p. 183 in *Organic Chemistry of Life: Readings from Scientific American,* Freeman, San Francisco, 1973. The determination of the amino acid sequence of insulin.

Wold, F.: *Macromolecules: Structure and Function,* Prentice-Hall, Englewood Cliffs, N.J., 1971. Good introduction to the structures of proteins and other large biological molecules.

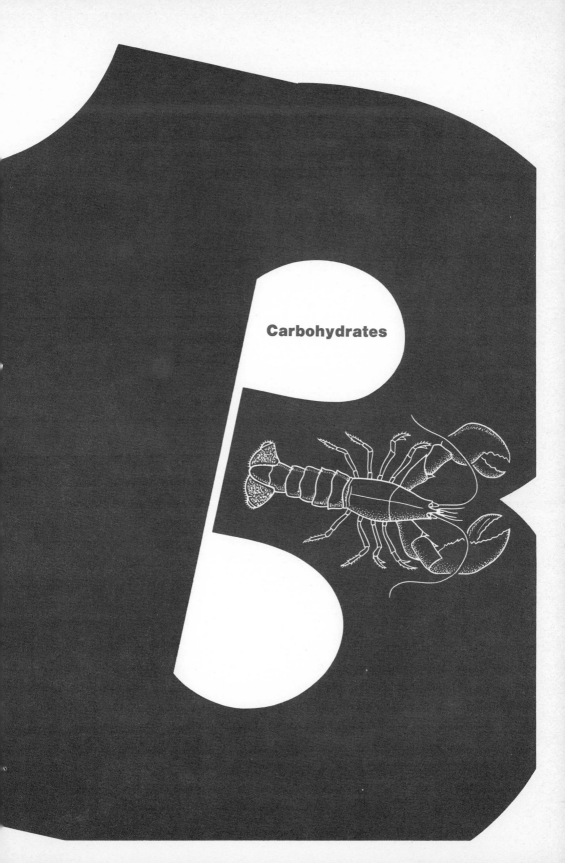

Carbohydrates

Carbohydrates are the most abundant organic compounds on earth. They serve many functions for living things—as sources of energy (for example, starch and sucrose), as structural materials (for example, cellulose), as control substances (for example, the phosphate esters of glucose), as starting points for syntheses (for example, syntheses of amino acids), and in many other ways.

The building blocks for the carbohydrates are the simple sugars, or *monosaccharides*. These compounds play much the same role in carbohydrate chemistry that amino acids play in protein chemistry. The properties of the monosaccharides and how they are put together into complex carbohydrates are the subjects of this chapter.

13-1 STRUCTURES OF MONOSACCHARIDES

Monosaccharides are the simplest carbohydrates. Typical monosaccharides contain a single straight carbon chain of three to eight carbon atoms. One of the carbon atoms is a carbonyl carbon; each other carbon has a hydroxyl group attached.

THE SIMPLEST SUGARS   There are two three-carbon sugars, called *trioses*, glyceraldehyde and dihydroxyacetone

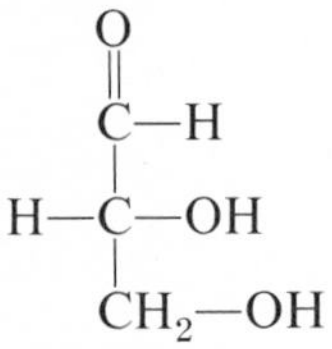

$$
\begin{array}{cc}
\begin{array}{c}
\overset{\displaystyle O}{\underset{\displaystyle |}{\overset{\displaystyle \|}{C}}}-H \\
| \\
H-C-OH \\
| \\
CH_2-OH
\end{array}
&
\begin{array}{c}
CH_2-OH \\
| \\
C=O \\
| \\
CH_2-OH
\end{array}
\\
\text{D-glyceraldehyde} & \text{dihydroxyacetone}
\end{array}
$$

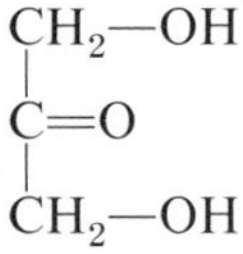

Sugars are frequently written, as here, in Fischer projections (section 4-4), with the carbon chain arranged vertically and the carbonyl group at or near the top.

Aldehyde sugars are called *aldoses*. The simplest aldose, glyceraldehyde, has one asymmetric carbon atom, and therefore has two stereoisomers (which are a pair of enantiomers). The D isomer shown here is the more common enantiomer in nature.

Ketone sugars are called *ketoses*. Dihydroxyacetone, the simplest ketose, is not chiral, although natural ketoses which contain more than three carbon atoms are chiral.

CLASSIFICATION   Aldoses and ketoses are classified by the number of carbon atoms they contain. A three-carbon sugar is called a *triose;* a four-carbon sugar is a *tetrose;* a five-carbon sugar is a *pentose;* a six-carbon sugar is a *hexose*. Thus, a four-carbon aldehyde sugar is called an *aldotetrose;* a five-carbon ketone sugar is called a *ketopentose*.

Aldoses containing four, five, and six carbon atoms occur throughout nature. There are two enantiomeric aldotrioses, D- and L-glyceraldehyde. Aldoses with more than three carbon atoms have additional asymmetric centers and therefore have more stereoisomers. There are four aldotetroses (two pairs of enantiomers), eight aldopentoses, and sixteen aldohexoses.

THE ALDOSE FAMILY

The D and L specifications for the aldoses are obviously not adequate to determine the configurations of all asymmetric carbons, except for glyceraldehyde. Instead, different diastereomers are given different names. The D or L refers to the configuration of the *lowest* asymmetric carbon in the Fischer projection.

The D-aldoses containing three to six carbon atoms are shown in figure 13-1. The compounds containing a particular number of carbon atoms differ in the configuration at one or more asymmetric carbons and are therefore diastereomers. One of each pair of enantiomers is shown.

Prove by use of molecular models that the aldopentoses shown in figure 13-1 are all different.

PROBLEM 13-1

Epimers are stereoisomers which have identical configurations at all asymmetric carbons except one. Identify all pairs of epimeric aldopentoses. Do not fail to consider both D and L isomers.

PROBLEM 13-2

Draw a Fischer projection of L-ribose.

PROBLEM 13-3

Ketoses containing three to seven carbon atoms occur commonly in nature. Nearly all naturally occurring ketoses have the carbonyl group one carbon away from the end of the carbon chain. A ketose has one fewer asymmetric centers than an aldose containing the same number of carbon atoms. The D-ketoses containing three to six carbons atoms are shown in figure 13-2.

THE KETOSE FAMILY

Sugars can be reduced with $NaBH_4$ to form sugar alcohols. Which of the compounds in figure 13-1 give meso compounds on reduction?

PROBLEM 13-4

Give the structure of a ketose that forms the same compound on reduction that glucose does.

PROBLEM 13-5

342

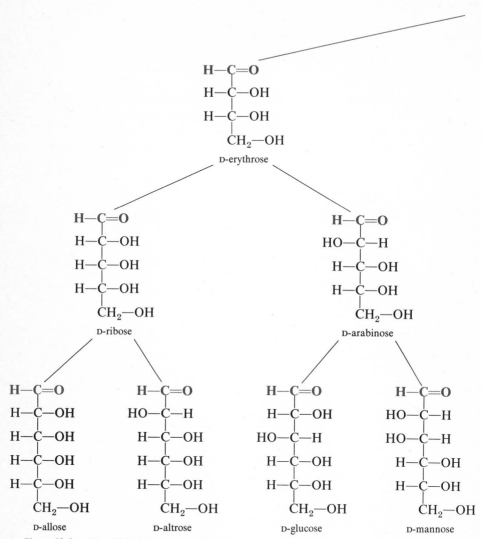

**Figure 13-1
Aldoses.**
The aldehyde sugars, or aldoses, can be
considered to be built up from
glyceraldehyde by addition of asymmetric
—CHOH— units. Only the D-aldoses are
shown.

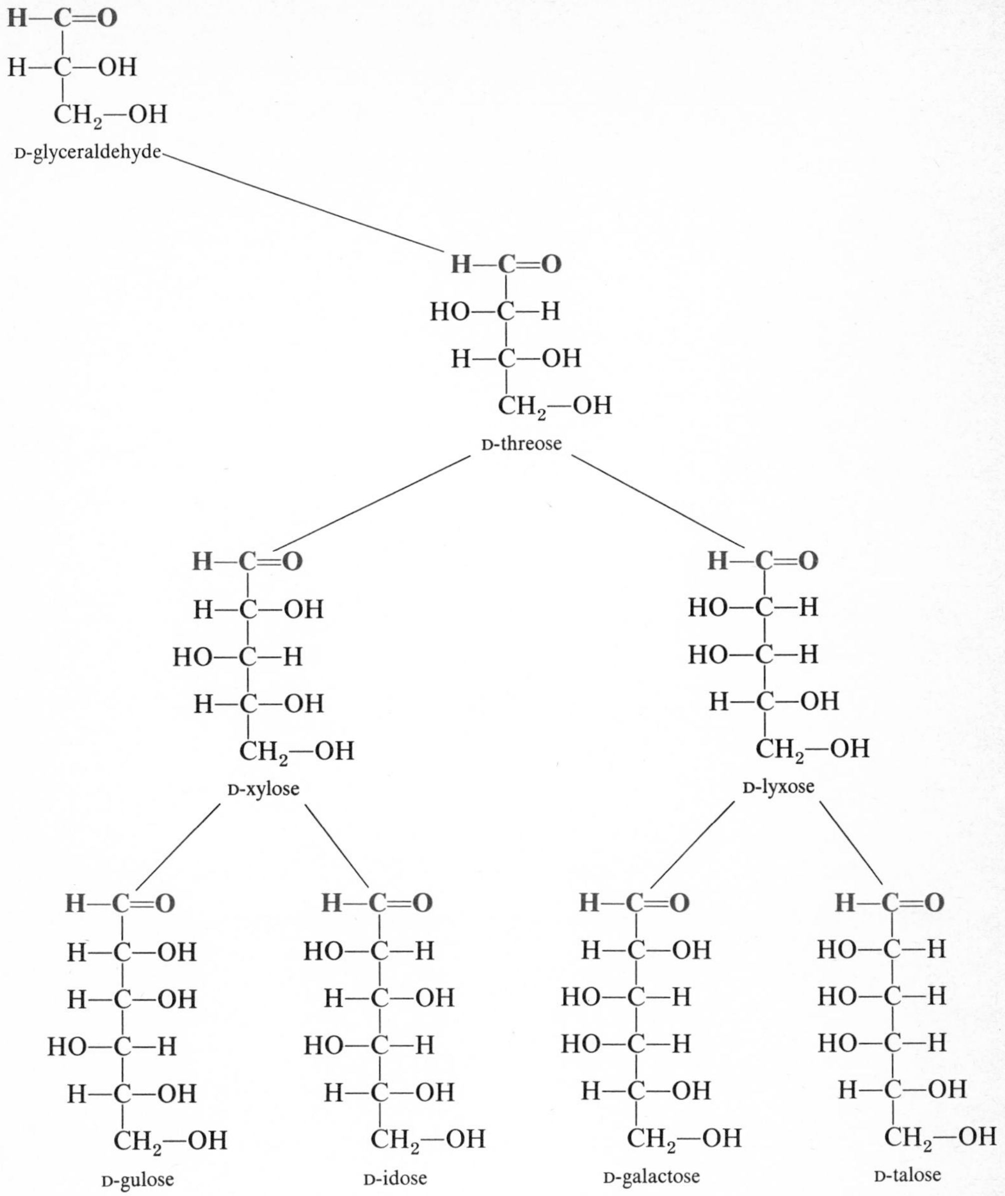

$$\begin{array}{l} H-C=O \\ H-C-OH \\ CH_2-OH \end{array}$$

D-glyceraldehyde

$$\begin{array}{l} H-C=O \\ HO-C-H \\ H-C-OH \\ CH_2-OH \end{array}$$

D-threose

$$\begin{array}{l} H-C=O \\ H-C-OH \\ HO-C-H \\ H-C-OH \\ CH_2-OH \end{array}$$

D-xylose

$$\begin{array}{l} H-C=O \\ HO-C-H \\ HO-C-H \\ H-C-OH \\ CH_2-OH \end{array}$$

D-lyxose

$$\begin{array}{l} H-C=O \\ H-C-OH \\ H-C-OH \\ HO-C-H \\ H-C-OH \\ CH_2-OH \end{array}$$

D-gulose

$$\begin{array}{l} H-C=O \\ HO-C-H \\ H-C-OH \\ HO-C-H \\ H-C-OH \\ CH_2-OH \end{array}$$

D-idose

$$\begin{array}{l} H-C=O \\ H-C-OH \\ HO-C-H \\ HO-C-H \\ H-C-OH \\ CH_2-OH \end{array}$$

D-galactose

$$\begin{array}{l} H-C=O \\ HO-C-H \\ HO-C-H \\ HO-C-H \\ H-C-OH \\ CH_2-OH \end{array}$$

D-talose

13-2 PHOTOSYNTHESIS

The respiration of animals (the process by which they make energy)
is characterized by oxidation. The ultimate products of this reaction
are carbon dioxide and water

$$\text{Food} + O_2 \longrightarrow CO_2 + H_2O + \text{energy}$$

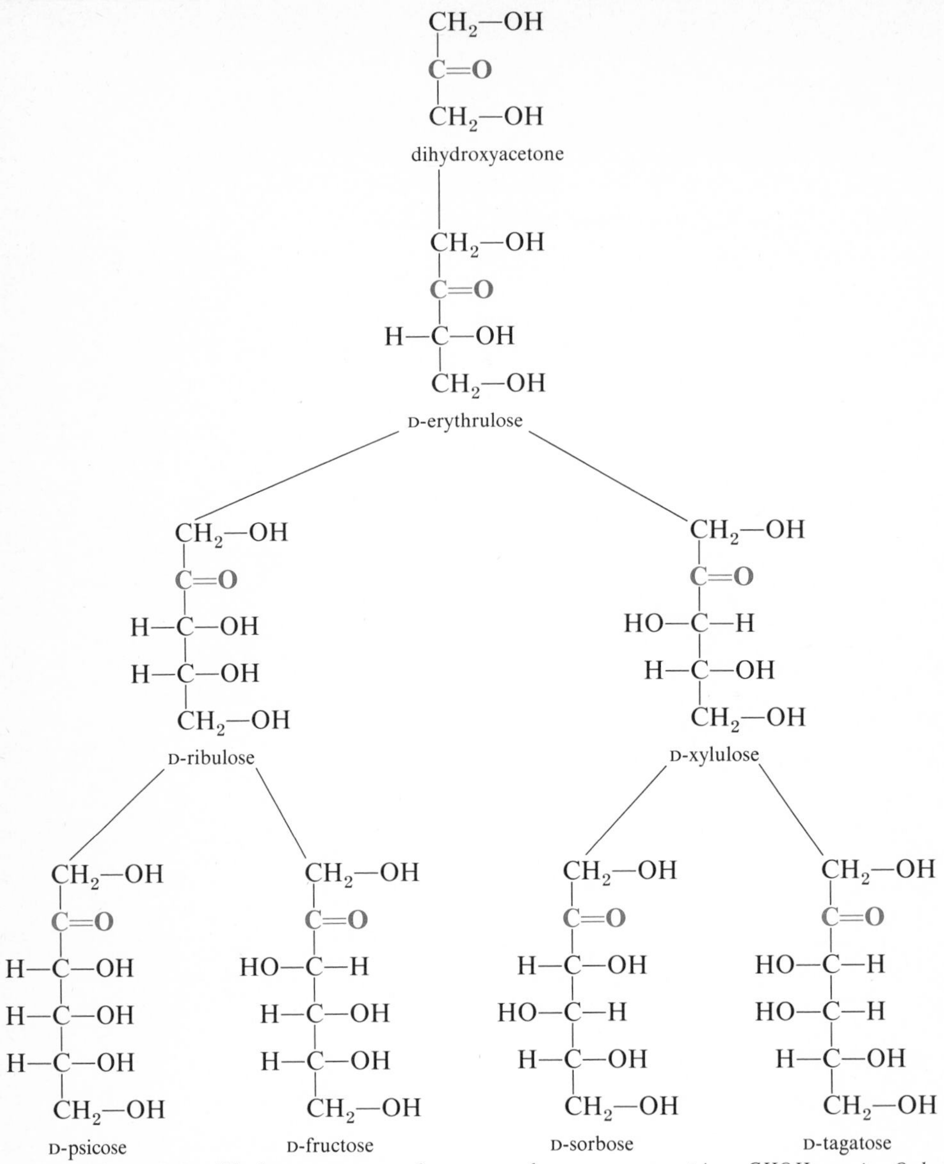

$$CH_2\!-\!OH$$
$$|$$
$$C\!=\!O$$
$$|$$
$$CH_2\!-\!OH$$

dihydroxyacetone

$$CH_2\!-\!OH$$
$$|$$
$$C\!=\!O$$
$$|$$
$$H\!-\!C\!-\!OH$$
$$|$$
$$CH_2\!-\!OH$$

D-erythrulose

$$CH_2\!-\!OH$$
$$|$$
$$C\!=\!O$$
$$|$$
$$H\!-\!C\!-\!OH$$
$$|$$
$$H\!-\!C\!-\!OH$$
$$|$$
$$CH_2\!-\!OH$$

D-ribulose

$$CH_2\!-\!OH$$
$$|$$
$$C\!=\!O$$
$$|$$
$$HO\!-\!C\!-\!H$$
$$|$$
$$H\!-\!C\!-\!OH$$
$$|$$
$$CH_2\!-\!OH$$

D-xylulose

$$CH_2\!-\!OH$$
$$|$$
$$C\!=\!O$$
$$|$$
$$H\!-\!C\!-\!OH$$
$$|$$
$$H\!-\!C\!-\!OH$$
$$|$$
$$H\!-\!C\!-\!OH$$
$$|$$
$$CH_2\!-\!OH$$

D-psicose

$$CH_2\!-\!OH$$
$$|$$
$$C\!=\!O$$
$$|$$
$$HO\!-\!C\!-\!H$$
$$|$$
$$H\!-\!C\!-\!OH$$
$$|$$
$$H\!-\!C\!-\!OH$$
$$|$$
$$CH_2\!-\!OH$$

D-fructose

$$CH_2\!-\!OH$$
$$|$$
$$C\!=\!O$$
$$|$$
$$H\!-\!C\!-\!OH$$
$$|$$
$$HO\!-\!C\!-\!H$$
$$|$$
$$H\!-\!C\!-\!OH$$
$$|$$
$$CH_2\!-\!OH$$

D-sorbose

$$CH_2\!-\!OH$$
$$|$$
$$C\!=\!O$$
$$|$$
$$HO\!-\!C\!-\!H$$
$$|$$
$$HO\!-\!C\!-\!H$$
$$|$$
$$H\!-\!C\!-\!OH$$
$$|$$
$$CH_2\!-\!OH$$

D-tagatose

*Figure 13-2
Ketoses.* The ketone sugars, or ketoses, can be considered to be built up from dihydroxyacetone by the addition of asymmetric —CHOH— units. Only the D-ketoses are shown.

In the course of this long series of reactions the oxidation numbers of the carbon atoms of foods are gradually increased from 1, 2, and 3 to 4.

All animals depend ultimately on plants for their food. Plants derive their energy from the sun via *photosynthesis,* the synthesis of carbohydrates from carbon dioxide, water, and sunlight

$$CO_2 + H_2O + sunlight \longrightarrow carbohydrate + O_2$$

This reaction is precisely the opposite of the reaction by which animals make energy: carbon dioxide (oxidation number 4) is reduced to oxidation number 1 or 2.

Photosynthesis has been studied by chemists, biochemists, and biologists, and many details of this important process are now understood. Two important light-dependent reactions are known. The first is the formation of adenosine triphosphate (ATP, section 10-11) from adenosine diphosphate (ADP), using light as a source of energy

$$ADP + phosphate + light \longrightarrow ATP + H_2O$$

The second reaction is the reduction of nicotinamide adenine dinucleotide phosphate (NADP$^+$, section 11-8) to the reduced form (NADPH), using light as a source of energy

$$NADP^+ + H_2O + light \longrightarrow NADPH + \tfrac{1}{2}O_2 + H^+$$

The incorporation of carbon dioxide into organic compounds occurs in a reaction which does not itself require light; instead, the ATP and NADPH produced in the light-dependent reactions shown above are used as sources of energy. In this reaction, ribulose diphosphate reacts with CO_2 to form two molecules of phosphoglyceric acid. As is clear from the equation, this reaction is extremely complex

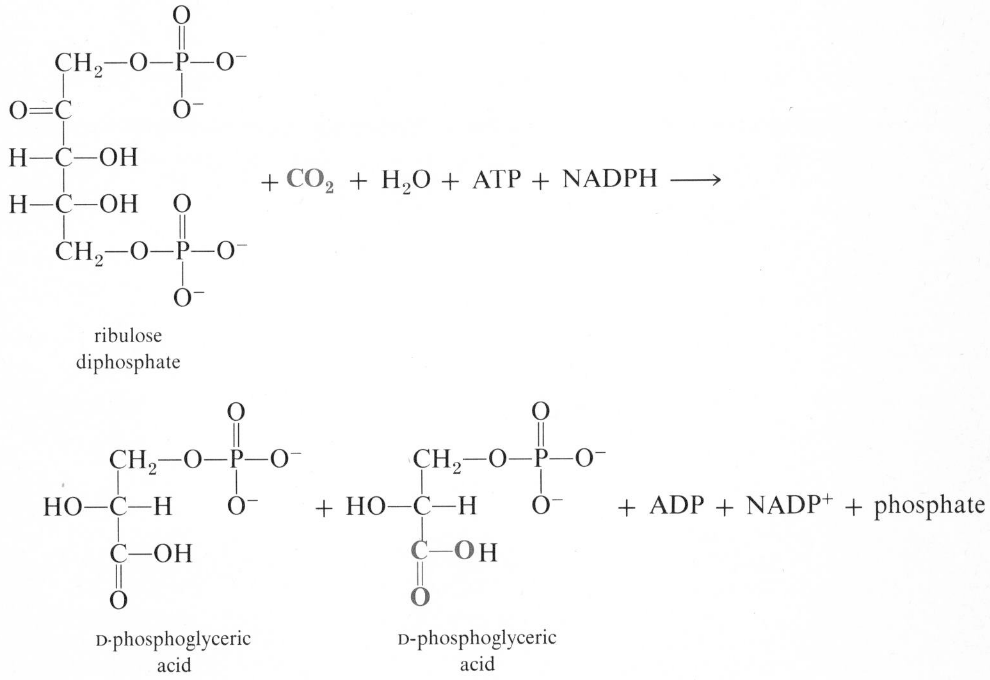

ribulose
diphosphate

$+ CO_2 + H_2O + ATP + NADPH \longrightarrow$

D-phosphoglyceric
acid

$+$

D-phosphoglyceric
acid

$+ ADP + NADP^+ + phosphate$

The carbon atom of the carbon dioxide becomes the carboxyl carbon atom of one of the two molecules of phosphoglyceric acid. In this reaction, the oxidation level of that carbon is reduced from 4 to 3.

Phosphoglyceric acid is the key starting compound for the synthesis of amino acids and fatty acids (figure 13-3). Further reduction of phosphoglyceric acid produces glyceraldehyde phosphate, which is the key compound in the synthesis of carbohydrates. In addition, glyceraldehyde phosphate is used to make glycerol for use in the synthesis of fats.

13-3 CYCLIC FORMS OF SUGARS

Aldehydes and ketones react with alcohols in the presence of either acid or base to form hemiacetals and hemiketals (sections 7-4 and 7-5),

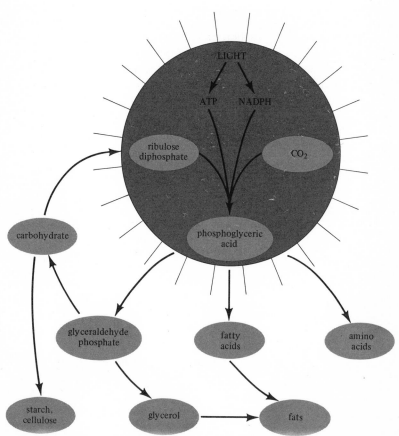

Figure 13-3
Photosynthesis and metabolism.
Phosphoglyceric acid, the primary product of photosynthesis, is the starting material for the synthesis of other compounds in plant metabolism. The general outline of these reactions is shown here. Each arrow may represent a number of reaction steps.

although the equilibrium usually does not favor hemiacetal or hemiketal formation

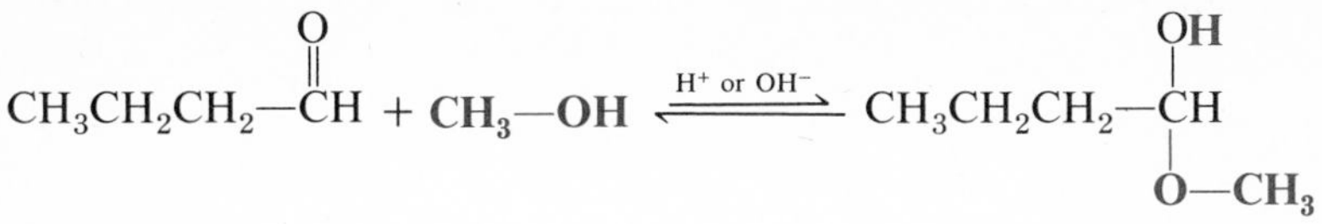

$$\underset{\text{aldehyde}}{CH_3CH_2CH_2-\overset{\displaystyle O}{\overset{\|}{C}H}} + CH_3-OH \underset{}{\overset{H^+ \text{ or } OH^-}{\rightleftharpoons}} \underset{\text{hemiacetal}}{CH_3CH_2CH_2-\underset{\displaystyle O-CH_3}{\overset{\displaystyle OH}{\underset{|}{\overset{|}{C}H}}}}$$

Sugars, being both carbonyl compounds and alcohols, are able to undergo this reaction all by themselves: the reaction occurs between a hydroxyl group and a carbonyl group *of the same molecule,* thus forming a cyclic hemiacetal or hemiketal. Unlike the situation with simple aldehydes and ketones, the equilibrium in the case of sugars favors hemiacetal or hemiketal formation.

The aldotetrose *erythrose* provides a simple example of hemiacetal formation (figure 13-4). The structure of erythrose is frequently written in the acyclic form, but this representation is not strictly correct. Instead, the hydroxyl group of the terminal carbon is combined with the carbonyl carbon to form a cyclic hemiacetal with a five-membered ring. Such a five-membered ring is called a *furanose* because of its

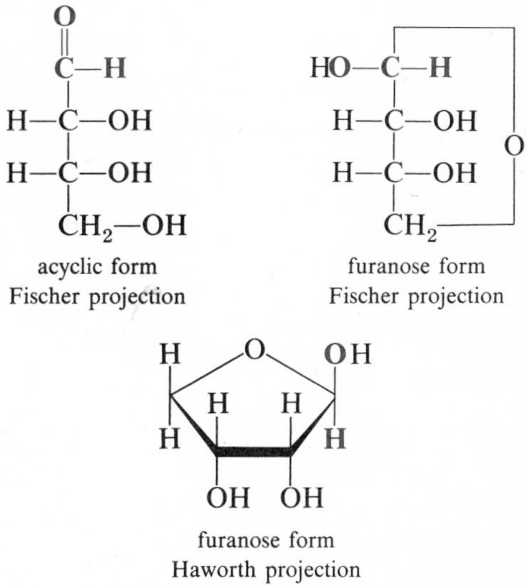

ERYTHROSE, A FURANOSE

acyclic form
Fischer projection

furanose form
Fischer projection

furanose form
Haworth projection

The Fischer projection of the acyclic form is the starting point for understanding this structure, even though the compound actually exists in nature almost entirely in a cyclic form. The Fischer projection of the furanose form is correct and is easily drawn, but the stereochemistry of the five-membered ring is difficult to see. The stereochemistry can be seen clearly in the Haworth projection.

Figure 13-4
D-*Erythrose.*

relationship to the heterocyclic compound *furan*

furan

THE ANOMERIC CARBON Sugars are capable of forming two cyclic forms which differ only in the configuration of one carbon atom, the carbon atom which was the aldehyde carbon in the acyclic form. These two forms, called *anomers,* are shown for erythrose in figure 13-5. The carbon atom which is the carbonyl carbon in the acyclic form is called the *anomeric carbon.* The two cyclic forms shown in figure 13-5 are identical except for the configuration of the hydroxyl group attached to the anomeric carbon. The hydroxyl group is down in the α form and up in the β form. The two structures in figure 13-5 are separate and distinct compounds; conversion of one compound into the other occurs slowly (page 351).

HAWORTH PROJECTIONS OF CYCLIC FORMS Strictly speaking, Fischer projections like those shown in figures 13-1 and 13-2 are not correct structures of sugars because they do not show the hemiacetal or hemiketal. The hemiacetal or hemiketal can be indicated in a Fischer projection, as in figure 13-4, and such structures are useful for correlating Fischer projections with Haworth projections. The transformation of a cyclic Fischer projection into a Haworth projection is made by the following rules:

1 In the Haworth projection the oxygen atom in the ring is placed at the back of a five-membered ring or at the back right of a six-membered ring.
2 The anomeric carbon is placed on the right.
3 Hydroxyl groups which are on the right in the Fischer projection are down in the Haworth projection; those on the left are up.
4 The substituent on the carbon whose hydroxyl group forms the hemiacetal or hemiketal follows the opposite rule.

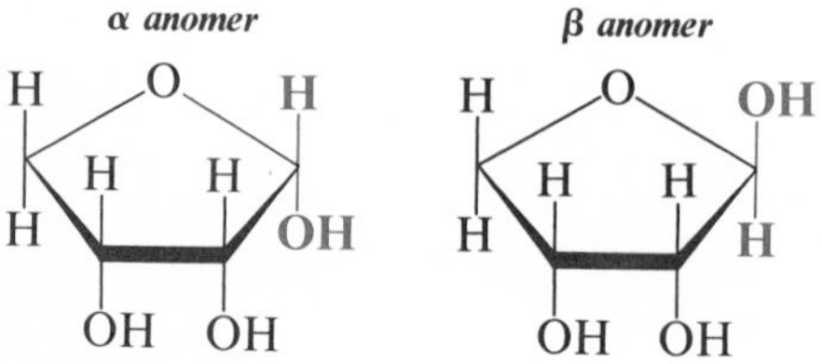

Figure 13-5 Two anomers of D-erythrofuranose. Anomers differ only in the stereochemistry of the hemiacetal or hemiketal carbon.

α anomer   β anomer

*acyclic
form*

CH$_2$—OH

C=O

H—C—OH

H—C—OH

CH$_2$—OH

α anomer   *β anomer*

HOCH$_2$—C—OH   HO—C—CH$_2$OH

H—C—OH   H—C—OH

H—C—OH   O   H—C—OH   O

CH$_2$ CH$_2$

Fischer projection   Fischer projection

Haworth projection   Haworth projection

The acyclic form is not observed in   *The two anomers can be interconverted,*   **Figure 13-6**
nature. Instead, two cyclic forms, an α   *but they are distinct and separate*   D-**Ribulose.**
anomer and a β anomer, are observed.   *compounds.*

Use molecular models to confirm the correctness of the representations   ***PROBLEM 13-6***
given in figures 13-4 and 13-5.

Formation of a furanose ring for the ketopentose ribulose is shown   *RIBULOSE*
in figure 13-6. Because ribulose is a ketose, the cyclic form is a hemiketal rather than a hemiacetal.

Glucose, the most common sugar of all, is an aldohexose which exists   *GLUCOSE, A*
in a six-membered ring form called a *pyranose,* after the heterocyclic   *PYRANOSE*
compound *pyran* (figure 13-7)

α-pyran

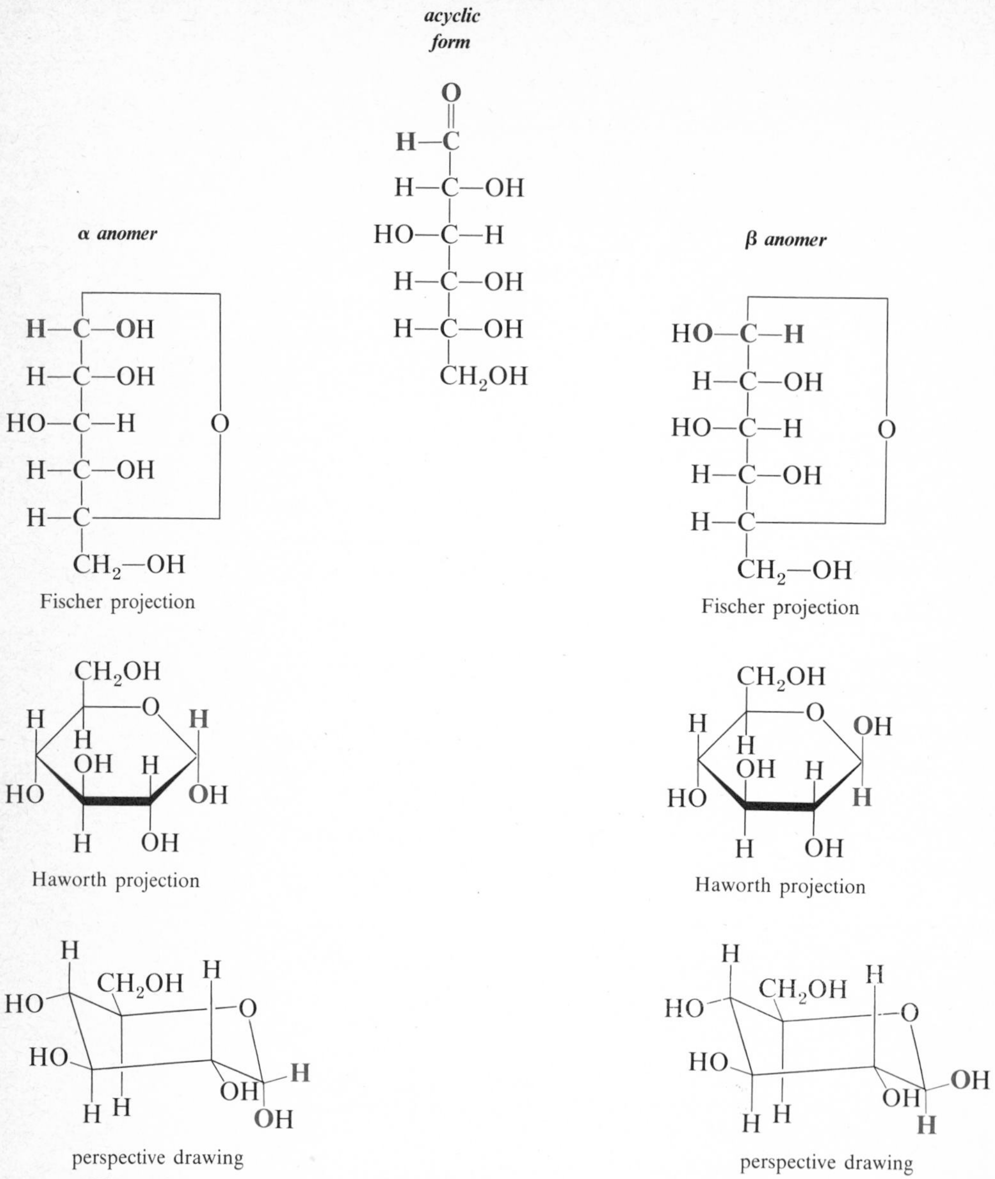

acyclic form

α anomer

Fischer projection

β anomer

Fischer projection

Haworth projection

Haworth projection

perspective drawing

perspective drawing

Figure 13-7
D-**Glucose.**
The two anomers can be shown as Fischer projections, as Haworth projections, or as perspective drawings. Only the last of these pictures correctly show the geometry of the six-membered ring. The acyclic form is not a stable form.

Glucose, also known as *dextrose,* occurs free in the juices of various fruits and in a combined form in sucrose, maltose, lactose, and other complex sugars. Glucose is the basic building block from which cellulose, starch, and many other complex polysaccharides are made, and it is important in the energy-producing metabolism of every living cell.

Furanose rings, like cyclopentane rings, are nearly flat, but pyranose rings, like cyclohexane rings, are puckered into a chair form (figure 13-7). It is significant that in glucose all the hydroxyl groups (with the possible exception of the anomeric hydroxyl) and the —CH$_2$OH group are in the more stable equatorial positions. Flipping the ring from one chair conformation to the other would make all these groups axial and would make that conformation of much higher energy than the conformation shown.

Because of the possibilities for pyranose and furanose formation, the complete description of the structure of a given anomer of a monosaccharide includes:

DEFINING THE STRUCTURES OF MONO-SACCHARIDES

1 The nature of the monosaccharide and whether it is D or L

2 Whether a ring is formed and whether it is a pyranose or a furanose

3 Whether the anomeric hydroxyl group is α or β

The only cyclic structures which are normally formed by sugars are those containing five- and six-membered rings. If a sugar is capable of forming both types, the pyranose form (six-membered ring) will predominate, although in some cases small amounts of furanose will be formed as well.

The equilibrium between the α and β anomers of most sugars is quite evenly balanced, and appreciable amounts of both anomers are present at equilibrium. Because the anomers interconvert only slowly, it is frequently possible to separate and study individual anomers.

The interconversion of α and β anomers is called *mutarotation.* If either anomer is dissolved in water containing a trace of acid or base, over a period of minutes or hours an equilibrium mixture of the two anomers is formed.

MUTAROTATION

The stereochemistry of mutarotation is very interesting (figure 13-8). The first step in the reaction is decomposition of the hemiacetal or hemiketal to form the free aldehyde or ketone. Rotation of the bond between the carbonyl carbon and the adjacent carbon through 180°, followed by formation of the hemiacetal or hemiketal, produces the other anomer. Although the free aldehyde or ketone form of the sugar is an intermediate in this reaction, its concentration is too low to be detectable.

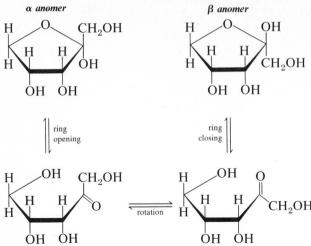

Figure 13-8
Mutarotation of
α-D-ribulose.

If the α anomer is dissolved in water containing a small amount of acid or base, it will be converted into an equilibrium mixture of the α and β anomers. The interconversion takes place by opening of the hemiketal to the ketone form, rotation of the carbon-carbon bond of the open-chain form, and reaction of the carbonyl with the same —OH as before but from the other side.

Mutarotation is a moderately slow reaction, requiring at least a few minutes to reach equilibrium. Most enzyme-catalyzed reactions of monosaccharides in nature are effective with only one of the two anomers. As a result, enzymes exist whose sole function is to increase the rate of mutarotation, thereby providing the proper anomer for further reactions.

PROBLEM 13-7 Draw Fischer and Haworth pictures of the two anomers of D-threo-furanose.

PROBLEM 13-8 Draw Fischer and Haworth projections of four cyclic forms of D-fructose.

13-4 CHEMISTRY OF MONOSACCHARIDES

The chemistry of monosaccharides is a composite of the chemistry of the carbonyl group, the hydroxyl group, and the hemiacetal linkage. Monosaccharides are puzzling because some of the time they appear to react like aldehydes or ketones and other times they react like hemiacetals or hemiketals.

ACETAL AND KETAL FORMATION

Simple aldehydes and ketones react with two molecules of an alcohol in the presence of an acid catalyst to form acetals and ketals (section 7-5). However, since most sugars are already in the hemiacetal or hemiketal form, reaction with a single molecule of alcohol in the presence of an acid catalyst converts them into acetals or ketals, which are called *glycosides*

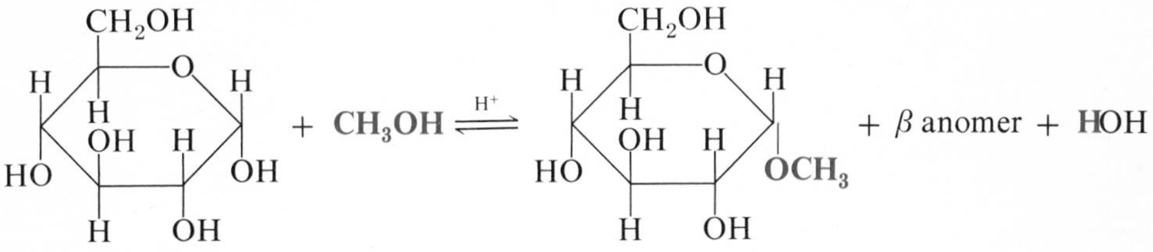

α-D-glucopyranose   methyl α-D-glucopyranoside (a glycoside)

The mechanism of this reaction has already been given (mechanism 7-3). Acetals and ketals are much more stable than hemiacetals and hemiketals. They decompose only in the presence of an acid catalyst or an enzyme catalyst. Like other cyclic sugar derivatives, they can exist as both α and β anomers.

PROBLEM 13-9

The above reaction forms both anomers of the acetal. Give the mechanism of this reaction and explain why both anomers are formed.

ALKYLATION AND ACYLATION

The hydroxyl groups of sugars can act as nucleophiles in a variety of displacement reactions. Such reactions have been widely used both for the synthesis of sugars and for the determination of their structures. For example, glucose reacts with methyl iodide to form a pentaether*

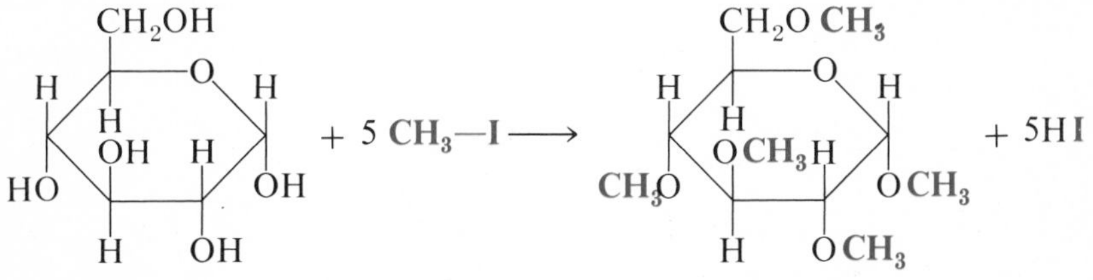

α-D-glucopyranose   pentamethyl α-D-glucopyranoside

*The hydroxyl groups of sugars, like those of simple alcohols, are not very nucleophilic. As a result, the reaction as written here works only poorly. In practice, the reaction is conducted in the presence of Ag_2O. The Ag^+ ions accelerate the reaction by exerting a pulling effect on the iodine during the displacement step. In this case, the product of the reaction is not HI but the insoluble AgI.

Hydrolysis of this compound in the presence of dilute aqueous acid converts the acetal into a hemiacetal but leaves all the other ether linkages intact

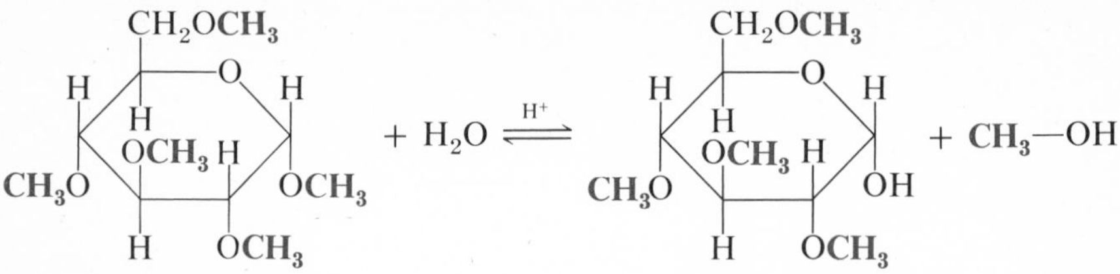

pentamethyl
α-D-glucopyranoside

tetramethyl
α-D-glucopyranoside

Acylation of the hydroxyl groups of sugars can be conducted with anhydrides or acid chlorides. All available hydroxyl groups react

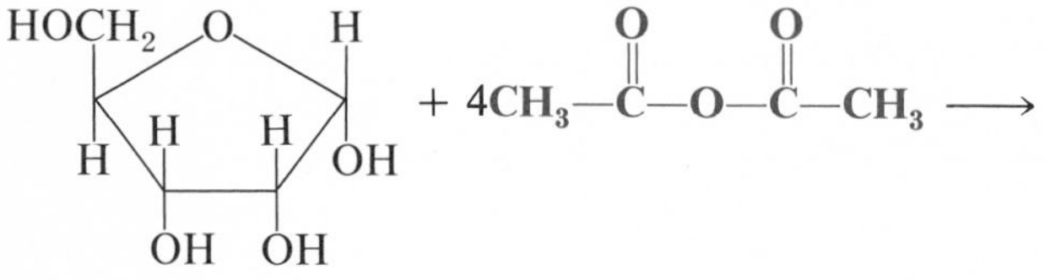

α-D-ribofuranose

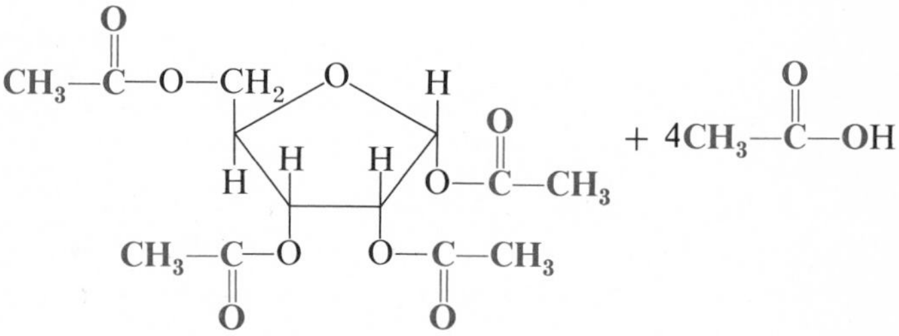

tetraacetyl α-D-ribofuranose

BIOLOGICAL INTERCONVERSION OF ALDOSES AND KETOSES

Enolization is an important reaction of the carbonyl groups of sugars. The synthesis of a sugar by means of an enolization reaction was shown in section 7-10. Another reaction in which enolization plays an important role is the interconversion of aldoses and ketoses (figure 13-9). Like many other reactions of sugars in nature, this reaction involves the phosphate esters of sugars rather than the free sugars themselves.

VITAMIN C

Scurvy is a disease which first became common in the fifteenth and sixteenth centuries on the long ocean voyages of such explorers as Columbus, da Gama, and Cartier. The connection between scurvy and

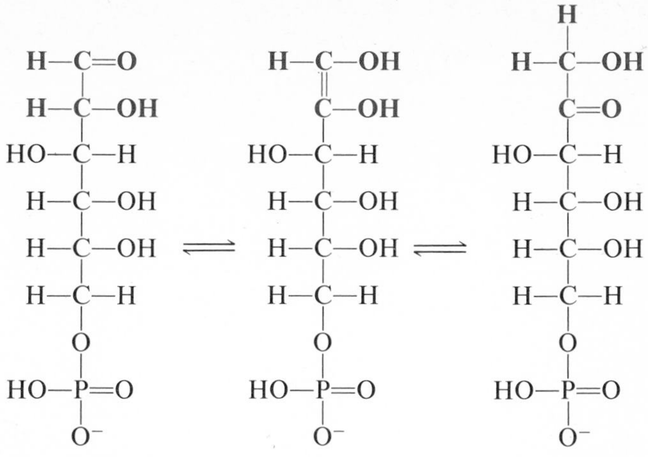

$$
\begin{array}{ccc}
\text{H—C}{=}\text{O} & \text{H—C—OH} & \text{H—C—OH} \\
\text{H—C—OH} & \text{C—OH} & \text{C}{=}\text{O} \\
\text{HO—C—H} & \text{HO—C—H} & \text{HO—C—H} \\
\text{H—C—OH} & \text{H—C—OH} & \text{H—C—OH} \\
\text{H—C—OH} \rightleftharpoons & \text{H—C—OH} \rightleftharpoons & \text{H—C—OH} \\
\text{H—C—H} & \text{H—C—H} & \text{H—C—H} \\
\text{O} & \text{O} & \text{O} \\
\text{HO—P}{=}\text{O} & \text{HO—P}{=}\text{O} & \text{HO—P}{=}\text{O} \\
\text{O}^- & \text{O}^- & \text{O}^-
\end{array}
$$

| glucose 6-phosphate | enol intermediate | fructose 6-phosphate |

This reaction takes place by formation of an enol and then formation of the carbonyl group at the position adjacent to that of the original compound. The configurations of other chiral centers in the molecule are not affected by this transformation. In nature these reactions usually involve phosphate esters of the sugars, rather than the free sugars. This reaction is catalyzed by the enzyme phosphohexose isomerase.

Figure 13-9
Aldose-ketose interconversion.

diet was recognized about 1750, and not long thereafter the British navy began to require a daily ration of lime juice for all sailors (hence the name "limey" for a British sailor).

The structure of vitamin C, the active ingredient of lime juice, was not elucidated until 1932. Vitamin C (also known as *ascorbic acid*) is a lactone derived from an L-ketohexose. The commercial synthesis of ascorbic acid is shown in figure 13-10. The synthesis of ascorbic acid in nature is similar, though not identical.

The role of vitamin C in human nutrition is still a subject of controversy. Unlike most vitamins, it seems to play only a limited part in metabolism. Nonetheless, ascorbic acid is vital to human health.

13-5 DISACCHARIDES

Just as polypeptides and proteins are constructed from amino acids by means of amide linkages, polysaccharides are constructed from monosaccharides by means of acetal and ketal linkages. The principle on which these structures are built is simple: the anomeric oxygen of one sugar unit is derived from a hydroxyl group of the adjacent sugar. This same principle applies to all complex sugars, from the disaccharides (two sugar units) to the high-molecular-weight polysaccharides.

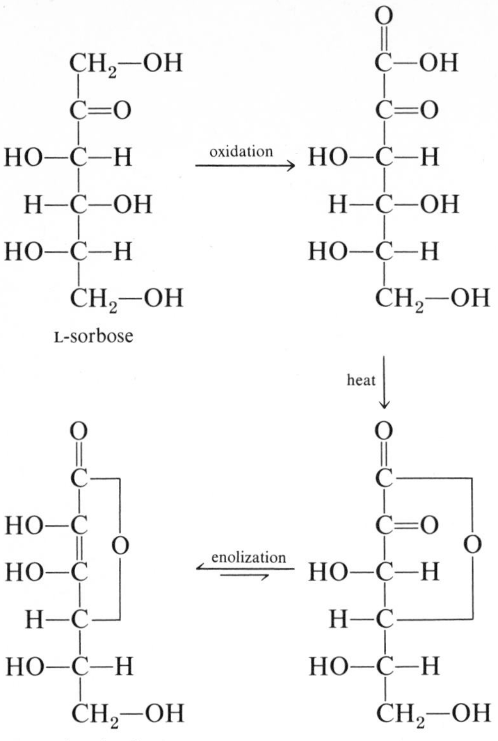

$$
\begin{array}{ccc}
\text{CH}_2\text{—OH} & & \text{C—OH (}\overset{\text{O}}{\|}\text{)} \\
\text{C=O} & & \text{C=O} \\
\text{HO—C—H} & \xrightarrow{\text{oxidation}} & \text{HO—C—H} \\
\text{H—C—OH} & & \text{H—C—OH} \\
\text{HO—C—H} & & \text{HO—C—H} \\
\text{CH}_2\text{—OH} & & \text{CH}_2\text{—OH}
\end{array}
$$

L-sorbose

heat ↓

$$
\begin{array}{ccc}
\text{(}\overset{\text{O}}{\|}\text{)C} & & \text{(}\overset{\text{O}}{\|}\text{)C} \\
\text{HO—C} & & \text{C=O} \\
\text{HO—C} \quad \text{O} & \xleftarrow{\text{enolization}} & \text{HO—C—H} \quad \text{O} \\
\text{H—C} & & \text{H—C} \\
\text{HO—C—H} & & \text{HO—C—H} \\
\text{CH}_2\text{—OH} & & \text{CH}_2\text{—OH}
\end{array}
$$

vitamin C

Figure 13-10 Commercial synthesis of vitamin C.   The naturally occurring sugar L-sorbose is oxidized by nitric acid to a carboxylic acid, which cyclizes under the influence of heat to a lactone. Enolization of this compound occurs easily, forming vitamin C.

β-MALTOSE   When starch is hydrolyzed by the enzyme β-*amylase*, the product is β-maltose. This compound is constructed from two glucose units (figure 13-11). The hydroxyl group at carbon atom 4 of one glucose unit has become the anomeric oxygen of the second glucose. Thus, β-maltose contains one hemiacetal linkage (in the right ring) and one acetal linkage (in the left ring). Hydrolysis of β-maltose results in the formation of two molecules of glucose.*

SUCROSE   Disaccharides, being made up of two monosaccharide units, must contain two carbonyl carbon atoms or their equivalent. In most disaccharides, one of these carbons is present as a hemiacetal or hemiketal

---

*One of the principal problems in sugar chemistry is the problem of recognizing significant features of these very complex structures. The first step is to identify all carbonyl carbon atoms (the only carbons in the molecule at oxidation level 2). Acetal and ketal linkages can then be recognized easily.

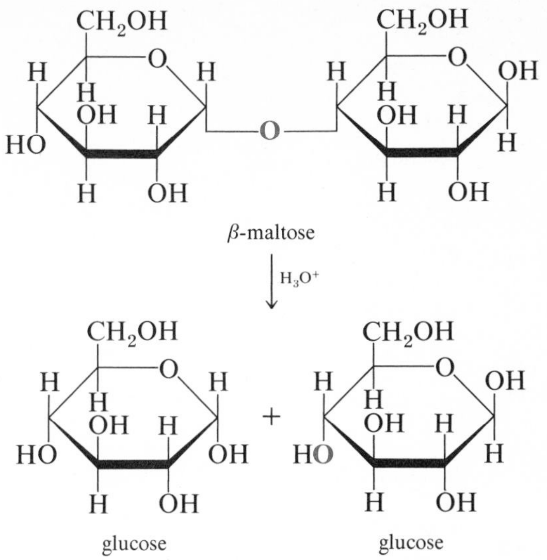

β-maltose

This disaccharide is composed of two glucose units connected by an acetal linkage. Hydrolysis of β-maltose produces two molecules of glucose.

Figure 13-11 β-Maltose.

and the other as an acetal or ketal, but this is not always so. Sucrose, for example, has an acetal linkage and a ketal linkage and no hemiacetals or hemiketals (figure 13-12). Hydrolysis of sucrose catalyzed by acid or by the enzyme *sucrase* gives one molecule of glucose and one of fructose.

Sucrose is by far the most abundant disaccharide in nature. It is the common sugar of commerce, and is produced (from sugarcane and sugar beets) in amounts exceeding 100 million tons per year. Honey contains very little sucrose but contains instead a 1:1 mixture of glucose and fructose. This mixture, called *invert sugar,* is sweeter than sucrose and occurs in honey because of the presence of the enzyme sucrase.

Because of the number of hydroxyl groups in most sugars, a variety of disaccharides can be formed from a particular pair of monosaccharides. The complete specification of the structure of a disaccharide requires four pieces of information:

DETERMINING THE STRUCTURES OF DISAC-CHARIDES

1 The nature of the two sugars
2 Whether each sugar is in a furanose or pyranose form
3 Whether the anomeric carbons are α or β
4 The position of the linkage between the two sugars

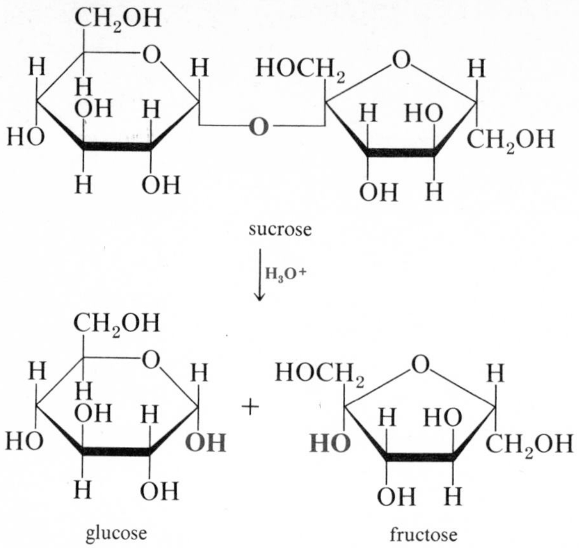

Figure 13-12
Sucrose.
This disaccharide is unusual because the two anomeric carbon atoms are connected to the same oxygen atom. Hydrolysis of sucrose forms a molecule of glucose and a molecule of fructose.

PROBLEM 13-10   Describe fifteen disaccharides that might be formed from two molecules of D-ribofuranose.

13-6 POLYSACCHARIDES

Polysaccharides are large molecules put together from many sugar units—often as many as several thousand. Unlike proteins, which have very complex sequences with little regularity, polysaccharides may have very regular structures, being composed of only one or a small number of different monosaccharide units. In addition, whereas proteins have a definite size, polysaccharides are always heterogeneous with respect to size and (unlike proteins) often contain branched chains.

Polysaccharides fall into two classes, the *structural polysaccharides* and the *energy-storage polysaccharides*. Both types may be used as food by other organisms. Cellulose, for example, is a structural polysaccharide for plants, but is used as food by snails and a number of microorganisms.

CELLULOSE   Cellulose is a common polysaccharide of tremendous economic and biological importance. Roughly 50 percent of the dry weight of wood and 97 percent of the dry weight of cotton are cellulose. About 10^{11}

tons of cellulose are produced by plants each year, making this the most abundant organic substance in the world.

Cellulose is a straight-chain polymer made up of a large number of units of β-D-glucose (figure 13-13). The linkage between adjacent units connects the anomeric carbon of one glucose unit to oxygen atom 4 of the adjacent glucose unit. The configuration of every anomeric carbon is β. Cellulose chains are typically several thousand glucose units long, with no chain branching.

Man and many other mammals are unable to digest cellulose; when ingested, it passes unchanged through the digestive system and is excreted. The apparent ability of cattle, goats, and other ruminants to digest cellulose is due to the presence in their digestive tracts of microorganisms which convert the cellulose into digestible compounds, principally fatty acids.

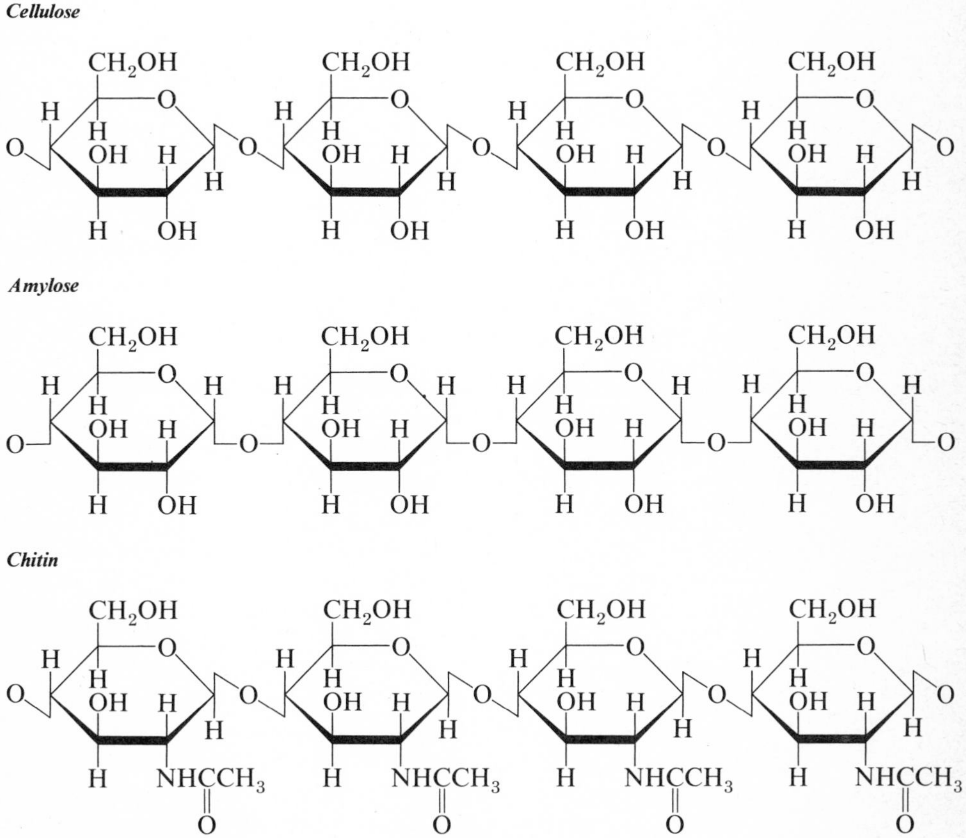

Cellulose

Amylose

Chitin

All three are very long, regular structures, only small segments of which are shown here. Cellulose and amylose differ only in the configuration of the anomeric carbon atoms.

Figure 13-13
Structures of cellulose, amylose, and chitin.

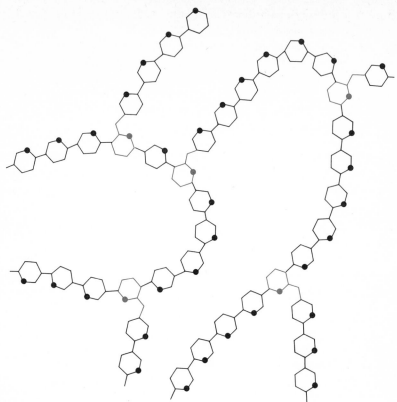

Figure 13-14
Glycogen.

Each hexagon represents a glucose unit linked 1,4 to two adjacent glucose units. Each colored hexagon represents a

glucose unit linked 1,4,6 to three adjacent glucose units. All anomeric carbons are α.

PROBLEM 13-11   When cellulose is partially hydrolyzed with acid, only one disaccharide is produced. What is its structure?

STARCH   Starch is a polysaccharide which is used as a nutritional reservoir in plants, thereby serving as food for animals. Starch is a mixture of two components, *amylose* and *amylopectin*. Amylose has almost the same structure as cellulose; it is composed of a large number of glucose units held together by acetal linkages between the anomeric carbon of one glucose unit and oxygen atom 4 of the adjacent unit (figure 13-13). The difference between cellulose and amylose is in the configuration of the anomeric carbons; they are β in cellulose but α in amylose.

Amylopectin is also composed of glucose units linked together like those in amylose, but amylopectin has branched chains, in which the hydroxyl group of carbon atom 6 becomes the anomeric carbon of another glucose unit and thus the beginning of another chain.

The difference in function between cellulose and amylose is due to the difference in configuration at the anomeric carbon. The enzymes that synthesize these materials and the enzymes that degrade them are specific for a particular configuration of this carbon atom and have no difficulty in distinguishing between the two configurations.

Whereas plants use starch as their energy-storage carbohydrate, ani- *GLYCOGEN* mals use glycogen. Like starch, glycogen is made up of very long chains of glucose units with the oxygen atom at carbon atom 4 of one glucose linked α to the anomeric carbon of the adjacent glucose. However, unlike those in starch, all chains in glycogen are branched; in a substantial fraction of the glucose units the hydroxyl group of carbon atom 6 becomes a branch point by forming an acetal to another glucose (figure 13-14). This glucose is the starting point for another chain.

The lobster shown at the beginning of this chapter has a shell contain- *CHITIN* ing a large amount of chitin, a polymer of the acylamino sugar *N-acetylglucosamine* (figure 13-13). Chitin occurs in a wide variety of lower forms of plant life, in fungi, and in invertebrates. It is probably the second most abundant organic substance on the earth.

13-12 Draw and name four cyclic forms of D-arabinose.   ***PROBLEMS***

13-13 Define and give an example of each of the following: (*a*) aldotetrose; (*b*) ketohexose; (*c*) furanose; (*d*) pyranose; (*e*) anomer; (*f*) disaccharide; (*g*) mutarotation.

13-14 Give the products of the reaction of α-D-threofuranose with (*a*) H_3O^+; (*b*) $H_3O^+ + CH_3OH$; (*c*) acetic anhydride; (*d*) excess $CH_3I + Ag_2O$.

13-15 Give the products of the reaction of sucrose with the reagents of problem 13-14.

13-16 Summarize the differences among cellulose, glycogen, and starch.

13-17 Which L-ketohexoses can be used to make ascorbic acid by the scheme shown in figure 13-10?

13-18 (*a*) From what two monosaccharides is β-melibiose constructed? (*b*) Identify the anomeric carbon atoms. (*c*) Identify any pyranose or furanose rings. (*d*) Are the anomeric carbons in the α or β configurations? (*e*) Label any acetal, ketal, hemiacetal, or hemiketal linkages.

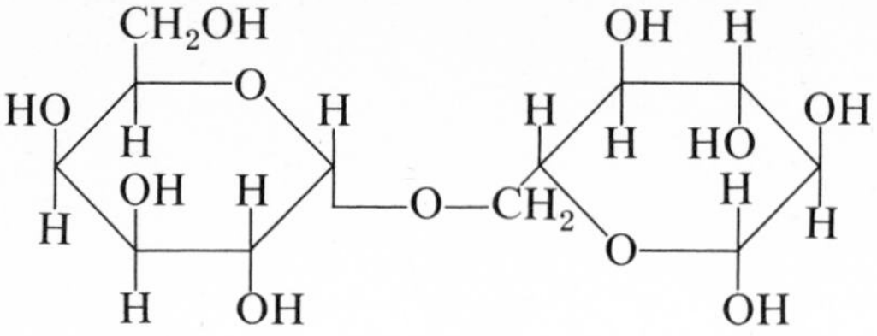

β-melibiose

SUGGESTED READINGS

Bassham, J. A.: The Path of Carbon in Photosynthesis, p. 383 in *Organic Chemistry of Life: Readings from Scientific American,* Freeman, San Francisco, 1973. The chemistry of photosynthesis.

Clayton, R. K.: *Light and Living Matter,* McGraw-Hill, New York, 1971. A comprehensive discussion of photosynthesis and related topics.

Preston, R. D.: Cellulose, *Sci. Am.,* September 1957, p. 157.

ANSWERS TO SELECTED PROBLEMS

2-1 (a) Ca^{2+} 2:Cl:⁻ (b) $2Na^+$:S:²⁻ (c) H^+ :Cl:⁻

2-3 PH_3: H:P:H
$\qquad\quad$ H

2-4 (a) H:N:H, + charge on nitrogen; (b) :O::N:O:⁻, − charge
$\qquad\;\;$ H
on right oxygen; (c) H:C⟨:O:⟩ , − charge on lower oxygen
$\qquad\qquad\qquad\qquad$:O:⁻

2-9 (c) H:C:Cl: The carbon is sp^3-hybridized. The carbon-
$\qquad\quad$:Cl:
$\qquad\qquad$ H
chlorine bonds are polar. (e) H:C:N:H The carbon-nitrogen bond
$\qquad\qquad\qquad\qquad\qquad$ H
$\qquad\qquad\qquad\qquad\qquad$ H H
is polar. The carbon is sp^3-hybridized.

2-12 (a) H—S—H (b) H—O—N—H (c) H—C—S—H
$\qquad\qquad\qquad\qquad\qquad\;$ |$\qquad\qquad$ |
$\qquad\qquad\qquad\qquad\qquad$ H$\qquad\qquad$ H
$\qquad\qquad\qquad\qquad\qquad\qquad\qquad\qquad$ H

2-14 The center carbon atom is sp-hybridized, and the carbon-carbon-carbon bond angle is 180°. The two end carbon atoms are sp^2-hybridized. The molecule is not planar. The hydrogens and the carbon at one end of the molecule form a plane perpendicular to that formed by the hydrogens and carbon at the other end.

2-15 There is no way to hybridize a carbon atom so that all four orbitals point in the same direction. Thus, a quadruple bond between two carbons cannot exist because four orbitals of one carbon cannot be made to overlap with four orbitals of the other.

2-18 (1) The nitrogen might be sp^3-hybridized, in which case the hydrogen-nitrogen-hydrogen angles would be 109°; or (2) the nitrogen might be sp^2-hybridized, in which case the angles would be 120° and the molecule would be planar; or (3) the nitrogen might be unhybridized, with an s and three p orbitals, in which case the angles would be 90°. (The first possibility is the one actually observed.)

CHAPTER 3

3-5 (a) 3-Methylpentane; (b) 3-ethyl-2,3-dimethylhexane; (c) 3-ethylhexane

3-9 (a) 3-Ethyl-2-pentene; (b) *trans*-3-methyl-2-pentene; (c) *trans*-3-isopropyl-2-hexene

3-11 There are two isomers, a cis isomer and a trans isomer:

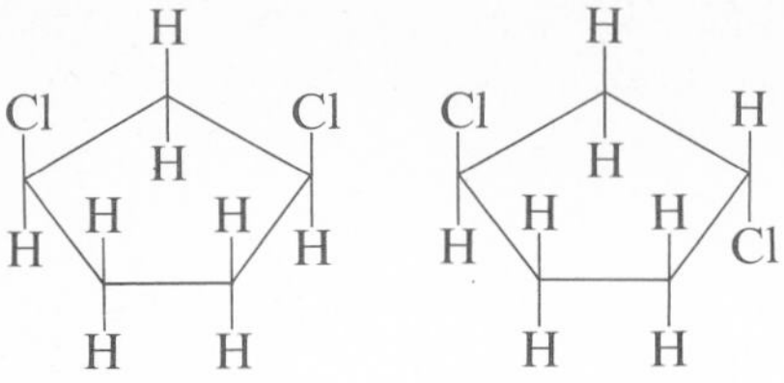

cis   trans

3-12 It is clear from molecular models that *trans*-cyclooctene is the smallest trans cycloalkene.

3-16 There are three isomers, 1,1-dimethylcyclopropane, *cis*-1,2-dimethylcyclopropane, and *trans*-1,2-dimethylcyclopropane.

3-17 (*a*) There are three isomers. (*b*) There are six isomers.

3-18 (*a*)

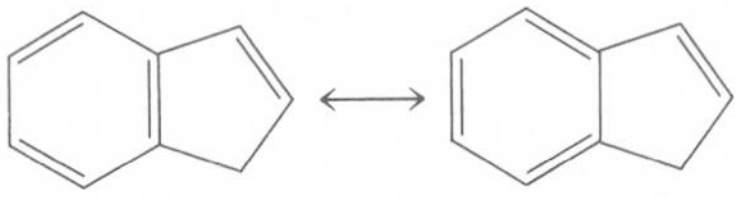

(*b*) There are four resonance forms.

3-21 Two

3-25

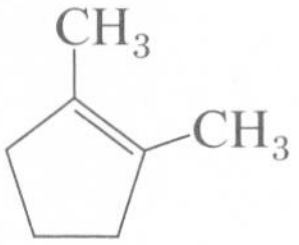

3-26 (*a*) Br—CHCH$_2$CH$_3$ CH$_3$CCH$_3$ Br—CH$_2$CHCH$_3$
          | | |
          Cl   Cl   Cl

Cl—CH$_2$CHCH$_3$ Cl—CH$_2$CH$_2$CH$_2$—Br   (*b*) There are twenty-
          |
          Br

three isomers of C$_5$H$_8$.   (*c*) There are five isomers of C$_3$H$_5$I.

3-30 There are nine possible products.

3-33 The most stable conformation is the anti conformation.

3-35 There are sixteen cyclic compounds with empirical formula C$_6$H$_{12}$.

3-36 There are only two isomers of methyladamantane.

3-42 There are seven isomers of dimethylspiropentane.

3-43 Only the cis isomer will be shown:

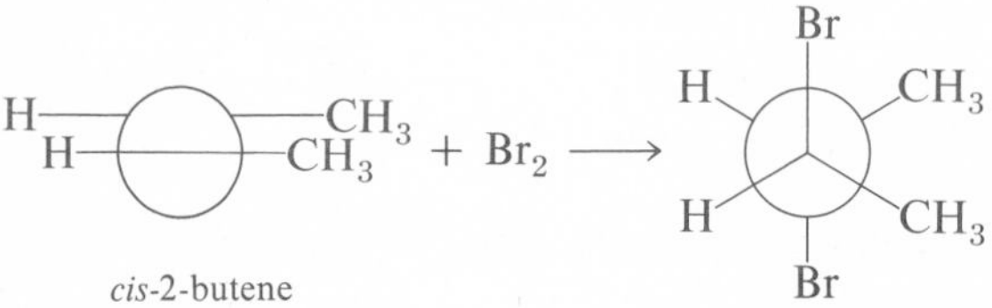

cis-2-butene

Molecular models will show that the trans isomer forms a different product.

3-45 (a)

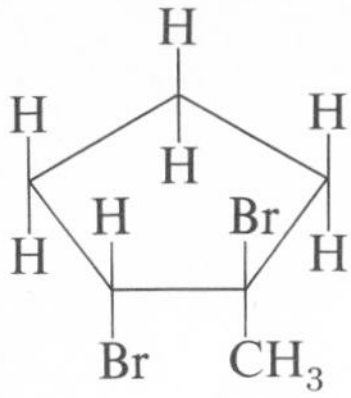

(b)

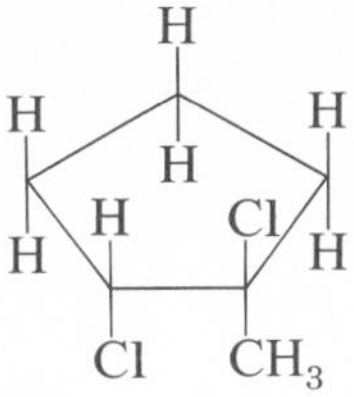

(c)

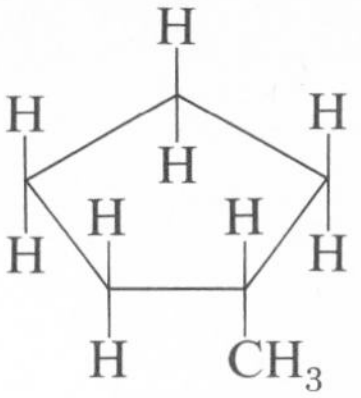

(d)

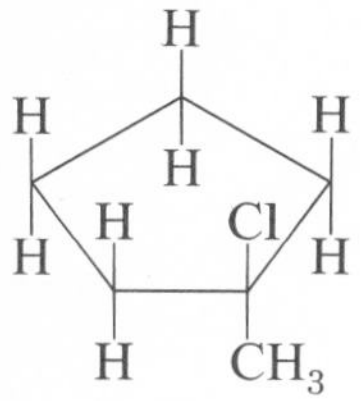

(e)

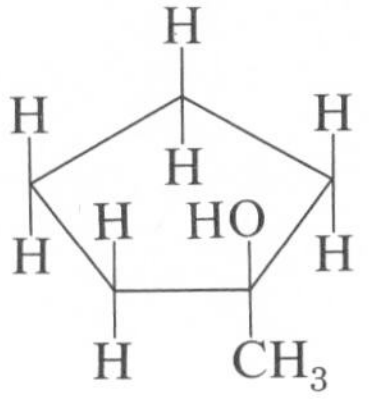

(f) $CH_3\overset{\overset{\displaystyle O}{\|}}{C}CH_2CH_2CH_2\overset{\overset{\displaystyle O}{\|}}{C}H$

(g)

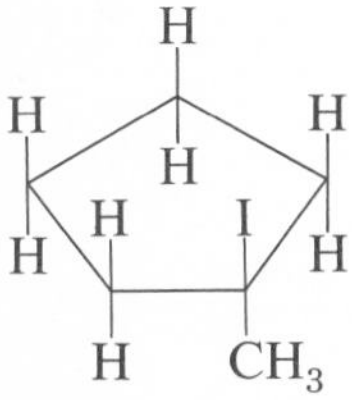

4-1 (a) Not chiral; (b) chiral; (c) chiral; (d) not chiral; (e) not chiral; (f) chiral   ***CHAPTER 4***

4-3 (b), (d), and (e) are chiral.

4-4 The microorganism consumed the half of the alanine which had the configuration normally found in nature. The unnatural enantiomer remained after growth was complete, and the optical activity observed at that time was due to this enantiomer. Thus, the unnatural enantiomer has a negative rotation, and the natural enantiomer has a positive rotation.

4-7 (a) $CH_3\overset{\overset{\displaystyle CH_2CH_3}{|}}{\underset{\underset{\displaystyle CH_2CH_2CH_3}{|}}{C}}H$

(b) $CH_3\overset{\overset{\displaystyle CH=CH_2}{|}}{\underset{\underset{\displaystyle CH_2CH_3}{|}}{C}}H$

(c)

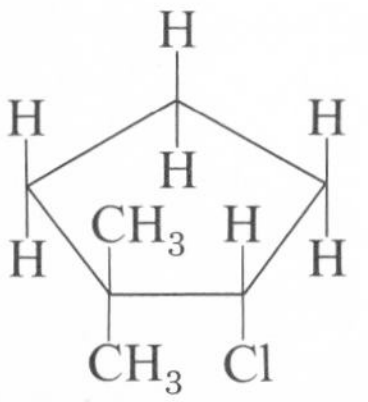

(d)

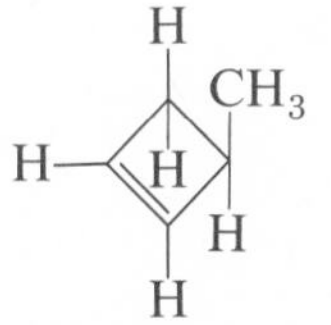

4-9 *Rule:* A compound with two asymmetric carbon atoms is meso if and only if (a) the substituent groups attached to one asymmetric carbon are identical with the substituent groups attached to the other and (b) the configuration of one asymmetric carbon is R and the other is S.

4-10 Only one isomer will be shown for each part:

(a)
$$\begin{array}{c} \text{S} \quad CH_3 \\ H-\overset{|}{\underset{|}{C}}-Cl \\ \text{R} \quad CH_2 \\ H-\overset{|}{\underset{|}{C}}-Cl \\ CH_2CH_3 \end{array}$$

(b)
$$\begin{array}{c} \text{S} \quad CH_3 \\ H-\overset{|}{\underset{|}{C}}-Cl \\ C_5H_{11} \end{array}$$

(c)
$$\begin{array}{c} \text{S} \quad CH_3 \\ H-\overset{|}{\underset{|}{C}}-OH \\ \text{R} (CH_2)_3 \\ H-\overset{|}{\underset{|}{C}}-OH \\ CH_3 \end{array}$$
(meso)

(d)
$$\begin{array}{c} \text{R} \quad CH_2Br \\ H-\overset{|}{\underset{|}{C}}-Br \\ \text{R} \quad CH_2 \\ H-\overset{|}{\underset{|}{C}}-Br \\ CH_3 \end{array}$$

4-12 The 1,1 isomer is not chiral. The cis-1,2, cis-1,3, cis-1,4, and trans-1,4 isomers are all meso isomers. The trans-1,2 and trans-1,3 isomers are chiral.

4-16 2-Methyloctane is not chiral. 3-Methyloctane and 4-methyloctane are chiral. In the case of nonane, the 2-methyl and 5-methyl isomers are not chiral; the 3-methyl and 4-methyl isomers are chiral.

4-17 $-1.4°$; $0°$

4-18 Only (d) is chiral.

CHAPTER 5

5-1 $109.5°$

5-3 (a) $HCl + CH_3CH_2O^- \rightleftharpoons Cl^- + CH_3CH_2OH$
(b) $CH_3OH_2^+ + NH_3 \rightleftharpoons CH_3OH + NH_4^+$

5-7 The intermediate is the carbonium ion

The center carbon of this carbonium ion is sp^2-hybridized. The intermediate is therefore not chiral, and the product of the reaction, though chiral, is racemic.

5-12 The oxygen atom should be sp^3-hybridized, and the C—O—C bond angle should be $109.5°$.

5-13 The mechanism involves protonation of the ether oxygen and displacement of an alcohol by bromide ion. Displacement occurs more easily at a primary carbon than at a secondary carbon; as a result, the products are 2-butanol and 1-bromobutane.

5-14

$$R-\overset{O}{\overset{\triangle}{CH-CH}}-R' + HI \longrightarrow R-\overset{OH}{\underset{I}{CH}}-\overset{}{\underset{}{CH}}-R' + R-\overset{OH}{\underset{}{CH}}-\overset{}{\underset{I}{CH}}-R'$$

disparlure

5-17 (a) 1-Hexanol; (b) 2-methyl-2-butanol; (c) 2-hexene (mixture of cis and trans isomers); (d) 1-iodohexane; (e) 2-hexene (mixture of cis and trans isomers); (*f*) $CH_3CH_2CH_2CH_2\overset{\overset{\displaystyle O}{\|}}{C}CH_3$

5-18 (*a*) 2(*S*)-iodohexane; (*b*) racemic 3-methyl-3-hexanol

5-19 (*a*) 1-Chloro-4-methylpentane, primary; (*b*) 2-iodobutane, secondary; (*c*) 5-ethyl-2-heptanol, secondary; (*d*) 1,2-ethanediol, primary; (*e*) 2-chloro-2-methylpropane, tertiary

5-21 $O=C=N^- \longleftrightarrow {}^-O-C\equiv N + CH_3CH_2Br \longrightarrow$
$CH_3CH_2-O-C\equiv N + CH_3CH_2-N=C=O$

5-23 *S*-Adenosylethionine is less reactive than *S*-adenosylmethionine because displacement in the former case must occur at a primary carbon, rather than at a methyl carbon.

5-25 The product of the reaction is

When the R,S isomer of the starting material is used, the product is cis and is a meso isomer. When the R,R or the S,S isomer is used, the product is trans and therefore chiral.

5-29 (*e*) $ClCH_2CH_2CH_2CN$; (*f*) $ICH_2CH_2CH_2CH_2I$;
(*g*) $CH_3CH_2CH_2SH$; (*n*) $2CH_2=C(CH_3)_2$; (*o*) $CH_3CH_2CH_2S^-$ Na^+

6-1 (*a*) $CH_3CH_2CH_2CH_2NH_2$; (*b*) $CH_3NH\overset{\overset{\displaystyle CH_3}{|}}{C}HCH_3$;   ***CHAPTER 6***

(*c*) $CH_3N^+(CH_2CH_3)_3$ Br^-; (*d*) $CH_3NHCH_2CH_2CH_3$

6-3 (*a*) $CH_3CH_2CH_2NH_2$ $CH_3CH_2NHCH_3$ $CH_3\overset{\overset{\displaystyle}{|}}{N}CH_3$
$\overset{}{}CH_3$

  n-propylamine   methylethylamine   trimethylamine

(*b*) There are eight isomers of $C_4H_{11}N$.

6-8 $HC\equiv N \longrightarrow CH_3NH_2$; $CH_2=CH_2 \longrightarrow CH_3-CH_3$

6-9 See if the bacteria will catalyze the reduction of hydrazine to ammonia. A second experiment would be to look for traces of hydrazine formed during nitrogen fixation, but the likelihood of success with such an experiment is small.

6-10 Amantadine is not chiral.

6-12 At pH 10, the following amines exist primarily as ammonium ions: methylamine, dimethylamine, piperidine, and pyrrolidine. At pH 7 all amines exist as ammonium ions except aniline.

6-14 (*a*)

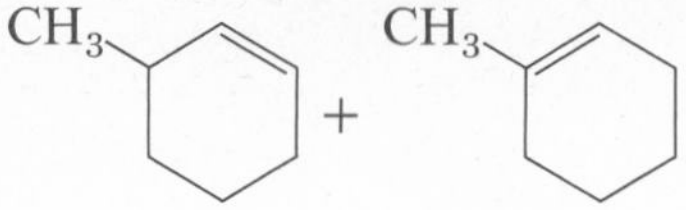

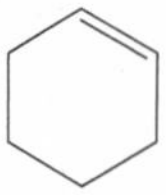

major product   minor product

(*b*) $CH_2{=}CH_2$ +

major   minor
product   product

6-19 The final product is $CH_2{=}CHCH_2CH{=}CHCH_2CH_2CH_3$, which would probably rearrange during the elimination to $CH_3CH{=}CHCH{=}CHCH_2CH_2CH_3$.

6-20 (*a*)

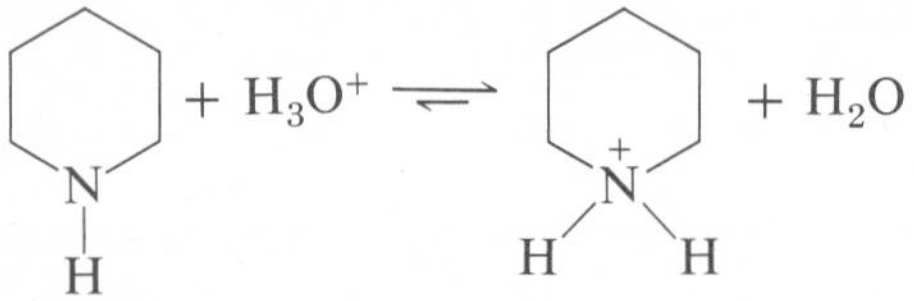

(*b*)

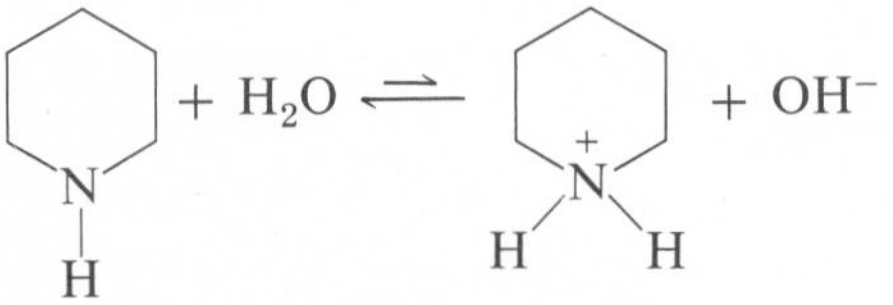

6-21 In acid, the amine is converted into an ammonium ion, which does not have an unshared pair of electrons and cannot act as a nucleophile.

6-22 Although in acidic solution the amine exists as an ammonium salt, even in the strongest acid a small fraction (perhaps only 1 molecule out of 10^{10} or fewer) of the amine is free amine. This small portion of free amine racemizes rapidly. Since the free amine is in equilibrium with the ammonium salt, racemization of the ammonium salt is the ultimate result of this process. Thus, the overall racemization mechanism is

(R)-salt $\rightleftharpoons$ (R)-amine + H$^+$ $\overset{fast}{\rightleftharpoons}$ (S)-amine + H$^+$ $\rightleftharpoons$ (S)-salt

6-24

(*a*)

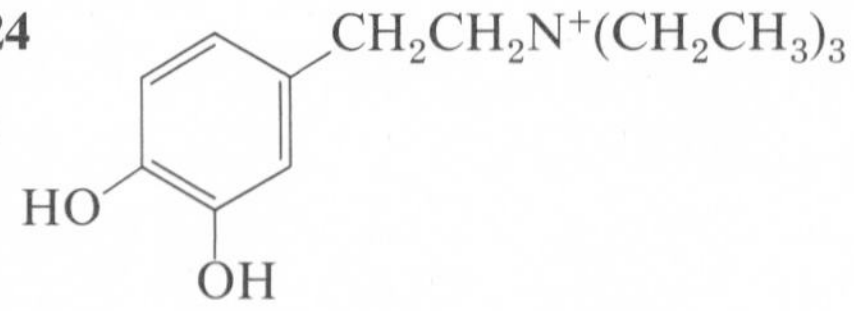

(*b*)

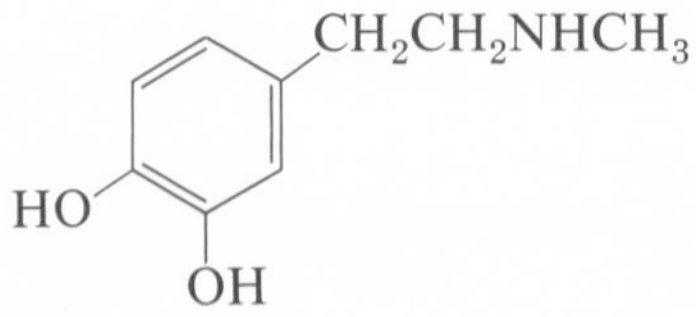

(*d*) $CH_2CH_2CH_3$ Br^- + HBr

7-1 $CH_3-\overset{\overset{O}{\|}}{C}-CH_3 \longleftrightarrow CH_3-\overset{\overset{O^-}{|}}{C^+}-CH_3$

The oxygen atom and all three carbon atoms must lie in the same plane.

7-3 This mechanism is exactly the same as mechanism 7-1 in reverse.

7-4

7-6 (*a*) $CH_3(CH_2)_4\underset{\underset{MgBr}{|}}{C}HCH_2CH_3$ (*b*) $CH_3(CH_2)_4\underset{\underset{CH_3CH_2CHOMgBr}{|}}{C}HCH_2CH_3$

(*c*) $CH_3(CH_2)_4\underset{\underset{CH_3CH_2CHOH}{|}}{C}HCH_2CH_3$

7-8 (*a*) $CH_3CH_2CH_2CH_2N=\underset{\underset{CH_3}{|}}{C}CH_2CH_3$

(*b*) $CH_3CH_2CH_2CH_2\underset{\underset{H_3B^-}{|}}{N}-\underset{\underset{CH_3}{|}}{C}HCH_2CH_3$

(*c*) $CH_3CH_2CH_2CH_2NH-\underset{\underset{CH_3}{|}}{C}HCH_2CH_3$

7-10 $CH_3CH_2CH_2-\overset{\overset{O}{\|}}{C}-CH_2^-$ and $CH_3CH_2\overset{}{CH}^--\overset{\overset{O}{\|}}{C}-CH_3$

7-12 $CH_3\overset{\overset{O}{\|}}{C}-CH^--\overset{\overset{O}{\|}}{C}CH_3 \longleftrightarrow CH_3\overset{\overset{O^-}{|}}{C}=CH-\overset{\overset{O}{\|}}{C}CH_3 \longleftrightarrow CH_3\overset{\overset{O}{\|}}{C}-CH=\overset{\overset{O^-}{|}}{C}CH_3$

7-16 (*a*) $CH_3\underset{\underset{CH_3}{|}}{C}H\overset{\overset{O}{\|}}{C}CH_2CH_2CH_3$

(*b*) $CH_3CH_2\underset{\underset{CH_3CH_2}{|}}{C}HCH_2CH_2CH_2\overset{\overset{O}{\|}}{C}H$

$$\overset{O}{\overset{\|}{CH_3CHCH}}$$

(c) $\overset{O}{\overset{\|}{HCH}}$ (d) [structure: phenyl ring attached to CH₃CHCH with C=O]   (e) [structure: seven-membered ring ketone with C=O]

7-17 Such an imine would have two resonance structures

$$\overset{R_2N^+}{\underset{RCH}{\|}} \longleftrightarrow \overset{R_2N}{\underset{R\overset{+}{C}H}{|}}$$

Like the protonated imine shown in mechanism 7-2 and the protonated aldehyde shown in mechanism 7-3, this imine would be very reactive and would readily decompose to an aldehyde on reaction with water.

7-18 $CH_3\overset{O}{\overset{\|}{C}}(CH_2)_5\overset{O}{\overset{\|}{C}}CH_3$ **7-21** [benzene ring]$-\overset{SCH_2CH_3}{\underset{SCH_2CH_3}{CH}}$

7-24 (a) [cyclohexene ring with $=N-NH\overset{O}{\overset{\|}{C}}NH_2$ substituent]

(b) $CH_3CH_2CH_2CH_2\overset{CH_3CH_2CH_2O}{\underset{CH_3CH_2CH_2O}{C}}CH_3$

(e) $CH_3CH_2CH_2\overset{OH}{\underset{CH_3}{C}}-$[benzene ring]   (f) $CH_3\overset{OH}{\underset{CH_3}{CH}}CH(CH_2)_4CH_3$

(g) $CH_3\overset{CH_3CH_2}{\underset{CH_3CH_2}{C}}-\overset{O}{\overset{\|}{CH}}$

CHAPTER 8 *Note:* In all syntheses, the student should write the structure of the compound formed in each step, together with the reagents required for each step.

8-1 (a) Step 1: reduction with $LiAlH_4$ followed by addition of water forms an alcohol. Step 2: reaction with concentrated sulfuric acid produces an olefin. Step 3: reaction of the olefin with HBr produces the desired halide.
(b) Same as steps 1 and 2 of part (a).
(c) Steps 1 to 3 as in part (a). Step 4: reaction with dimethylamine forms the desired tertiary amine.

(*d*) Steps 1 and 2 as in part (*a*). Step 3: reaction with Cl_2 forms the desired halide.

(*e*) Steps 1 and 2 as in part (*a*). Step 3: reduction of the olefin with H_2 in the presence of a platinum catalyst produces the desired alkane.

(*f*) Steps 1 to 3 as in part (*a*). Step 4: reaction of the bromide with $NaSCH_3$ produces the desired sulfide.

(*g*) Step 1: reaction of cyclohexanone with ammonia produces an imine. Step 2: reduction of the imine with $NaBH_4$ produces the desired amine.

8-3 (*b*) Step 1: reaction of the chloride with hydroxide ion forms an alcohol. Step 2: oxidation of the alcohol with dilute $KMnO_4$ produces the desired aldehyde. The aldehyde is removed from the reaction by distillation as it is formed in order to prevent oxidation to benzoic acid.

(*e*) Step 1: reaction of the bromide with KOH forms an olefin. Step 2: reaction of the olefin with bromine produces the desired dibromide.

(*h*) Step 1: reaction with HCl produces a secondary halide. Step 2: reaction with magnesium in ether produces a Grignard reagent. Step 3: reaction of the Grignard reagent with formaldehyde, followed by addition of water, forms a primary alcohol. Step 4: reaction of the alcohol with concentrated sulfuric acid produces an olefin. Step 5: reaction of the olefin with hydrogen in the presence of a platinum catalyst produces the desired hydrocarbon.

9-1 120°

9-2 (*a*) $CH_3CH_2CH_2\overset{\overset{\displaystyle CH_3}{|}}{C}HCH_2\overset{\overset{\displaystyle O}{\|}}{C}OH$ (*b*) $CH_3(CH_2)_4\overset{|}{C}H—\overset{\overset{\displaystyle O}{\|}}{C}HCOH$
$\qquad\qquad\qquad\qquad\qquad\qquad\qquad\qquad\qquad CH_3CH_2\ \ CH_2CH_3$

(*c*) $CH_3(CH_2)_5\overset{\overset{\displaystyle Br}{|}}{C}HCH_2CH_2\overset{\overset{\displaystyle O}{\|}}{C}OH$

9-3 (*a*) $CH_3\overset{\overset{\displaystyle O}{\|}}{C}OH + Na^+\ {}^-OCH_3 \rightleftharpoons CH_3\overset{\overset{\displaystyle O}{\|}}{C}O^-\ Na^+ + HOCH_3$

(*b*) $H\overset{\overset{\displaystyle O}{\|}}{C}OH + CH_3NH_2 \rightleftharpoons H\overset{\overset{\displaystyle O}{\|}}{C}O^- + CH_3NH_3{}^+$

9-4 This compound is not a β-keto acid and cannot form an enol by decarboxylation.

9-5 (*a*) $CH_3CH_2\overset{\overset{\displaystyle O}{\|}}{C}OH$;   (*b*) and (*c*) $CH_3CH_2\overset{\overset{\displaystyle O}{\|}}{C}O^-$

9-6 The final product is $HO\overset{\overset{\displaystyle O}{\|}}{C}CH_2CH_2\overset{\overset{\displaystyle O}{\|}}{C}—\overset{\overset{\displaystyle O}{\|}}{C}OH$.

9-10 (*a*) Step 1: reaction of the halide with magnesium in ether forms a Grignard reagent. Step 2: reaction of the Grignard reagent with CO_2, followed by addition of water forms the desired product.

(*b*) Step 1: reaction of the bromide with NaOH will form a primary alcohol. Step 2: reaction of the primary alcohol with $K_2Cr_2O_7$ in the presence of acid will produce the desired product.

CHAPTER 10 **10-1** $CH_3(CH_2)_6\overset{\displaystyle O}{\overset{\displaystyle \|}{C}}OCHCH_2CH_3$
$\underset{\displaystyle |}{}$
CH_3

2-butyl octanoate, or
sec-butyl octanoate

10-2 $CH_3CH_2CHCNCH_2CH_3$
$\underset{CH_3}{|}\ \underset{CH_3}{|}$

N-ethyl-*N*-methyl-2-methylbutyramide

10-3 $CH_3CH_2CH_2\overset{O}{\overset{\|}{C}}O\overset{O}{\overset{\|}{C}}CH_2CH_2CH_3$ $CH_3CH_2CH_2\overset{O}{\overset{\|}{C}}Cl$

butyric anhydride   butyryl chloride

10-4 $CH_3CH_2CH_2CH_2C\equiv N$

pentanonitrile

10-6 Note that there are two monoglycerides, two diglycerides, and one triglyceride.

10-7 Step 1: reaction of *n*-hexanol with sodium forms an alkoxide. Step 2: reaction of an excess of the alkoxide with ethyl acetate forms *n*-hexyl acetate.

10-9 $CH_3\overset{O}{\overset{\|}{C}}NHOH$

10-11 Reaction of an anhydride with an amine results in formation of equal amounts of amide and carboxylic acid. As a result, only half of the total carboxylic acid used as starting material is converted into amide.

10-12 Step 1: reaction of *n*-propyl bromide with magnesium in ether forms a Grignard reagent. Step 2: reaction of ethyl acetate with an excess of this Grignard reagent, followed by addition of water, forms the desired alcohol.

10-13 $CH_3CH_2CH_2CH_2NH_2$

10-18 Enolates can only be formed in basic solution. Under those conditions carboxylic acids are converted into carboxylate anions. Because of unfavorable charge-charge repulsions, this carboxylate anion is unable to lose a proton and form a doubly charged enolate anion.

10-22 $NH_2\overset{\overset{\displaystyle O}{\|}}{C}NH_2$ + $HO\overset{\overset{\displaystyle O}{\|}}{C}\overset{\overset{\displaystyle O}{\|}}{C}OH$ $\xrightarrow{\text{heat}}$ barbital

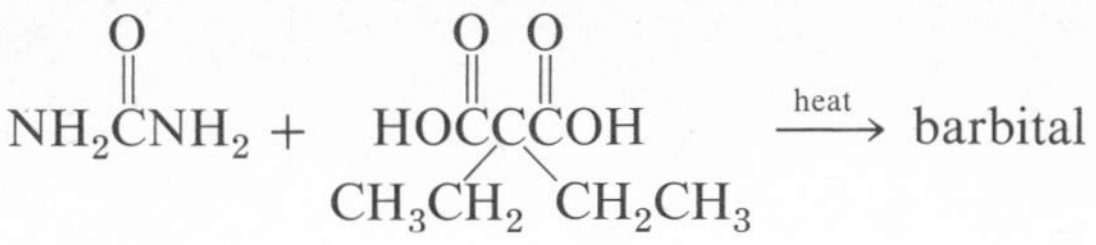

$\qquad\qquad\qquad CH_3\overset{|}{C}H_2 \quad CH_2CH_3$

10-27 $HO\overset{\overset{\displaystyle O}{\|}}{-}P-O-CH_2CH_2-\overset{\overset{\displaystyle CH_3}{|}}{N^+}-CH_3$

$\qquad\qquad \overset{|}{O^-} \qquad\qquad\quad \overset{|}{C}H_3$

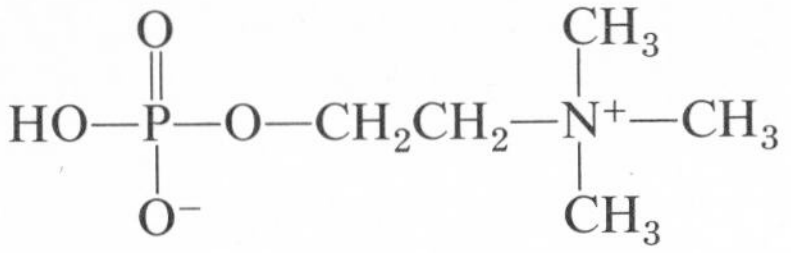

10-29 Attack of the nucleophile on the carbonyl carbon atom forms ethyl acetate and phosphate. Attack of the nucleophile on phosphorus forms ethyl phosphate and acetate.

10-31 The anion formed by loss of a proton from phthalimide is resonance-stabilized like the anion formed by loss of a proton from a β-diketone (see problem 7-12).

10-32

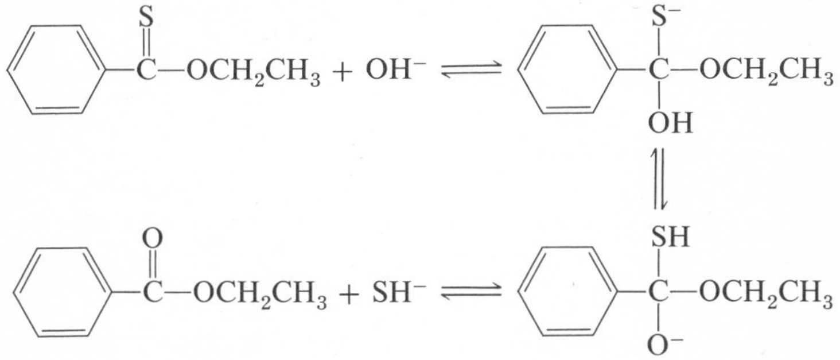

10-37 (*a*)

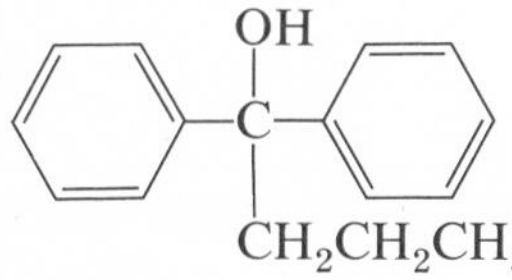

(*b*) $CH_3CH_2CH_2CH_2OH + CH_3OH$

(*c*) $CH_3CH_2CH_2\overset{\overset{\displaystyle O}{\|}}{C}\overset{|}{C}H\overset{\overset{\displaystyle O}{\|}}{C}OCH_3$

$\qquad\qquad\qquad\quad \overset{|}{C}H_3CH_2$

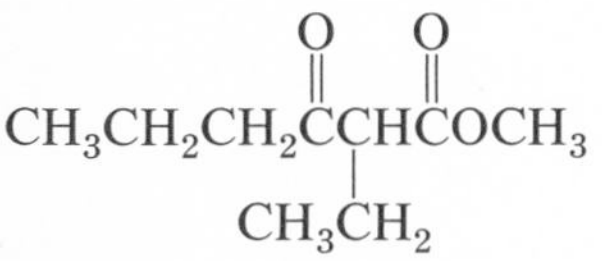

(*d*) $CH_3CH_2CH_2\overset{\overset{\displaystyle O}{\|}}{C}O^-\ Na^+ + CH_3OH$

11-1 (*a*) Two isomers; (*b*) ten isomers; (*c*) three isomers

CHAPTER 11

11-2

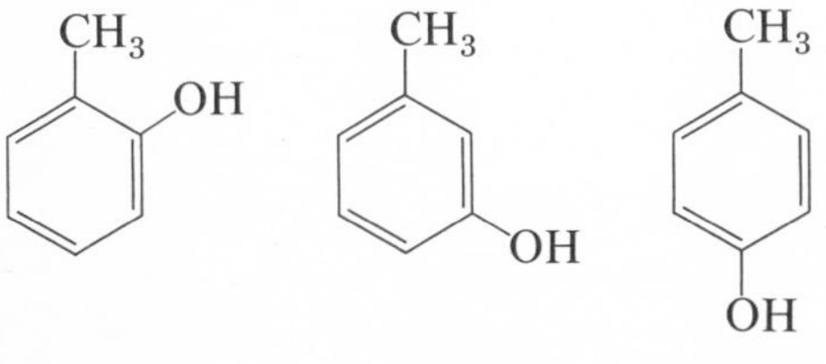

o-cresol   *m*-cresol   *p*-cresol

11-5

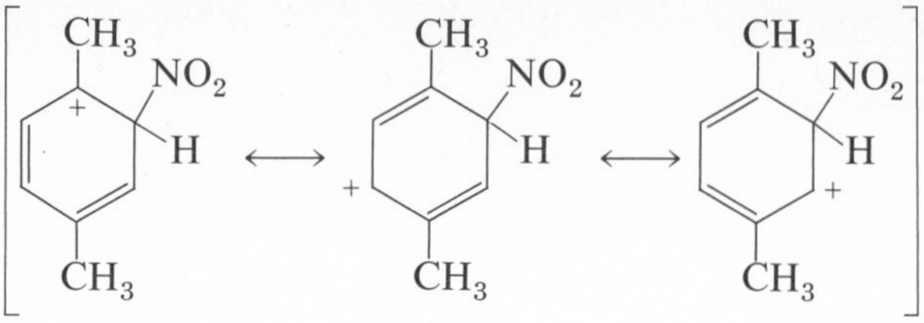

$$\left[\underset{}{\underset{}{}} \longleftrightarrow \longleftrightarrow \right]$$

11-8 $CH_3\overset{+}{-}\overset{}{C}=O \longleftrightarrow CH_3-C\equiv O^+$

11-9 $\longrightarrow$ $CH_2CH_2CH_2\overset{+}{C}=O$ $\longrightarrow$ $\left[\underset{+\ H}{\overset{}{}} \overset{}{\underset{O}{}} \right] \longrightarrow$

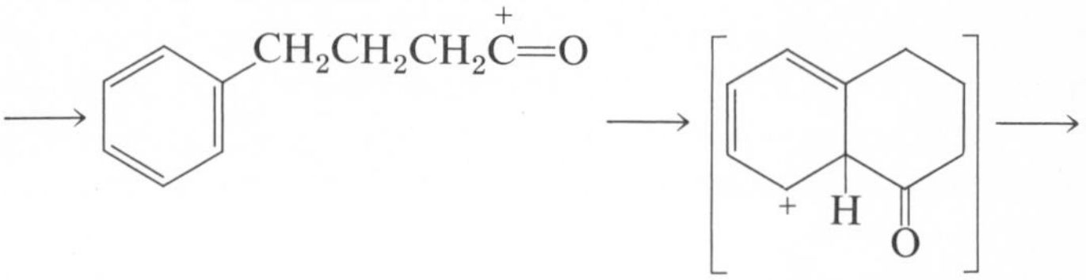

resonance-stabilized
intermediate

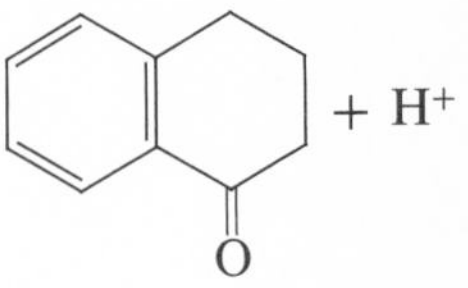

$+\ H^+$

11-10 (a) $\overset{OCH_3}{\underset{}{}}SO_3H$ $+$ $\overset{OCH_3}{\underset{SO_3H}{}}$ (b) $\overset{OCH_3}{\underset{}{}}NO_2$ $+$ $\overset{OCH_3}{\underset{NO_2}{}}$

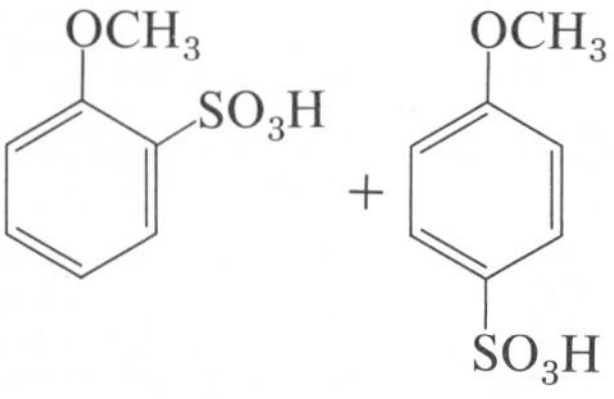

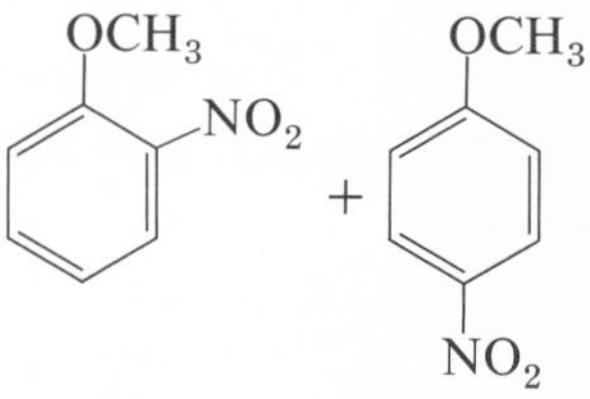

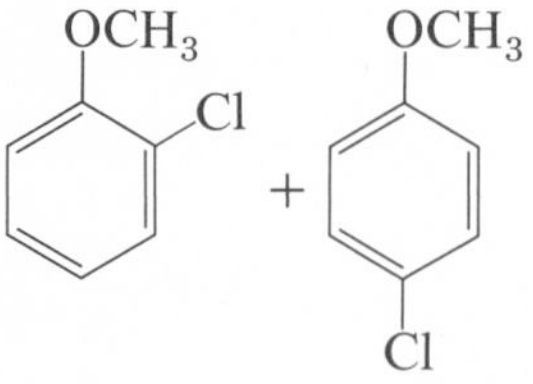

(c) $\overset{OCH_3}{\underset{}{}}Cl$ $+$ $\overset{OCH_3}{\underset{Cl}{}}$

(d) $\overset{NO_2}{\underset{SO_3H}{}}$ $\overset{NO_2}{\underset{NO_2}{}}$ $\overset{NO_2}{\underset{Cl}{}}$

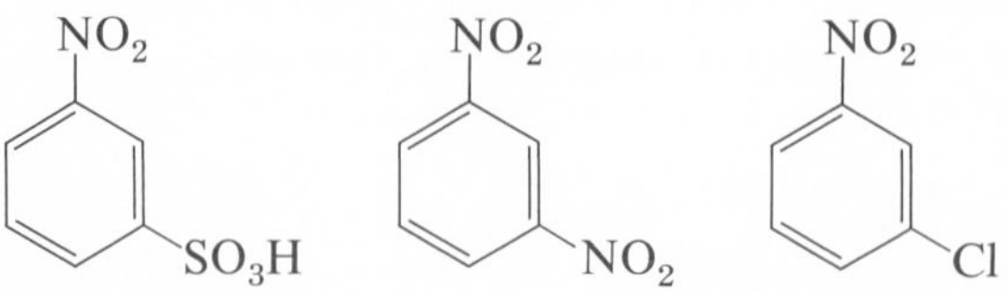

11-11 *p*-Nitrophenol is more acidic than phenol. The most important resonance form is

The negative charge is stabilized because it is adjacent to the positively charged nitrogen.

11-12 As in the comparison of pyridine with pyrrole, the important question is whether or not the electron pair which is involved in acid-base reactions is part of the π electron system. In imidazole that electron pair is not part of the π system, and so imidazole is considerably more basic than pyrrole.

11-14 The analog would not work because the conversion of the oxidized to the reduced form of this compound cannot be made by addition of a single hydride ion, H^-.

11-19 $\xrightarrow{\text{oxidation}}$ $\xrightarrow{\text{hydrolysis}}$ $+ NH_3$

11-22 The carbonium ion is extensively stabilized by resonance. There are a total of forty-four resonance forms for the carbonium ion.

12-2 Phenylalanine is converted into tyrosine in the course of normal metabolism (see figure 11-6). Thus, if there is sufficient phenylalanine in the diet, tyrosine is not required. *CHAPTER 12*

12-5 (*a*) Step 1: 3-methylbutanal + NH_3 + HCN forms an amino-nitrile. Step 2: hydrolysis of the nitrile with aqueous acid forms leucine.

12-6 The bromoacid method would require synthesis of the compound $NH_2(CH_2)_4\overset{Br}{C}H-\overset{O}{C}OH$. This compound is unstable; the amino group would immediately displace the bromine, forming a cyclic amino acid.

12-7 (*a*) Step 1: β-phenylpropionic acid + Br_2 in the presence of phosphorus forms a bromoacid. Step 2: reaction of the bromoacid with ammonia forms phenylalanine.

12-8 Allylglycine can be made by the Strecker method but not by the bromo acid method, which requires the reaction of bromine with

$$CH_2\text{=}CHCH_2\overset{\overset{\displaystyle O}{\|}}{C}OH.$$ Besides the usual bromination of the α carbon, addition of bromine to the double bond would occur.

12-9 If the diketopiperazine is made from (R)-phenylalanine, the product is a cis isomer of the R,R configuration, which is chiral. If the diketopiperazine is made from racemic phenylalanine (an equal mixture of the R and S enantiomers), the product is a mixture of cis and trans diketopiperazines. The cis isomer is chiral but racemic. The trans isomer is not chiral; this is one of those rare compounds which lacks a plane of symmetry but is nonetheless identical with its mirror image.

12-10 (a) $$(CH_3)_3N^+(CH_2)_4\overset{\overset{\displaystyle O}{\|}}{\underset{\underset{\displaystyle {}^+N(CH_3)_3}{|}}{C}}HCOH$$

(b) $$CH_3CH_2O\overset{\overset{\displaystyle O}{\|}}{C}CH_2CH_2\overset{\underset{\underset{\displaystyle {}^+NH_3}{|}}{}}{C}H\overset{\overset{\displaystyle O}{\|}}{C}OCH_2CH_3$$

(c) $$CH_3\overset{\overset{\displaystyle O}{\|}}{C}OCH_2\overset{\underset{\underset{\displaystyle O}{\underset{\displaystyle \|}{NHCCH_3}}}{|}}{C}H\overset{\overset{\displaystyle O}{\|}}{C}OH$$

(d)

(e)

12-14 A total of 8! = 40,320 possible sequences

12-15 (a) Glycine for valine would probably be well tolerated. (b) Substitution of glycine for glutamic acid would cause a large change in polarity and a change in charge and so would probably not be well tolerated. (c) Substitution of glutamine for glutamic acid causes no significant change in polarity but a large change in charge. This substitution might be possible, depending on the role of that particular amino acid in the protein structure. (d) Aspartic acid for glutamic acid should be satisfactory. (e) Because of the change in charge, substitution of lysine for glutamic acid is questionable. (f) Methionine for lysine would probably not be satisfactory.

12-17 The side chain on the single D-amino acid would point in the wrong direction (figure 12-7). Unless the side chain were very bulky, it probably would not disrupt the helix.

13-2 D-ribose and D-arabinose; D-ribose and D-xylose; D-arabinose and D-lyxose; D-xylose and D-lyxose; all the corresponding L pairs; D-ribose and L-lyxose; L-ribose and D-lyxose; D-arabinose and L-xylose; L-arabinose and D-xylose

13-3

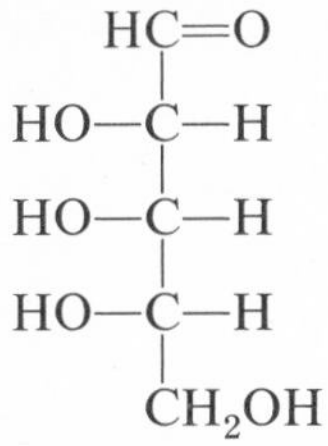

$$
\begin{array}{c}
HC{=}O \\
HO{-}C{-}H \\
HO{-}C{-}H \\
HO{-}C{-}H \\
CH_2OH
\end{array}
$$

13-4 Glyceraldehyde, erythrose, ribose, xylose, allose, and galactose

13-5 Fructose

13-7 The α anomer:

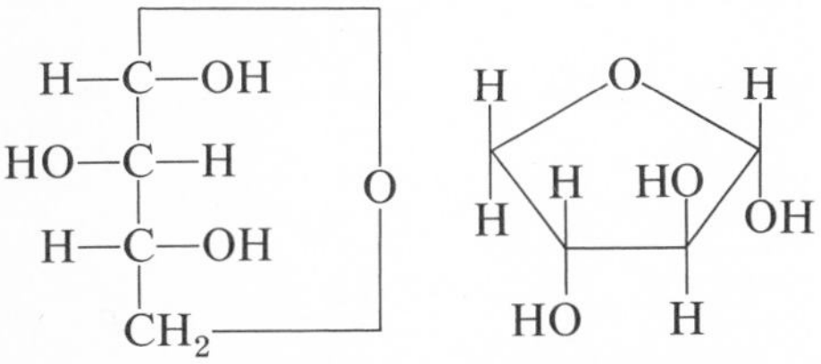

13-9 See mechanism 7-3. The cation formed by loss of water can be attacked by methanol from either above or below the ring. The first attack results in formation of the β anomer, the second, the α anomer.

13-11

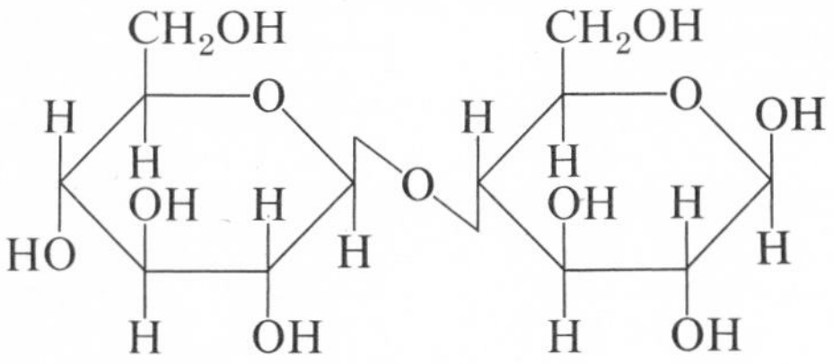

13-17 L-Sorbose and L-tagatose

Index